STRATEGIC DECEPTION

Rhetoric and Public Affairs Series

Eisenhower's War of Words: Rhetoric and Leadership
Martin J. Medhurst, editor

The Nuclear Freeze Campaign: Rhetoric and Foreign Policy in the Telepolitical Age
J. Michael Hogan

Mansfield and Vietnam: A Study of Rhetorical Adaptation
Gregory A. Olson

Truman and the Hiroshima Cult
Robert P. Newman

Post-Realism: The Rhetorical Turn in International Relations
Francis A. Beer and Robert Hariman, editors

Rhetoric and Political Culture in Nineteenth-Century America
Thomas W. Benson, editor

Frederick Douglass: Freedom's Voice, 1818–1845
Gregory P. Lampe

Angelina Grimké: Rhetoric, Identity, and the Radical Imagination
Stephen Howard Browne

STRATEGIC DECEPTION

Rhetoric, Science, and Politics in Missile Defense Advocacy

Gordon R. Mitchell

Michigan State University Press
East Lansing

∞ The paper used in this publication meets the minimum requirements of ANSI/NISO Z39.48–1992 (R 1997) (Permanence of Paper).

Michigan State University Press
East Lansing, Michigan 48823–5202

Printed and bound in the United States of America.

07 06 05 04 03 02 01 00 1 2 3 4 5 6 7 8

LIBRARY OF CONGRESS CATALOGING-IN-PUBLICATION DATA

Mitchell, Gordon R., 1967-
Strategic deception: rhetoric, science, and politics in missile defense advocacy / Gordon R. Mitchell.
p. cm.— (Rhetoric and public affairs series)
Includes bibliographical references (p.) and index.
ISBN 0-87013-557-0 (clothbound: alk. paper)
ISBN 0-87013-558-9 (pbk.: alk. paper)
1. Ballistic missile defenses—United States. 2. United States—Armed Forces—Appropriations and expenditures. 3. United States—Politics and government—1945-1989. 4. United States—Politics and government—1989- 5. Rhetoric—Political aspects—United States. I. Title. II. Series.
UG743 .M58 2000
358.1'74'097309045—dc21

00-008705

Book and cover design by Michael J. Brooks, Royal Oak, Michigan.

Visit Michigan State University Press on the World Wide Web at:
www.msu.edu/unit/msupress

In memory of Grandma Who-Who's
passion for words
and for Nana's smiling spirit

Contents

ACRONYMS AND ABBREVIATIONS

ABM	anti-ballistic missile
ACDA	Arms Control and Disarmament Agency
AEC	Atomic Energy Commission
ASAT	antisatellite
AVT	Analysis Validation Team
BMD	Ballistic Missile Defense
BMDO	Ballistic Missile Defense Organization
BUR	bottom-up review
BW	biological weapon
CIA	Central Intelligence Agency
CNN	Cable News Network
CRS	Congressional Research Service
CTBT	Comprehensive Test Ban Treaty
DCIS	Defense Criminal Investigative Service
DIS	Defense Information Service
DOD	Department of Defense
DOE	Department of Energy
DOT&E	Director of Testing and Evaluation
DVG	Data Validation Group
EKV	exo-atmospheric kill vehicle
EU	European Union
FACA	Federal Advisory Committee Act
FAS	Federation of American Scientists
FOIA	Freedom of Information Act
GAO	General Accounting Office
GBI	ground based interceptor
GBR	ground based radar
GOP	Grand Old Party
GPALS	Global Protection Against Limited Strikes
GPO	Government Printing Office
HEL	high energy laser

HF	Hertitage Foundation or High Frontier
HOE	Homing Overlay Experiment
HTK	hit-to-kill
ICBM	intercontinental ballistic missile
IDDS	Institute for Defense and Disarmament Studies
IDF	Israeli Defense Force
IFT	integrated flight test
ISIS-Europe	International Security Information Service (Europe)
ISIS-U.K.	International Security Information Service (United Kingdom)
IPCC	Intergovernmental Panel on Climate Change
KFE	Kalman Feature Extractor
LLNL	Lawrence Livermore National Laboratory
MAD	Mutual Assured Destruction
MARV	Maneuvering Re-Entry Vehicle
MEADS	Medium Altitude Air Defense
MED	Manhattan Engineer District
MIRACL	Mid-Infrared Advanced Chemical Laser
MIRV	multiple independently targeted reentry vehicle
MIT	Massachusetts Institute of Technology
NASA	North American Space Association
NATO	North Atlantic Treaty Organization
NAVSTAR	navigation satellite tracking and ranging
NGO	nongovernmental organization
NIMBY	"Not In My Back Yard"
NMD	National Missile Defense
NPT	Nonproliferation Treaty
NSC	National Security Council
NSDD	National Security Decision Directive
OSRD	Office of Scientific Research and Development
OTA	Office of Technology Assessment
PAC-3	Patriot Advanced Capability-3
PAT	probability of assigned target

PGRV	Precision Guided Re-Entry Vehicle
PLV	payload launch vehicle
R&D	research and development
RMA	revolution in military affairs
ROK	Republic of Korea
SALT	Strategic Arms Limitation Talks
SAMs	surface-to-air missiles
SBIRS	Space-based Infrared Tracking System
SDI	Strategic Defense Initiative
SDIO	Strategic Defense Initiative Organization
SMDC	Space and Missile Defense Command
SSBMs	nuclear powered ballistic missile submarine
START II	Strategic Arms Reduction Talks II
TBM	theater ballistic missile
THAAD	Theater High Altitude Air Defense
THEL	Thermal High Energy Laser
TLAM	Tomahawk Land Attack Missile
TMD	theater missile defense
UEWR	upgraded early warning radar
UN	United Nations
USAF	United States Air Force
VDU	video display unit
WHO	World Health Organization

Demonstrators march on the Menwith Hill Signals Intelligence Centre near Harrogate, North Yorkshire (U.K.). Peace movement activists organized this 4 March 2000 rally to protest integration of the Menwith Hill radar station into the proposed U.S. National Missile Defense system. Photograph reprinted courtesy of J. Brierly/Yorkshire CND.

Preface

On 4 March 2000, *Star Wars* characters Darth Vader, Princess Leia and Chewbacca marched with hundreds of friends down a road near Harrogate, North Yorkshire in the United Kingdom. On first glance, this parade had the appearance of a George Lucas fan club gathering or a sci-fi movie convention. Yet messages written on placards revealed that marchers had donned the costumes to make a political point; slogans such as "Say No to Star Wars" and "Keep Space for Peace" marked this gathering as a demonstration against missile defense. Some 400 activists from political groups such as the Campaign for Nuclear Disarmament had gathered outside the 13th USSA Field Station to protest conversion of the U.S. National Security Agency surveillance facility at Menwith Hill into a National Missile Defense (NMD) radar tracking station.

News of this protest reached me when I was in the final stages of updating material for this book. Subsequently, I took great interest in the Menwith Hill facility and its possible role in NMD. For background research, I retrieved information about Menwith Hill's radar capabilities from the Internet, then phoned the Public Affairs Office of the U.S. Ballistic Missile Defense Organization (BMDO) to confirm that the facility was indeed being considered as a possible radar tracking station to be used in the proposed NMD system.

After I explained my query to the BMDO employee who fielded my phone call, she responded that she couldn't help me, because information on the possible NMD role for the Menwith Hill facility was "sensitive." I replied that such information couldn't be *that* sensitive, because I had just found a bevy of Internet websites discussing the subject, including one that contained pictures of Star Wars characters demonstrating against use of the facility for missile defense! Our short conversation ended with her remarkable statement: "We don't have to give out information like that to the public; *that's the great thing about this!*"

I hung up the phone and reflected on this exchange. It was hard for me to believe that the information I requested was secret. Sure enough, after some digging, I eventually found it listed in several BMDO Fact Sheets available in the public domain. Upon further reflection, what I found especially revealing was the way the BMDO employee rejected my routine query so gleefully, punctuating her refusal to help me with the giddy comment: *"that's the great thing about this!"* Later, I wrote a letter to her boss at BMDO, Public Affairs Director Lt. Col. Richard Lehner. In this letter, I described the phone conversation and suggested that the "cavalier attitude regarding information classification" exhibited by his staff "does not reflect well on your organization, whose reputation for democratic accountability is eroded by such insouciance."

Shortly thereafter, I received an apologetic phone call from Lehner, who explained that my misfortune was the result of a mix-up. As Lehner clarified, BMDO's Office of Public Affairs *should* have fielded my query; information on the Menwith Hill facility's possible NMD role *was not* classified secret; and the only reason such information was withheld from me was because my call was fielded by a "very young and inexperienced" staffer at BMDO, who was not even supposed to be working in the Office of Public Affairs. Lehner said that he had confronted this staffer about the mistake, and that when he asked her why she had celebrated her power to withhold information with the ebullient outburst *"that's the great thing about this!"* she replied: "I don't know."

New York Times journalist William J. Broad won the Pulitzer Prize for his groundbreaking analysis of how the seductive power of secrecy beguiled and transformed missile defense scientists working on Edward Teller's x-ray laser project in the 1980s. Broad found that Teller's scientists experienced a sort of intoxication by working on classified projects in tightly compartmentalized work spaces, cut off from the rest of the world. Their language changed. The isolation literally went to their heads, producing a bizarre strain of scientific megalomania.

Perhaps the opening vignette describing my recent phone conversation with a BMDO junior staffer is a poignant reminder that a culture of excessive secrecy still intoxicates members of the Pentagon's missile defense enterprise. According to retired TRW senior engineer and missile defense insider Roy Danchick, this enterprise "has been running on intellectual fumes for decades." What makes the missile defense program a fascinating site of rhetorical study is that despite occasional sputters and stalls resulting from failed projects and disclosures of testing deception, the program has been revived perennially in episodes of rhetorical re-invention.

Today, we find ourselves in the midst of such a re-invention episode, with missile defense advocates pressing proposals for a dizzying array of BMD systems including Airborne Laser, National Missile Defense, Navy Theater Wide, Patriot Advanced Capability-3, Space Based Laser, Theater High Altitude Air Defense, and Thermal High Energy Laser. This technological cornucopia stirs memories of the Reagan Star Wars era, when missile defense advocates trumpeted a similar bevy of gadgets. In mid-1980s SDI promotion campaigns, BMD enthusiasts touted chemical lasers, free electron lasers, rail guns, neutral particle beams, and X-ray lasers as devices that could deliver on Reagan's vision to render nuclear weapons "impotent and obsolete."

I first studied these devices, and the controversy surrounding Reagan's Star Wars proposal, when I was assigned to research the national policy debate topic as an undergraduate member of Northwestern University's intercollegiate debating team in 1987. That year, the topic chosen for debate was: "Resolved: That the United States should reduce substantially its military commitments to NATO member states." During the tens of thousands of intercollegiate contest rounds held across the country at numerous debate tournaments that season, members of some 150 different debating societies grappled with this topic and argued about the proper U.S. military posture in the late stages of the cold war. Reagan's Star Wars proposal came up frequently in such discussions, and I recollect always facing an uphill battle when called upon to refute debate cases that argued for termination of SDI. My understanding of missile defense was honed by the challenge of competing against such affirmative cases, which were advocated by outstanding debaters including Dean Groulx, Aaron Hawbaker, Martin Loeber, J. Daniel Plants, Ken Schuler, and Scot Thomson. Other teammates, competitors and judges who contributed substantially to my grasp of military strategy and security logic that year included Timothy Alderete, Casey Anderson, Julie Arthur, Ben Attias, Jon Bruschke, Terry Check, Joel Christie, David Coale, Rodger Cole, Rex Copeland, Alan Coverstone, Chris Decker, Steve Dolley, Erik Doxtader, Todd Flaming, Mark Friedman, Jon Garcia, Michael Green, Sherry Hall, David Hingstman, Scott Hodges, Judd Kimball, Stacey Kinnamon, Kevin Kuswa, Madison Laird, Alexander Lennon, Denise Loshbough, Frank Lowrey, Scott Maberry, Steve Mancuso, T. A. McKinney,

Shaun Martin, Cameron Murray, Bill Newnam, Dallas Perkins, Barry Pickens, Gus Puryear, Gus Ramsey, Dan Reiter, Calvin Rockefeller, Marc Rubinstein, Andrew Schrank, Roger Solt, James Speta, Fred Sternhagen, Joseph Thompson, Robert Wick, and Spencer Zuzulo.

The world of intercollegiate policy debate is an odd and magical place, where a keen spirit of competition drives debaters to amass voluminous research in preparation for tournaments, and where the resulting density of ideas spurs speakers to cram arguments into strictly timed presentation periods during contest rounds. Expert judges trained in policy analysis keep track of such contests as they unfold at breakneck speed, with speakers routinely delivering intricate argumentation at over 300 words per minute. To the uninitiated onlooker, this style of debate reveals itself as an unintelligible charade, something like a movie-length Federal Express commercial or an auctioneering competition gone bad. But there are rich rewards for participants who master policy debate's special vocabulary, learn its arcane rules, and acclimate themselves to the style of rapid-fire speaking needed to keep up with the flow of arguments. The rigorous dialectical method of debate analysis cultivates a panoramic style of critical thinking that elucidates subtle interconnections among multiple positions and perspectives on policy controversies. The intense pressure of debate competition instills a relentless research ethic in participants. An inverted pyramid dynamic embedded in the format of contest rounds teaches debaters to synthesize and distill their initial positions down to the most cogent propositions for their final speeches. In many ways, the style of inquiry and argument I learned through intercollegiate debate is performed in the pages of this book, so sincere thanks are due to the debate mentors who have shared their wisdom with me through the years: Shelley Clubb, Cori Dauber, Erik Doxtader, Eric Gander, Scott Deatherage, Doug Fraleigh, Scott Harris, Charles Kauffman, Allan Louden, Arnold Madsen, Shirley Markle, Cate Palczewski, Carla Robinson, Ross Smith, Walter Ulrich, and Henry Weissman. I also owe a debt of gratitude to the many debate students I have been privileged to coach and teach in a variety of pedagogical settings, who have given back more than they know in argumentative wisdom, especially Kevin Ayotte, Tara Beichner, Daryl Burch, John Butler, Rick Fledderman, Bryon Gill, Mark Grant, Marcy Halpin, Kelly Happe, Terry Johnson, Brian Lai, Brian Lain, James Lyle, Holly Lewis, Sean McCaffity, Mason Miller, Timothy O'Donnell, Joseph Peery, Mike Ridge, Jeff Pierson, Almas Sayeed, Maxwell Schnurer, Andrew Stangl, Marcia Tiersky, Jody Terry, Laura Veldkamp, Ron Von Burg, Bill Ziegelmueller and members of the 1998 Cherub Study Group on Missile Defense (Clint Burr, Melissa Gainey, Theo Schweitz, David Summers, Tony Todero, Hester Tsui, Barbara Tunkis, and Adam Wyatt).

Perhaps the most idiosyncratic aspect of the contemporary intercollegiate policy debate community is that, by and large, it keeps to itself. Contrary to the populist tradition of debate as the quintessential genre of public discourse, contemporary intercollegiate policy debate is an insular and specialized academic activity. The research products generated by thousands of debaters nation-wide are generally put toward a singular end: winning tournament competitions. Sometimes this insularity appears absurd to those who stumble across a slice of the debate community for the first time. In the summer of 1990, Madison Laird (then captain of the Loyola University debate squad) was assigned the task of entertaining Earth Day organizer Bill Keepin during Keepin's visit to the Loyola campus in Los Angeles, California. After Keepin delivered a speech on nuclear power to the student body, Laird led him on a campus tour that ended up in the debate squad room, where yards and yards of argument briefs were stowed away in filing drawers. When Keepin asked to see the files containing research on nuclear power, Laird pulled open one file drawer stuffed to the gills with high-quality research. Keepin was stunned, asking incredulously: "How long have you folks kept this stuff locked up?!" In a small way, this story illustrates the folly associated with the intercollegiate debate community's insular nature. Indeed, it would not be surprising to find countless other Bill Keepins out there who could make tremendous use of the research and knowledge generated out of intercollegiate policy debate competition. To reach them, debaters need only to realize that they can make vital contributions to public arguments swirling beyond the rarefied confines of debate tournament sites.

My good fortune of having a helpful and supportive doctoral dissertation committee composed of former intercollegiate debaters turned communication professors (Thomas Farrell, Thomas Goodnight, and David Zarefsky) was instrumental in helping me translate successfully the fruits of my competitive policy debate research into a 1997 Northwestern University thesis. Professor Goodnight deserves particular recognition here for his outstanding advising and timely help in finding a sympathetic publisher for a politically provocative work. The task of converting a sprawling thesis into a into a book suitable for publication took place at my current academic home in the Department of Communication at the University of Pittsburgh, where my work was buoyed in myriad ways by colleagues Thomas Kane, John Lyne,

Janet McCarthy, Lester Olson, Regina Renk, Debra Siegel, Carol Stabile, and Ted Windt. Completion of the book would not have been possible without significant institutional backing from the University of Pittsburgh, which provided crucial research leave time and grant support. Financing from the Richard D. and Mary Jane Edwards Endowed Publication fund made the addition of several high quality photographs in the book and a top-notch index possible.

A research project focusing on secrecy and deception poses special challenges regarding access to information. In this regard, I was lucky to receive assistance from numerous sources who helped me navigate through the complex and shadowy world of missile defense science and politics. John Pike of the Federation of American Scientists opened up his cavernous vault of space policy archives in Washington, D.C. Robert Bowman, Lucas Fischer, Ted Gold, Rick Goodman, Laura Lee, Michael Nacht, John Pike, Theodore Postol, Carol Rosin and Stephen Schwartz gave generously of their time during interviews and background conversations. Erik Doxtader and Patricia Mische forwarded key documents and also provided important clues marking the trail to other research finds.

My strategy of contributing research findings to public debate as they emerged and stabilized, rather than waiting for the entire book to be completed, was made possible by helpful editors such as Mike Moore, Lauren Spain, and Linda Rothstein at the *Bulletin of the Atomic Scientists;* Stephen Pullinger and John Ziman at the International Security Information Service (U.K.); Catriona Gourlay and Russell Picard at the International Security Information Service (Europe); John Pike at the Federation of American Scientists' Space Policy Project; Andrew King at *The Quarterly Journal of Speech;* Ellsworth Fuhrman at *Science, Technology, and Human Values;* and John Allison at the *Pittsburgh Post-Gazette.*

Robert Bowman, Anthony Cave Brown, Karlyn Kohrs Campbell, Thomas Farrell, Thomas Goodnight, Henry Krips, John Lyne, Timothy O'Donnell, David Helwich, John Pike, Carol Rosin, Nira Schwartz, David Zarefsky, John Ziman, and three anonymous reviewers provided excellent feedback after reading parts or all of the manuscript.

I am indebted to Martin J. Medhurst, senior editor of the Michigan State University Press series on Rhetoric and Public Affairs, for taking stock in my work and then following up to push it toward its potential. Julie Loehr, Julie Reaume, and Annette Tanner at MSU Press deserve plaudits for guiding the project through turbulent political waters created by election-year politics as well as the production manager's nightmare of an avalanche of last-minute updates.

Portions of chapter three published previously in *The Quarterly Journal of Speech* appear here courtesy of the National Communication Association. Stretches of chapter four published previously in *Science, Technology and Human Values* appear here courtesy of Sage Publications. Parts of chapter five were published previously in *The Bulletin of the Atomic Scientists, The Pittsburgh Post-Gazette,* and briefing paper series produced by the International Security Information Service (UK and Europe). Thanks are due to these organizations for releasing material to be published here.

Hal Williamson's trenchant systems analysis contributed substantial kernels of insight. I built up a great head of steam for this project from a lifetime of intellectual collaboration with my mother, Sharon Eakes. It is difficult to distill Melissa Butler's contribution to this project, since her wisdom and loving energy have suffused every aspect of research, writing, and revision.

Finally, I dedicate this book to my grandmothers, the late Ruth Morgan Mitchell (Who-who) and Weedie Louise Eakes (Nana). As a lover of language, Who-who used to summon me to her work space and drill me on grammar. When I grew weary, she would enchant me with tales of how the best radio interviewers would use the same principles to explore pertinent issues of the day. It was during these episodes that I developed profound respect for language as an agent of change in society. Only recently have I become aware of Nana's profound influence on my political outlook. It is easy to slip into the vortex of cynicism while immersing oneself in the study of weapons, waste, and deception. But the resilience of Nana's smiling spirit renews my faith in the possibility that a hopeful outcome could emerge at any time.

The Stanley R. Mickelsen Safeguard complex in Nekoma, North Dakota. Safeguard was the only operational ballistic system ever deployed by the United States. It was deactivated in 1976 after being operational for only four months. This aerial view shows Safeguard's pyramid-shaped radar bunker in the background, and the missile field containing thirty Spartan and sixteen Sprint interceptor missiles in the foreground. U.S. Army photo. Photograph is courtesy of the U.S. Army.

CHAPTER ONE

Missile Defense Advocacy in Rhetorical and Historical Perspective

A "secret government report" titled *Report From Iron Mountain* first hit the shelves of American bookstores in 1967. In the opening pages, readers of this document learn that it is the work product of a fifteen-member "Special Study Group," commissioned by high-level officials from the Kennedy administration. The group was charged to "determine, accurately and realistically, the nature of the problems that would confront the United States if and when a condition of 'permanent peace' should arrive, and to draft a program for dealing with this contingency."[1] To put it mildly, the report's analysis and recommendations are shocking. In what journalist John Leo calls "a cold, flat style—described by some readers as 'perfect bureaucratese,'"[2] the report argues that since society has become so economically, politically, and culturally dependent on war, any transition to a state of "permanent peace" would entail wrenching upheaval and dislocation. To lessen the shocks of such a turbulent transition, the report prescribes the reintroduction of slavery and the invention of new ritual blood sports.[3] For population control in the absence of Malthusian death checks from military conflict, the report proposes a program of computer-engineered eugenics.[4] The report also recommends a deliberately wasteful and lavish space program, designed to reach unreachable points in space, as a mechanism to replace the war system's valuable function as the "balance wheel" of the economy.[5]

For five years after the publication of *Report From Iron Mountain*, scholars and pundits debated the significance, origin, and authorship of the mysterious report.[6] Was the U.S. government secretly considering the bizarre schemes recommended by the Special Study Group? Why was the report suppressed? How were the report's contents leaked to the public? Who were the fifteen shady authors who claimed to have unmatched expertise in dealing with issues of such massive socio-political significance? In a 1972 *New York Times* op-ed piece, Leonard C. Lewin put many of these questions to rest by disclosing that the report was a hoax, and that he, not the "Special Study Group," was the sole author. The idea behind the extravagant parody, Lewin explained, was to "caricature the bankruptcy of the think-tank mentality by pursuing its style of scientistic thinking to its logical ends."[7] "Authors" of the fictitious report wrote as if they had "cornered the market on objectivity,"[8] but as foreign policy analyst Kenneth Boulding explains, the piece is "clever pinchbeck, operating at the level of popularized folk science."[9] Delivering prose that commentator Felix Kessler calls at once "authoritatively bureaucratic but quietly preposterous,"[10] Lewin sought to highlight the dangers and absurdities of scientism run amok.

Like many think-tank treatises on military policy,[11] *Report From Iron Mountain* is chiseled in such a way that as commentator Marc Pilisuk explains, "the product can have a ring of scientific jargon that sounds good, or at least incomprehensibly authoritative, when presented in Congressional testimony or in special White House reports."[12] Indeed, pages of the report are cluttered with sprinklings of statistical analysis, healthy doses of austere value-free language, signposts of methodological rigor, and repeated denials of personal interest or bias by the authors. These markers lend a certain credibility to each individual fragment of analysis contained within the report. However, when the outrageous argument of the report is considered in toto, Lewin's disturbing suggestion emerges quite clearly: uncritical, wholesale reliance on the idiom of scientific objectivity can lead to preposterous conclusions in military analysis.

By labeling his parody a "secret and suppressed" government report, Lewin added an additional layer of depth to this warning. Because most military initiatives are cloaked in secrecy, the scientific reasoning undergirding major projects sponsored by the war system is often exempt from outside scrutiny. While some military projects are shielded from external inspection on the grounds that release of sensitive information would constitute, in the words of the Supreme Court, a "grave, immediate and direct" threat to national security, many other military initiatives are kept under wraps simply because disclosure would be inconvenient for what President Dwight Eisenhower called the "military-industrial complex (MIC)."[13]

During the cold war, the precarious balance of nuclear deterrence dictated that in controversies over secrecy and classification of scientific information, the burden of proof should rest on critics of the MIC calling for open disclosure and public scrutiny. This burden was justified with arguments either that imprudent release of information would aid the USSR materially, thus upsetting the balance of deterrence, or that the amplification of controversy and dissent itself would undermine Soviet perceptions of American will and resolve, also jeopardizing deterrence. With the unraveling of the Warsaw Pact and the fall of the Berlin Wall, the strictness of the deterrence equation has eased. The specter of an opportunistic Soviet nuclear first strike in the face of American weakness has been dismissed and no longer functions credibly as an argumentative trump card in public discussion of military policy. Accordingly, one might expect a presumption of openness to prevail in post–cold war controversies regarding classification and secrecy of scientific research, with the burden of proof resting on those calling for secrecy and the fettered flow of information.

However, this presumption of openness has failed to materialize, and the cold war legacy lingers on today, more than a decade after the fall of the Berlin Wall. This legacy features a systematic pattern of threat inflation, a relentless stream of surplus weapons development, and a predilection for a brand of secret science that serves the interests of defense contractors and hawkish politicians, but frays the fabric of democracy. There is perhaps no military project more representative of this legacy than ballistic missile defense (BMD). The basic idea behind BMD is simple enough—to protect against enemy missile attack, build a system that sends up interceptor missiles to shoot down incoming rockets in midflight. However, the reality is more complex. Not only does this task involve notoriously difficult "rocket science," but the margins for error in BMD physics are razor-thin, given the level of engineering precision demanded to "hit a bullet with a bullet," each traveling thousands of miles per hour.

Politically seductive but scientifically elusive, the notion of missile defense has given rise to waves of runaway rhetoric featuring technical claims that have outstripped the supporting scientific data. "Vast sums of money, a wall of secrecy that limits peer review, and strong political commitment to deployment" have created what international relations scholar Gary Guertner calls "a clear and present danger that [missile defense] scientists on both the inside and outside may become more interested in advocacy than in proof."[14] Guertner's concerns about the dangers of excessive secrecy and scientific fraud materialized dramatically in the late stages of the cold war, when senior officials and scientists in the Reagan administration turned to overclassification, deception, and intimidation to keep their pet BMD programs afloat. Although promises made since the fall of the Berlin

Wall have supposedly signaled a fresh start for U.S. BMD programs, deception schemes still continue to distort public perceptions of post–cold war BMD projects such as Patriot, Theater High Altitude Air Defense (THAAD), and National Missile Defense (NMD).

This spotty record of scientific integrity led the editors of *New Scientist* magazine to single out missile defense as an area of defense policy particularly deserving of close, searching scrutiny. "When looking for programmes to place under the microscope, the SDI should be at the very top of the list. No programme in the Pentagon has been surrounded with such political salesmanship and such pressures to produce wildly optimistic scenarios," they argue.[15] Critical examination of missile defense advocacy campaigns promises to shed light not only on the technical aspects of missile defense science, but also on the strategies used by missile defense advocates to harness the ethos of the scientific enterprise as they negotiate their way through the scientific, political, mythological, and even theological currents of public argument.

On the one hand, an intense sense of psychological vulnerability to enemy missile attack among civilians and military officials alike has swelled a significant reservoir of political support for missile defense. On the other hand, persistent engineering problems dogging BMD systems have frustrated efforts to translate this political will into a technically reliable missile defense program. The friction arising from this devilish double bind has worked to transform the missile defense issue into a lightning rod for controversy. In scientific quarters, expert disagreement over BMD feasibility has given rise to technical controversies that pivot around esoteric and speculative aspects of physics.[16] In policy-making circles, defense planners have clashed in controversies concerning the political desirability of BMD systems. The high degree of scientific uncertainty built into missile defense engineering and the substantial political stakes associated with BMD deployment are factors that have combined to charge BMD controversies with crackling argumentative energy.

A survey of the historical record of missile defense debates shows scientists stretching technical claims to serve political interests, and politicians stretching political claims to serve scientific interests. Sometimes, these persuasive strategies have unfolded in relatively obscure venues such as think-tank boardrooms and scientific laboratories. Other times, BMD discussions have spilled into public spheres, where wider audiences have picked up the currents of argumentation. As these dialogues have evolved, the strategies of advocates and reactions of audiences have determined the character, direction, and significance of the U.S. BMD program. In the process, BMD discussions have also served much wider precedent-setting purposes by shaping social norms about broader issues such as the authority of science in society, the role of the public in national security decision making, the power of the military as patron of the scientific enterprise, the meaning of scientific consensus, and the very definition of legitimate scientific inquiry.

My purpose in this study is to approach the recurrent BMD controversies from a rhetorical perspective that opens lucid sight lines into the technical and political dimensions of these disputes. This approach foregrounds analysis of scientific practice, locates the significance of such practice in public argument, and evaluates the normative dimensions of such arguments as they play out in extended controversies. The remainder of this chapter sets the stage for such analysis with a two-part discussion. Part one contains a historical account of the origins of U.S. BMD programs, while part two explains the critical approach this book takes to explore the entwinements of rhetoric, science, and politics in missile defense advocacy.

HISTORY OF U.S. BALLISTIC MISSILE DEFENSE

In the late stages of World War II, Germany unveiled the V-2 rocket, a new weapon that fundamentally altered the nature of strategic warfare.[17] Unlike relatively slow and bulky bombers carrying explosive payloads, the V-2 rocket could deliver devastating conventional warheads at speeds overwhelming enemy radar and interception capabilities.[18] Luckily for allied forces, World War II ended before the German V-2 rocket program had reached fruition. However, postwar discoveries of the ambitiousness of the German missile program demonstrated the massive potential lethality of this new ballistic missile technology.[19] Such bleak assessments were darkened further by the postwar realization that ballistic missiles could be outfitted with atomic bombs. According to historian Steven Guerrier, the prospect of nuclear-tipped intercontinental ballistic missiles (ICBMs) "changed everything."[20] Donald Baucom, another historian, points out that this frightening development "indicated clearly that the near-absolute security Americans had enjoyed during the war was becoming a thing of the past."[21]

DAWN OF THE U.S. BMD PROGRAM (1944–1966)

The realization that nuclear bombs could be delivered by rockets prompted the U.S. Army to initiate Project Thumper, the nation's first BMD research program, begun in 1944.[22] In less than a year, the air force took over this line of BMD research and merged it with its own Ground-to-Air Pilotless-Aircraft (GAPA) program, to become Project Wizard.[23] As the Wizard program progressed, it faced competition from the Nike BMD initiative, a new program launched by the army in early 1945. Nike grew out of a joint effort by Bell Telephone Laboratories and the Douglas Aircraft Company to develop long-range Surface-to-Air Missiles (SAMs).[24]

Between 1945 and 1958, the air force's Project Wizard and the various generations of the army's Nike program[25] vied for ascendancy as the nation's premier BMD system. Interservice rivalry, which would become a perennial feature of American BMD politics, flared up as the respective service branches pointed to a variety of events (e.g., the first Soviet atomic bomb test in 1949,[26] the outbreak of the Korean War in 1950,[27] the launch of *Sputnik I* in 1957[28]) to justify increased appropriations for their favored strategic defense programs.[29] In 1958, Secretary of Defense Neil H. McElroy quieted this rivalry by naming the army as the service branch in charge of the nation's fledgling BMD program.[30] However, despite this show of support, the army's third-generation Nike system (Nike-Zeus) continued to be plagued by technical difficulties, and its political backing eroded as several high-profile scientific reports emerged in 1958–59 that questioned the feasibility of the system.[31]

The 1960 election of President John F. Kennedy provided a new opportunity for the army to lobby for increased support for Nike-Zeus. Army advocates rode the crest of President Kennedy's "missile gap" election rhetoric in full-scale Nike-Zeus promotion campaigns launched in Congress and the media.[32] While these campaigns succeeded in generating increased enthusiasm for BMD in Congress,[33] this momentum was blunted by the efforts of new Secretary of Defense Robert McNamara, who emerged quickly as a formidable opponent of missile defense early deployment.[34] McNamara succeeded in holding off pressure for fast deployment of Nike-Zeus in the first three years of the new administration, proposing in its place an exploratory basic research program designed to search for long-term BMD solutions utilizing "exotic" technologies.[35]

By 1963, basic research conducted under the aegis of Project Defender had yielded technical advances in phased array radars, high acceleration rockets, and advanced data processing.[36] These technical improvements served as the basis for the fourth-generation Nike BMD system, "Nike-X." While McNamara continued to insist on "passive" civil defense measures as prerequisites to BMD, several countervailing factors swelled budget appropriations for Nike-X, including China's first detonation of an atomic bomb in October 1964 and Russia's deployment of the "Galosh" antiballistic missile (ABM) system[37] around Moscow in 1966.[38]

The "Great ABM Debates" and the Road to the ABM Treaty (1967–72)

In the late 1960s, BMD proponents began to point out that design features of the Nike-X system addressed many of the technical concerns raised earlier against Nike-Zeus (e.g., radar jamming, decoys, countermeasures). In light of these new technical advances, McNamara found that in order to stem the tide of support for quick BMD deployment, he had to rely much more on "political" arguments, namely the "action-reaction" thesis that U.S. BMD deployment would trigger a spiraling superpower arms race.

For students of BMD's political dynamics, the eventual insufficiency of McNamara's argumentative strategy in this regard is very instructive. McNamara found that BMD advocates such as the Joint Chiefs of Staff were not persuaded by his "action-reaction" hypothesis.[39] BMD advocates based worst-case assessments of enemy military capability on state-of-the-art American technology as a likely index of Soviet power and equated security with technological advancement. Use of this "mirror imaging" logic by BMD advocates to argue for hasty deployment of strategic defenses triggered what physicist Herbert York calls in his book *Race to Oblivion* an "internal arms race."[40] As historian Ernest Yanarella argues, this regressive and self-referential reasoning created an impetus for the U.S. BMD program that came more from "American R & D laboratories and engineering offices within the Defense Department" than from the external threat environment.[41]

Losing traction against this "technological imperative"[42] paradigm for national security, McNamara delivered what Yanarella identifies as "one of the most momentous speeches in the history of ABM" on 18 September 1967.[43] Addressing members of the San Francisco Press Club, McNamara "pleaded passionately against the feasibility, utility, and wisdom of an anti-Soviet" BMD system, alluding at one point to the "mad momentum" of the arms race.[44] However, acknowledging the depth of recently formed pro-BMD sentiment, McNamara ended his speech by calling for a limited, "thin" system of defense to be deployed primarily as a hedge against China's small new nuclear arsenal. Although this system was to be based primarily on existing Nike-X technology, McNamara attempted to assert a measure of rhetorical control over BMD dialogue by giving the system a new name, "Sentinel."

Although the Sentinel program achieved the desired short-term effect of blunting the army's zealous call for quick deployment of a "thick" BMD system,[45] its announcement brought a host of new voices into the debate about strategic defense. Prior to the "great ABM debates" of the 1960s, the level of public understanding of missile defense and the degree of participation in missile defense discussions was extremely low. "Throughout the 1950s and 1960s, the extent of public knowledge and interest in BMD was virtually nil. In fact, a poll taken in 1965 showed that fully two-thirds of the American people believed that the United States was already protected by an ABM system," Yanarella explains.[46]

After McNamara's speech, however, members of the general public became involved actively in discussions. Citizens in cities where the nuclear-armed Sentinel ABM batteries were slated to be built (primarily Seattle, Chicago, and Boston) organized vocal demonstrations that attracted national attention. In protests that would foreshadow the current era's "Not In My Backyard" (NIMBY) movements organized to fight hazardous waste dump siting, vocal members of these communities resisted the siting of Sentinel ABM batteries in their neighborhoods, because the interceptor missiles for the Sentinel system were equipped with nuclear warheads. "While the wider public quietly supported or remained indifferent to the administration's policy on ABM deployment," Yanarella notes, "a small but increasingly vocal segment of the public began to press its views on this issue into the political arena."[47] As *New York Times* writer Jack Finney describes, "[f]rom Boston to Honolulu, city councils, church groups, peace groups, conservationists, union leaders, real estate developers, and scientists banded together to protest the planned emplacement of nuclear-tipped missiles in the backyards of suburban America."[48] Additionally, numerous physicists were brought into the national limelight during high-profile congressional debates on Sentinel and Safeguard (President Nixon's scaled down version of BMD, introduced in 1970).[49] According to historian David Schwartz, "a broad spectrum of the U.S. scientific community now became embroiled in the controversy, and major figures carried on bitter debates before senators."[50]

Ultimately, Safeguard survived the great congressional debates by slim margins, and Nixon was able to utilize the newly approved system as a bargaining chip in arms control negotiations with the Soviet Union, which began in early 1972. On 26 May 1972 the SALT I Agreement was signed by Nixon and Soviet general secretary Leonid Brezhnev in Moscow. Part two of this agreement, "Limitation of Anti-Ballistic Missile Systems," became known commonly as the "ABM Treaty." The comprehensiveness of the ABM Treaty led commentators such as Gerard Smith to state that "[t]he Antiballistic Missile Treaty is the most fundamental arms agreement yet reached between the United States and the Soviet Union—a critical component of U.S. security policy."[51] Historian Douglas Lackey notes that after the ABM Treaty went into effect, "technicians on both sides were probably relieved."[52]

From the ABM Treaty to Star Wars (1972–83)

The narrow exceptions built into the language of the ABM Treaty[53] appeared to provide little room for vigorous pursuit of BMD deployment, and as Guerrier observes, "policy-makers showed little interest in ballistic missile defense in the decade following the ABM Treaty."[54] In political scientist Edward Reiss's account, after 1972, "BMD returned to the state of relative obscurity which it had enjoyed before the 'ABM debate' of the 1960s."[55] However, this lack of general interest did not translate into complete abandonment of the U.S. BMD program. A loophole in the ABM Treaty permitted each side to deploy two limited missile defense systems, one to protect the national capital and one to protect a selected ICBM field. Taking advantage of this loophole, the United States deployed a limited "Safeguard" system at Grand Forks Air Force Base in Nekoma, North Dakota. The system was completed in 1974 and it became fully operational on 1 October 1975.[56]

However, this instance of limited deployment did not reflect the broader priorities of the United States' post-ABM Treaty missile defense program. "In the wake of the ABM Treaty," explains Baucom, "the emphasis in the army's BMD program began to shift from developing deployable systems to research and development aimed at maintaining America's technological edge in the area of missile defenses to hedge against a possible Soviet breakthrough."[57] Although decreased appropriations reflected this new and relatively modest mission, advanced research in the areas of computing, optical sensors, kinetic-kill interceptors, and exotic space-based interceptors continued throughout the 1970s and into the early 1980s.[58] While earlier missile defense efforts were oriented primarily toward developing "off the shelf" technology for near-term deployment, post-ABM Treaty missile defense research was mainly exploratory in nature. As historians Gary Guertner and Donald Snow observe, "the effect of allowing R & D to continue was to permit the supporters of BMD to retire from the congressional and diplomatic battlefields bloodied but unbroken to the sanctity and privacy of their laboratories to prepare for a better day."[59]

The "better day" anticipated by Guertner and Snow turned out to be 23 March 1983, when President Ronald Reagan altered fundamentally the course of missile defense history with a dramatic televised speech to the nation. In what came to be known as the "Star Wars" address, Reagan outlined his vision of a futuristic missile defense system that would render nuclear weapons "impotent and obsolete."[60] Breaking with the tradition of missile defense advocacy that treated basic BMD research as an insurance hedge (e.g., Project Defender) or a pragmatic program to develop systems to defend limited ground areas (e.g., Nike, Safeguard, Sentinel), Reagan embraced the Strategic Defense Initiative (SDI), a much more ambitious vision of an impenetrable shield that would protect the entire United States from missile attack.

Reagan's rhetorical tack changed the dynamics of BMD discussions in a profound way. Previous missile defense debates (such as the struggle over the Safeguard system in the late 1960s) hinged on questions of whether science could be relied upon to shift the offense-defense equation marginally in favor of the defense. In the SDI era, the stakes were raised, as Reagan suggested that scientists could transform the very nature of strategic nuclear warfare by engineering an "Astrodome"-type leakproof defense. The rhetorical cornerstone of the Star Wars address was Reagan's claim that missile defense could transcend nuclear deterrence and emancipate the United States from the Mutual Assured Destruction (MAD) framework, which he felt was immoral. As Senator Jesse Helms (R-NC) noted on the floor of the U.S. Senate on the day after the Star Wars address, "[f]or the first time, a president of the United States has turned away from the incongruous doctrine known as MAD—Mutual Assured Destruction—and chosen to develop systems that can defend the American people."[61]

The seemingly impossible task of inventing a 100 percent reliable, space-based BMD system that could intercept all incoming enemy missiles did not deter SDI advocates from promoting the Star Wars vision; the daunting nature of this technical feat simply forced advocates to resort to the bluff as a routine argumentative strategy. Reagan administration officials found that by projecting the illusion of SDI feasibility, they could win budget increases for the program and achieve their geopolitical objectives. In the Star Wars era, official claims of missile defense effectiveness "were greatly magnified by the lack of serious science advice in the White House," as G. A. Greb explains.[62] However, as political scientist Erik Pratt observes, "this lack of input was not because of any system failure—it came about by design."[63] As one State Department official explained the typical bureaucratic reaction to statements about SDI, "We had gotten quite used to believing six impossible things before breakfast."[64] Recently declassified documents such as NSDD 172[65] and an April 1987 Joint Chiefs of Staff Memorandum[66] reveal that strategic deception was codified officially as a component of the SDI program at the highest level of government. "This misinformation altered, perhaps profoundly, the course of and expenditures for SDI," explains Kevin O'Neill of the Institute for Science and International Security.[67] The Star Wars address set in motion what the Union of Concerned Scientists' John Tirman calls an "entirely new and massive technocracy,"[68] a regime of scientific deception that legitimated itself by churning out technical misinformation overstating SDI's scientific feasibility. As Robert Hughes of the National War College observes, since Reagan's initiative was unveiled, "there has unfortunately sometimes been an almost 'rent-a-scientist' character to what has served for a public debate about the technical and technological merits of this SDI research."[69] Further discussion of this phenomenon of politicized science continues in chapter two, where consideration of strategic deception in the Star Wars case serves as the first case study in this book.

Critical Approach

Previous rhetorical studies of missile defense advocacy have focused primarily on textual analysis of Ronald Reagan's "Star Wars" address, and have tended to take government assessments of SDI's scientific validity at face value.[70] By deferring exploration of the scientific aspects of missile defense research, rhetorical critics have failed to treat adequately the prominent role of strategic deception in missile defense advocacy. The pages that follow sketch the outlines of a critical approach supporting an account of missile defense advocacy that places strategic deception at the center of attention and elucidates the interweaving dynamics of rhetoric, science, and politics. First, historiographical material on strategic deception in military contexts is consulted to develop context and theoretical insight useful for understanding institutionalized deception as a military strategy. An account of rhetoric is then given in part two that lays the foundation for criticism of key texts featured in episodes of public controversy over missile defense issues. Finally, the chapter's concluding remarks provide an overview of the three case studies to be examined in this book.

Strategic Deception

Writing in the context of military affairs, national security affairs scholar Michael Handel defines deception as "a purposeful attempt by the deceiver to manipulate the perceptions of the target's decision makers in order to gain a competitive advantage."[71] Sun Tzu's ancient maxim that "all warfare is based on deception"[72] is a reminder that military trickery has a legacy as old as the history of war itself. However, institutionalized strategic deception is a relatively new phenomenon. Prior to World War II, military deception was largely a tactical concern, as individual commanders were left to invent cunning battlefield maneuvers in an ad hoc fashion. "Since deception was not a systematically continued activity," explains Handel, "it required little or no co-ordination."[73] The Germans changed this trend by creating an official Disinformation Service in the 1920s, thereby institutionalizing an "official deception" program, designed to oversee and coordinate strategies of perception manipulation in military affairs.[74]

The prominence of strategic deception as a military instrument after World War II stimulated a burgeoning corpus of scholarly literature on the subject. Historians, international relations scholars, and political scientists generated a host of theoretical concepts and empirical generalizations helping to explain the dynamics of this new form of institutionalized lying. One important concept in this regard is Handel's distinction between "passive" deception (primarily based on secrecy and camouflage) and "active" deception (involving a calculated policy of disclosing half-truths).[75] The classic example of active strategic deception is Operation Barbarossa, an elaborate (and successful) Nazi strategy designed to mislead the Soviets into believing that they were safe from German aggression in 1941.[76] On the other hand, typical campaigns of passive strategic deception include those designed to hide military assets during arms control verification procedures.[77]

One old principle has emerged as a key to the success of the new deception strategies: "Tell them what they want to believe."[78] This Leninist slogan served as the basis of successful Soviet strategic deception campaigns waged against the United States during the cold war. "[T]he best Soviet strategic deception efforts are geared toward exploiting existing Western tendencies of thought. Principal among these is our tendency to engage in mirror-image perceptions, that is, where we tend to attribute to the Soviets such behavior as we would attribute to ourselves," explains former National Security Council member John Lenczowski.[79] One dramatic example of "mirror imaging" deception during the cold war is described by former Czechoslovakian disinformation officer Ladislav Bittman. In *The Deception Game,* Bittman describes how Soviet officials conspired with the "Bureau of Black Propaganda" in Prague to feed inflated assessments of Czech military forces to the West: "The general staff supplied Czechoslovak media with purposely distorted information on the Czechoslovak military, assuming that NATO analysts would pick it up."[80] This scheme used "mirror imaging" logic to play on Western perceptions regarding the reliability of local Czech media sources.

A similar case involved deception during Soviet military parades. For example, the Soviets mobilized every one of their Bison jet bombers at the 1955 Air Force Day ceremonies, and even though they "flew many of them by the reviewing stand twice," American intelligence agents were so primed to believe that the Soviets desired to massively build up their bomber forces, U.S. agents took the display "at face value," assessing "Soviet heavy bomber strength at four times the number of modern B-52 bombers then available to the U.S. Strategic Air Command," as historian Richards Heuer documents.[81] These examples of successful deception schemes are not aberrations. A quantitative study of strategic deception schemes conducted by military analysts Ronald Sherwin and Barton Whaley shows that most often, well-designed trickery works to manipulate audience perceptions about military issues.[82] However, there are notable exceptions of failure, including dramatic instances of "self-deception,"[83] "runaway deception,"[84] and other instances where deception has backfired on deceivers.

Normative discussions addressing the value of strategic deception often pivot on a distinction between wartime and peacetime deception. Thus, the generally accepted norm that strategic deception is a legitimate wartime strategy supports realpolitik justifications for virtually unlimited propaganda measures conducted in the heat of battle.[85] On the other hand, commentators such as Handel observe that "[i]n contrast to strategic and operational deception in war, deception in peacetime is quite likely to be counterproductive, even self-defeating."[86] In theory, this distinction appears to support a clear-cut basis for normative judgment regarding the appropriateness of strategic deception: "Deception of an enemy in wartime is certainly legitimate," proposes historian Trevor Barnes, but "[w]hen peace returns, all these doubtful means are supposed to be set aside."[87] However, as an odd mixture of war and peace, the cold war threw this bifurcated normative framework into disarray, and posed what Barnes calls "awkward ethical as well as practical problems for the West."[88] With case studies drawn from episodes of strategic deception pursued during the cold war, during the Persian Gulf War, and in peacetime, this book provides an opportunity to reflect on Barnes's observations in more detail. The following pages elucidate how consideration of such normative issues is possible from within a critical framework that highlights the interconnections between rhetoric, science, and politics in missile defense advocacy.

Rhetorical Criticism of Scientific Controversy

Trevor Melia points out that given the myriad ambiguities built into the terms "rhetoric" and "science," it would be prudent to sharpen these terms before turning to them as conceptual tools of analysis.[89] Melia's suggestion is particularly appropriate in the present context, because recently the very idea that rhetoric and science are intimately connected has spurred hotly contested and often esoteric disputations in the academic field of communication. Although participants in these scholarly disputes tend to share Aristotle's general belief that rhetoric involves "see[ing] the available means of persuasion in each case,"[90] they tend to part company over divergent interpretations of the proper scope of Aristotle's notion of rhetoric.

Specifically, so-called rhetoric of science critics such as Alan Gross hold that the art of rhetoric is used to invent persuasive proofs in all reaches of human affairs, including even science, that mode of inquiry traditionally thought to be territory off-limits for rhetoric.[91] In opposition to this view, Dilip Gaonkar argues that by locating rhetoric everywhere, Gross and others "globalize" Aristotle's conceptual framework, in the process stretching it so thin as to empty it of any meaningful content.[92] Rather than follow the path of scholars who have hacked bravely through the dense thickets of these academic arguments, I take a different critical route. Specifically, I focus on the particular genre of argument termed "deliberative rhetoric" by Aristotle.[93] For Aristotle, deliberative rhetoric is the art of inventing persuasive arguments in public debate over political affairs.

In Aristotle's day, the nascent tradition of scientific inquiry had not yet taken root in the popular imagination as an authoritative form of proof in deliberative argument. Instead, Aristotle explained that scientific explanations should remain confined to the province of formal education, with the contingent proofs of rhetoric (in the form of "topics," or "commonplaces") comprising the material for public argument in deliberative forums.

> [E]ven if we were to have the most exact knowledge, it would not be very easy for us in speaking to use it to persuade some audiences. Speech based on knowledge is teaching, but teaching is impossible [with some audiences]; rather, it is necessary for pisteis and speeches [as a whole] to be formed on the basis of common [beliefs], as we said in the Topics about communication with a crowd.[94]

To state a truism, things have changed mightily since the time that students studied science in Aristotle's Lyceum. Contemporary society's dependence on science as a source of useful technology has led to a blurring of science and technology (as well as basic and applied research) as distinct categories. What has emerged instead is what sociologist Bruno Latour calls the hybrid category of "technoscience."[95] Because the development and use of technology so often has direct political, social, and economic impacts that affect a wide variety of stakeholders, Thomas Farrell's observation that "many of the great controversies of our time are essentially public controversies about science—nuclear power, AIDS research, toxic waste disposal" should come as no surprise.[96]

In today's technologically complex world, "the discourse of science is now so much a part of civic discourse," as John Lyne explains, "that the study of it can no longer be seen as marginal to rhetorical studies."[97] This observation jibes with Philip Wander's suggestion that rhetoric of science scholars should prioritize "public policy deliberation" as a research site,[98] as well as with Steve Fuller's insight that the most fruitful area for science studies investigation is "not in the laboratory or the study, but in the policy forums where research is initially stimulated and ultimately evaluated."[99] These comments reflect the fact that in contemporary times, science itself has become a topos of argument in deliberative rhetoric. *Topos* is used here in the Aristotelian sense to mean a line of argument (or "topic" deployed commonly in the service of persuasion).[100]

The perceived epistemic (and political) authority of science as a deliberative topos in public argument is rooted ultimately in the concept of objectivity.[101] Advocates who can claim the mantle of scientific objectivity successfully gain the upper hand in public disputes, by virtue of their ability to exploit the ethos of scientific research and to tie their arguments to favorable cultural assumptions about scientific practice. This maneuver is accomplished by drawing upon the tradition of science as a practice that produces knowledge out of the "view from nowhere."[102] By following the rigorous rules of scientific method, it is suggested that any appropriately trained practitioner can attain the perspective of a detached, neutral observer and thus be afforded a direct cognitive window into the workings of nature.[103]

When rhetors advance truth claims and bolster the status of such claims by asserting that their statements deserve the status of scientific objectivity, they utilize the rhetoric of science as a topos of persuasion. "In our time, masters of the rhetoric of science command the most formidable rhetorical ethos. [T]heirs is a chaste rhetoric that pretends not to be rhetorical. . . . The intense conviction of testifying experts that they are not acting rhetorically casts an aura of authenticity over their performance," explain Michael McGee and John Lyne.[104] McGee and Lyne's observations tend to hold fast in high-stakes deliberative forums, since, as Chandra Mukerji observes, "the voice of science is authoritative to the extent that it seems objective and above politics even when applied to policy."[105]

Seen in this light, the rhetoric of science works enthymematically; it exploits the ethos of scientific research and development by appealing to commonly held ideas about the nature of science. There are a number of prevailing cultural presumptions that constitute a ready reserve of persuasive resources available to rhetors seeking to utilize the rhetoric of science in their public appeals. First, it is assumed widely that science is a neutral way of knowing, one that exhibits an "immunity to all kinds of local distorting factors like nationality, language, personal interest, and prejudice," as Theodore Porter explains.[106] Second, there is a common expectation that when it runs its course, scientific inquiry will yield progress and breakthroughs.[107] Third, scientific research is viewed generally as a cumulative enterprise, yielding a steadily growing knowledge base that is connected temporally in meaningful ways.[108] This notion reinforces the belief that scientific research progresses steadily over time, and it helps to reinforce the idea that scientific knowledge can advance in an incremental fashion.[109] Fourth, as the popular metaphor of "cutting edge" research suggests, the "advancement" of scientific knowledge is believed widely to push back an endless frontier, and in this sense, science is taken to be an open-ended enterprise.[110] Fifth, it is presumed by many that the process of scientific inquiry is accountable. The existence of peer review, the principle of the reproducibility of results, and the enforcement of scientific norms by professional organizations are taken by many to be adequate to ensure that renegade work will be ferreted out, assessed, and purged from the corpus of legitimately valid scientific knowledge. This presumption of accountability lends an automatic credibility to knowledge claims that have survived the scrutiny of science's system of institutional checks and balances.

As a topos of argument, rhetoric of science owes much of its efficacy to these prevailing cultural presumptions about the scientific enterprise. The fact that as sociologist Sheila Jasanoff explains, "almost anyone who claims to speak with the voice of science in the United States can find a place at the national policy table"[111] is a testament to the strength of these presumptions. Because science is viewed widely as neutral, progressive, cumulative, pregnant with infinite possibility, and accountable, arguments couched in the rhetoric of science have considerable persuasive purchase in public spheres of deliberation.

Overlaying this set of general cultural presumptions about science writ large is an additional series of beliefs about military science in particular that bolster attempts by military advocates to enlist the rhetoric of science as a persuasive tool in public dialogue. Many of these beliefs are entrenched due to popular celebration of the Manhattan Project. Early accounts of the Manhattan Project reified atomic scientists as "masters of the universe" and described the successful engineering of controlled nuclear fission as the "myth of second creation."[112] These celebrations endowed military practitioners with special latitude and authority to invoke the topos of science in public argumentation.[113]

When truth claims packaged in the parlance of scientific objectivity meet resistance in deliberative forums, and open public debate ensues, the stage is set for public controversy. In this situation, argumentation can take a variety of forms and can occur in a multitude of different forums. In strictly scientific forums such as laboratories and conferences, competing experts voice disagreements and debate topics such as experimental test design, interpretation of data, and research methods.[114] In deliberative forums such as congressional hearings and popular journals, a broader array of advocates struggle over questions bearing on the proper status and purpose of the scientific enterprise itself: Is a particular piece of research "legitimate" science? Is a particular social problem appropriate for resolution by scientific means? Should a certain branch of basic scientific research be supported with public funds?

Because the development and use of technology so often has direct political, social, and economic impacts that affect a wide variety of stakeholders, it is no surprise that many kinds of scientific controversies center on issues relating to the application and use of technology. However, other controversies have origins in disagreements that crop up in the murky interstices separating the spheres of basic research and applied technology. Just as grass pushes its way up through cracks in sidewalk concrete, scientific controversies tend to sprout in the points of cleavage and uncertainty that are scattered throughout the science / technology continuum. The ties that bind basic science, applied technology, and politics together cannot be established with formulaic precision. For example, indeterminacy in the experimental process often renders extrapolation of basic research results to the realm of technological engineering problematic. Likewise, the complexity of technological change often makes it difficult to anticipate the political significance that given engineering advances might hold for society. The realms of science, technology, and politics are linked together loosely, and the inherent slack in such connections provides opportunities for advocates to dispute the relevance, reliability, and meaning of scientific truth claims, particularly when issues of power, equity, and ethics are at stake.

With the foundational norms of science at stake in argumentation, scientific controversies become sites where prevailing opinions about the proper characteristics and purpose of the scientific enterprise can be aired, contested, and revised. In Farrell's formulation, rhetoric works here as a "generator" of "social knowledge."[115] The generative function of rhetorical practice ensures that the results of these types of debates establish social precedents for future controversies and, in the process, come to establish the nature of science as a social activity, as well as the accepted role of science in society.[116] H. Tristram Engelhardt Jr. and Arthur L. Caplan explain in the preface of their book-length anthology, *Scientific Controversies,* "[a] better understanding of the interplay between knowledge-directed and value- or politically directed forces in scientific disputes is required for an appreciation of the very character of culture."[117] Efforts to pursue such studies from a rhetorical horizon should be informed by knowledge of the pitfalls involved in treating science as nothing but rhetoric (Gaonkar's critique), as well the dangers inherent in overlooking rhetorical aspects involved in scientific controversies (Alan Gross's reminder). Located between these two critical cues is an approach that treats scientific facts as both objective and situated, having material essences and meanings constituted by social symbols. "It is possible," explains Celeste Condit, "to see facts as both objective and situated—both faithful to material realities and responsive to social conditions."[118]

Effective analysis in this regard requires getting to know scientific practice on its own terms, developing a working knowledge of what Latour calls "science in action."[119] Skeptics Paul Gross and Norman Levitt assert that such working knowledge can be acquired only by certified practitioners trained in the techniques of the natural sciences. "A serious investigation of the interplay of cultural and social factors with the workings of scientific research in a given field is an enterprise that requires patience, subtlety, erudition, and a knowledge of human nature. . . . A scholar devoted to a project of this kind must be, inter alia, a scientist of professional competence, or nearly so," they explain.[120] Paul Gross and Levitt's pronouncement on this point overlooks an impressive corpus of work produced by science studies scholars not certified professionally in technical fields, yet still capable of producing valuable commentary on scientific questions. These scholars have been able to overcome their lack of technical training, collaborating with learned scientists to generate accurate and often poignant insights regarding the role of science in society, as well as the significance of social factors as drivers of scientific practice. The result of such collaboration is often fine-grained, "patient," "subtle," and "erudite" work that evinces a level of understanding clearly reflective of "professional competence, or nearly so."[121] Outfitted with a working understanding of the origins and dynamics of scientific facts (one that includes a firm grasp on this dimension of materiality), rhetorical critics can interpret the ways in which such facts are utilized for persuasive ends, by examining public texts that cite the facts in support of political claims. This involves explaining how the interplay between fact and claim enhances the audience appeal of the argumentation under consideration, thus unpacking the persuasive work that is performed when the fact is included as a constituent of the public argument. When the overall argument is challenged in public debate, controversy erupts, and additional interpretive vistas open up for the rhetorical critic.

Configuring the critical act as an interpretation of controversy, scholars can then locate the stases of the dispute,[122] clarify the interests at stake by situating the disagreement in the wider sociocultural context, and identify which norms are being tested as the controversy unfolds. In controversies where the locus of dispute rests on what Charles Taylor calls "demarcation" issues, that is, "the relationship between scientific practice and rhetorical efforts at defining what is and what is not science,"[123] the broader significance of the controversy as constitutive of the definition of "science" itself is often expressed clearly on the surface of the controversy's textual artifacts.[124] For Trevor Pinch, controversies thus represent profound moments of opportunity for scholars to gain valuable knowledge about the nature of science, as previously submerged aspects of scientific practice come into view.

> It has been argued persuasively that scientific controversies form a strategic research site for studying science. During a controversy, social processes not normally visible within science can become unusually explicit. What counts as a repeatable experiment, the relationship between theory and experiment, the types of scientists who can legitimately contribute to the production of scientific knowledge, what counts as bias, impropriety and breaches of the scientific method, and the role of the media are just a few of the matters which are given concerted attention during a controversy. Under the lens of a scientific controversy, the good, the bad and the ugly within science come into focus as never before.[125]

Controversies not only serve as sites where the character of "science" as an institutional enterprise is revised and updated; broader social norms are also often at stake in such disputes. For example, in their reading of the controversy surrounding the Three Mile Island nuclear accident, Thomas Goodnight and Thomas Farrell show how the rhetoric of nuclear power experts, politicians, and members of the media reproduced "visions of the public" that constructed citizens as helpless spectators, vulnerable to manipulation by scientific experts in cyclical patterns of technical breakdown and catastrophe.[126] Goodnight and Farrell add normative purchase to their criticism by elucidating how the "accidental rhetoric" of nuclear power experts worked to reproduce patterns of public dialogue that erode possibilities for democratic control of high-risk technologies.

My approach in this book follows a similar path. In three case studies, I ascertain characteristics of strategic deception in missile defense advocacy, then unpack the consequences of strategic deception as an institutionally codified practice. To do so, I draw upon my expertise as a rhetorical critic in order to examine public speeches, media coverage, expert commentary, scientific reports, internal memoranda, government documents, popular films, novels, and interview transcripts that relate to distinct but connected episodes of missile defense advocacy featured in the case studies. This analysis will set up judgments based on normative assessments of consequences flowing from strategic deception. Some consequences include the way missile defense advocates' communicative practices affect the integrity of their own validity claims and impact the quality of scientific research in missile defense research and engineering projects. Since scientific discourse assumes that its technical claims will be tested in the crucible of debate, those arguments claiming that sanction, while strategically subverting the possibility of debate itself, are especially open to criticism in this regard. Other consequences involve issues of necessity and expediency in terms of wartime operations or peacetime deterrence. Spinning out even further are attendant consequences of strategic deception as the horizon of normative assessment expands. These consequences include systematic foreclosure of democratic deliberation in public spheres of discussion, victimization of vulnerable research subject populations, massive financial waste, vitiation of the credibility of science as an enterprise, and ripple effects of secrecy on the legitimacy of governmental institutions.

The three case studies featured in this book examine debates over the SDI, Patriot, and Theater High Altitude Air Defense (THAAD) missile defense systems. By looking at episodes of controversy spanning a seventeen-year period straddling the end of the cold war, I open up critical spaces for insight to be generated about strategic deception as a general phenomenon, as well as for understanding to be pursued about how deception works idiosyncratically in particular instances. The case studies chosen for scrutiny are not only richly textured controversies that support multiple interlocking levels of rhetorical analysis; they also represent important moments in the evolutionary history of missile defense argumentation. For example, in the disputes over the effectiveness of the Star Wars and Patriot missile defense systems, far more than questions of technical feasibility were at stake. Ground rules for discussion, future expectations of missile defense performance, the role of public input, the proper scope of classification, and the very reputation of science itself were implicated in arguments that ostensibly covered such arcane topics as laser brightness and post-war ground damage assessments.

My findings are offered not only as critical interpretations based on rhetorical readings of key texts; they are also deliberately positioned as interventions in present-day missile defense controversies yet to be resolved. This double gesture draws on the polyvalent nature of the rhetorical tradition as a simultaneously interpretive and productive art.[127] "In guiding the creation of discourses, and not just the interpretation of existing texts," Lyne explains that "the work of rhetoric is to invent language strategies that bring about change."[128] In the context of scientific criticism, Condit explains that a "broad" rhetoric of science approach invites critics to move beyond the (interpretive) role of rhetorical critic and embrace the (productive) role of transformative rhetor.[129] This sort of interventionist stance is facilitated when scholarly work is anchored in the investigation of public controversies. According to Thomas Brante, "[a]nother argument for studying modern controversies is that [scholars] . . . ought to confront problems carrying profound social and political relevance. The study of controversies in and around science provides a lever by which researchers in these fields may bootstrap themselves out of their academic ivory towers."[130] Brian Martin extends this point further to suggest that the ideals of researcher neutrality and detachment are called into question by the very choice to study public controversies. "Separation of the researcher and the researched may work in some cases, but practical experiences show that it often cannot be sustained in dealing with contemporary controversies with strong public involvement. To some extent, the social researcher is inevitably involved in the controversy being studied," he argues.[131]

While this manuscript was being prepared, my modes of critical engagement alternated between reflective stints involving research and interpretation of texts, on the one hand, and active interventions geared toward production of public arguments, on the other.[132] These public arguments amplified the basic findings of my research to multiple audiences, as such research findings emerged and stabilized. Specifically, I pursued a method of intervention that emphasized direct engagement with parties to the controversies under investigation, creation of networks linking my work to other actors, public amplification of my critical commentary beyond the specialized audience of my disciplinary peers, and concerted efforts to transform the dialogic and policy frameworks of public discussion in constructive ways. To reach professional audiences interested primarily in the technical and policy aspects of strategic deception, I published material with the *Bulletin of the Atomic Scientists* and the Federation of American Scientists Space Policy Project, while also filing briefs with congressional staff investigating missile defense research programs.[133] On an international level, I published briefing papers with the London-based International Security Information Service (ISIS-U.K.), and the International Security Information Service in Brussels, Belgium (ISIS-Europe).[134] Packaging my findings differently for academic audiences more interested in the rhetorical implications of strategic deception, I published articles in the *Quarterly Journal of Speech* and *Science, Technology, and Human Values.*[135] With co-author Anthony Todero, I presented a paper analyzing Japan-U.S. Theater Missile Defense co-operation at the First Tokyo Argumentation Conference in Tokyo, Japan.[136] Finally, as a strategy of intervention designed to convey my ideas to popular audiences, I published several op-ed pieces, including a collaborative effort with the Cherub Study Group, a research collective made up of high school debaters.[137]

Although the general tone of my public texts evinces a partisan anti–missile defense orientation, the broader telos underwriting my interventions is a vision of open and freewheeling public discussion. I used the opportunity to cycle between acts of interpretation and production to refine the critical perspective underwriting this telos. According to Martin, just as normative judgments reached during the interpretation of texts provide the bases for active intervention, intervention can in turn yield insight that helps critics revise earlier interpretive or normative judgments. "Generally speaking, action researchers see the process of gaining knowledge and changing society as interlinked, even inseparable. Intervention to change society produces understanding—including new perspectives of fundamental theoretical significance—which in turn can be used to develop more effective intervention," he explains.[138] Steve Woolgar characterizes this synergistic interplay among dimensions of inquiry as the "dynamic of iterative reconceptualization," a process whereby "practitioners in social studies of science from time to time recognize the defects of their position as an occasion for revising its basic assumptions."[139]

Conclusion

After Leonard Lewin disclosed that he was the author of the fictitious *Report From Iron Mountain,* commentator Irving Horowitz wrote that publication of this devilish parody of a "secret and suppressed" government report "should become an occasion for a new public demand for a penetrating examination and evaluation of Government reports on strategic planning for disarmament and peace."[140] One significant consequence of the dominance of the "technological imperative" paradigm of national security during the early days of U.S. missile defense was the fact that members of the public were mute in the vast majority of discussions regarding the proper nature, pace, and direction of the BMD program, and that the critical scrutiny called for by Horowitz was notably absent. With a few rare exceptions (e.g., the NIMBY protests of the Sentinel system), the authority to shape the course of U.S. missile defense rested with elite military planners. "To the degree that the issues of defense decision-making were construed as technical problems requiring information and instrumental strategies produced by scientific experts," explains Yanarella, "they were removed from political debate without being politically resolved."[141] This historical perspective reveals that there was more at stake in early missile defense debates than strategic nuclear stability; the very notion of a democratic politics of science was challenged by the primacy of operational systems analysis as a mechanism to steer the incipient missile defense programs. This tension between systems logic and democratic decision making remains a key element of today's BMD debates. This book offers analysis of contemporary episodes of missile defense advocacy highlighting many of the normative stakes implicated by this tension, a tension that has existed ever since the genesis of Project Thumper, the nation's first BMD program, in 1944.

Chapter two explores the experimentation surrounding the X-ray laser, since this laser was ultimately touted as the technological foundation for an "Astrodome" defense as outlined in Reagan's 1983 "Star Wars" speech. By tracking post–"Star Wars" address correspondence, such as letters and memoranda by Edward Teller's Livermore staff to policy makers in the executive and legislative branches, a more complete picture of the supporting network that buoyed the fortunes of Star Wars during much of the decade will come into focus.

Chapter three considers the controversy over Patriot missile accuracy in the Persian Gulf War. President Bush delivered a nationwide address from Raytheon, Inc., headquarters during the height of the Persian Gulf War, claiming that at that point in the war, Patriot was "41 for 42" in the "Scudbusting" department. Controversy erupted in 1992, however, when MIT physicist Theodore Postol published an article in the journal *International Security* alleging that no reliable data existed to establish Patriot's effectiveness. Because the ensuing scientific controversy prominently featured argumentation regarding the quality, status, and proper role of classified information in public debate about missile defense, the Patriot case is particularly appropriate for a study of strategic deception in missile defense advocacy.

Chapter four turns to the controversy over THAAD, the Clinton administration's favored missile defense system. This controversy involves a debate over the issue of whether the THAAD system violates the ABM Treaty.

In the final chapter, I summarize the study's relevance to contemporary missile defense debates by distilling perennial patterns of argumentation found in the case studies. This involves tracing the common threads that run through the X-ray laser, Homing Overlay Experiment (HOE), Patriot, and THAAD promotion campaigns, assessing the ways in which these threads have evolved over time, and speculating about what future developments seem likely, given the current trajectory of events.

NOTES

1. Leonard C. Lewin, *Report From Iron Mountain: On the Possibility and Desirability of Peace* (New York: Free Press, 1996), 1. The title of the report was taken from the name of a large nuclear fallout shelter in Hudson, New York, built to store the absolutely vital documents of the nation's biggest corporations in the event of a nuclear attack. Apparently, the Special Study Group met in the inner sanctum of this facility from 1963 to 1967 to produce *Report From Iron Mountain* (see Lewin, *Report,* 1–8).
2. John Leo, "'Report' on Peace Gets Mixed Views: Some See Book as Hoax, Others Take It Seriously," *New York Times,* 5 November 1967, 1.
3. Lewin, *Report,* 84.
4. Ibid., 100.
5. Ibid., 77–78.
6. For reprints of some of the key commentaries from 1967–1972, see Kenneth Boulding, ed., *Peace and the War Industry* (New Brunswick, N.J.: Transaction Press, 1973), 55–83.
7. Leonard C. Lewin, "The Guest Word," *New York Times,* 19 March 1972, 47.
8. Marc Pilisuk, "Comment," in *Peace and the War Industry,* ed. Kenneth Boulding (New Brunswick, N.J.: Transaction Press, 1973), 68.
9. Kenneth Boulding, "Comment," in *Peace and the War Industry,* ed. Kenneth Boulding (New Brunswick, N.J.: Transaction Press, 1973), 74.
10. Felix Kessler, "Who Wrote It? A Fad in Political Comment Is Using Pseudonyms," *Wall Street Journal,* 13 November 1967, 1.
11. For examples, see Victor Navasky, introduction to *Report From Iron Mountain,* ed. Leonard C. Lewin (New York: Free Press, 1996), v–xvi.
12. Pilisuk, "Comment," 70–71.
13. Dwight Eisenhower, "Farewell Address to the Nation," reprinted in *Bureaucratic Politics and National Security,* ed. David C. Kozak and James M. Keagle (Boulder, Colo.: Lynne Rienner, 1988), 278–82. Indeed, this was the logic that swayed Lewin's fictitious fifteen-member Special Study Group to recommend that their study be hidden from the public. In their letter of transmittal to the Kennedy administration, the "authors" of the *Report From Iron Mountain* explained their justification for secrecy: "Because of the unusual circumstances surrounding the establishment of this Group, and in view of the nature of its findings, we do not recommend that this Report be released for publication. It is our affirmative judgment that such action would not be in the public interest. The uncertain advantages of public discussion of our conclusions and recommendations are, in our opinion, greatly outweighed by the clear and predictable danger of a crisis in public confidence which untimely publication of this Report might be expected to provoke. The likelihood that a lay reader, unexposed to the exigencies of higher political or military responsibility, will misconstrue the purpose of this project, and the intent of its Participants, seems obvious" (Lewin, *Report,* 26).
14. Gary Guertner, "What is Proof?" in *The Search for Security in Space,* ed. Kenneth E. Luongo and W. Thomas Wander (Ithaca, N.Y.: Cornell University Press, 1989), 191–92.

15. "Defending the Indefensible: U.S. Defense Spending," *New Scientist,* 28 March 1992, 11.
16. Robert Hughes of the National War College explains that such debates crossed the Atlantic Ocean and spurred scientific disputes in Europe as well. "Like the exchanges in the United States between Hans Bethe and Richard Garvin [*sic*] of the Union of Concerned Scientists and SDI proponents such as General Abrahamson, Robert Jastrow, and James Fletcher, in Germany Hans Ruehle (head of the planning division of the Ministry of Defense) squared off in the popular magazine *Der Spiegel* in 1985 against Hans-Peter Duerr (director of the Max-Planck-Institute for Physics and Astrophysics), who had written the 'credo' of German scientists against SDI" (Robert C. Hughes, *SDI: A View From Europe* [Washington, D.C.: National Defense University Press, 1990], 109). For additional commentary on SDI from a European perspective, see Jonathan Alford, ed., *Arms Control and European Security* (Hampshire, England: Gower/ International Institute for Strategic Studies, 1984); Ian M. Cuthbertson, *The Anti-Tactical Ballistic Missile Issue and European Security* (New York: Institute for East-West Security Studies, 1990); and Lawrence Freedman, "Strategic Defence in the Nuclear Age," Adelphi Paper 224 (Oxford: Oxford University Press/International Institute for Strategic Studies, 1987). On SDI's implications for Asian-Pacific security, see Michael J. Mazarr, *Missile Defenses and Asian-Pacific Security* (London: Macmillan Press, 1989).
17. See Charles Donnelly, Report, in *United States Guided Missile Programs,* 86th Cong., 1st sess., Senate Hearing, Subcommittee on Preparedness Investigation of the Committee on Armed Services, 1959 (Washington, D.C.: GPO, 1959), 1–9.
18. Unprepared for the contingency of a rocket attack, Britain was unable to respond to such aggression before the rockets had detonated on the ground. As Baucom describes, "British scientists knew that England had been struck for the first time by German V-2 ballistic missiles traveling so fast that the sound of their approach was not heard until after their warheads had exploded" (Donald R. Baucom, *The Origins of SDI, 1944–1983* [Lawrence: University Press of Kansas, 1992], 3).
19. After the war, analyses of the German missile program revealed that the Nazis had been developing a two-stage rocket (the V-10). With an initial booster providing up to two hundred tons of thrust, the V-10 could have accelerated to a velocity of 3,360 miles per hour. "Had the war continued into 1946, the Germans might well have made good their plans to bombard New York City" (Baucom, *Origins of SDI,* 4).
20. Stephen W. Guerrier and Wayne C. Thompson, *Perspectives on Strategic Defense* (Boulder, Colo.: Westview, 1987), 7. While strategic defense deployed against conventional bombers needs to achieve only roughly 10–15 percent effectiveness to shift the offense-defense cost equation to the defense's favor, even a 90 percent effective defense against nuclear-armed ICBMs is unlikely to be acceptable to the defending side, given the enormous destructive force of even a single atomic explosion (see Guerrier and Thompson, *Perspectives,* 7; George Rathjens and Jack Ruina, "BMD and Strategic Instability," *Daedalus* 114 [1985]: 239–44).
21. Baucom, *Origins of SDI,* 4.

22. Ernest J. Yanarella, *The Missile Defense Controversy: Strategy, Technology, and Politics 1955–1972* (Lexington: University of Kentucky Press, 1977), 32–33. Project Thumper involved basic research of high altitude air defense that could be used against enemy bombers and V-2 generation rockets (see Guerrier and Thompson, *Perspectives,* 10). Inspiration for the program was drawn from the stunning success of Britain's air defense system in the closing stages of World War II. In the early 1930s, when Germany undertook an aggressive campaign to build up its strategic bombing forces prior to World War II, British military planners faced a stark choice. With the "indelible mark" of German World War I bombing still fresh in their minds, Britons knew that it was vital to counter Germany's buildup, but they were split over the proper response. Some military officials favored matching Germany's buildup by rushing more British bombers into production to ensure that the Royal Air Force would be able to deliver devastating retaliatory strikes against the German homeland in any future conflict. In contrast to this offense-oriented approach, other British planners (including Winston Churchill) urged pursuit of a defensive strategy that aimed to intercept and neutralize German bombers as they crossed into British airspace.

The technical aspects of this program included vigorous research and development of advanced technologies in "early warning, detection, and threat identification; kill mechanisms; command, control, and communication to tie the system together," as well as strategic planning in "the selection of targets for defending; and operational and organizational concepts to manage the system" (Benson Adams, "An Early SDI That Saved Britain," *Naval War College Review* 38 [November/December 1985]: 50–58). This latter option of strategic defense was eventually selected by British military planners. At the behest of Churchill and other strategic defense advocates, the British parliament formed the Tizard Committee for the Scientific Survey of Air Defense in 1934. "From this committee's work emerged the possibility of using radio waves for aircraft detection (radar), and from that the possibility of building an air defense system that could work. As a result . . . the British government decided in 1936–1937 to allocate more of its limited resources to fighter production, at the cost of additional bomber production, in order to build a system of air defenses" (Adams, "Early SDI," 13–14). This program has been widely credited by historians to be the basis of England's triumph in the 1940 "Battle of Britain" (see Guerrier and Thompson, *Perspectives,* 6; Herbert F. York, "Nuclear Deterrence and the Military Uses of Space," *Daedalus* 114 [1985]: 30). Hitler's air force triggered the Battle of Britain when a renegade German bomber crew bombed London in August 1940. Interpreting this as the prelude to a widespread German campaign to bomb British cities, Churchill ordered retaliatory strikes against Berlin, and this in turn prompted Hitler to direct a large portion of the Luftwaffe to bomb London in broad daylight (Adams, "An Early SDI," 16). However, the newly developed British air defense radar facilitated interception of approximately 10 percent of the incoming German planes, and Hitler abandoned the campaign (Guerrier and Thompson, *Perspectives,* 6).

It should come as no surprise that contemporary advocates of ballistic missile defense (BMD) such as Newt Gingrich cite the Battle of Britain as a historical precedent demonstrating the prudence and effectiveness of strategic defense as a military option (see Newt Gingrich, statement, *Congressional Record,* 17 March 1997, H1023–H1030; Joseph Cirincione, "The Persistence of the Missile Defense Illusion," paper presented at the Conference on Nuclear Disarmament, Safe Disposal of Nuclear Materials or New Weapons Development, Como, Italy, 2–4 July 1998, Internet. Carnegie Endowment for International Peace Website, Online at http://www.ceip.org/people/cirincio.htm). Churchill's 1934 call to "acquire some means or methods of destroying sky marauders" and Britain's subsequent success in developing and deploying radar to defeat the German Luftwaffe stand as powerful examples of prudent foresight and prescient military planning (see Winston Churchill, Speech from the House of Commons Debate, "The Need for Air Defense Research," 28 November 1934, reprinted in *Promise or Peril: The Strategic Defense Initiative,* ed. Zbigniew Brzezinski, [Washington, D.C.: Ethics and Public Policy Center, 1986], 5–7).

23. Guerrier and Thompson, *Perspectives,* 10; see also Baucom, *Origins of SDI,* 6.
24. Yanarella, *Missile Defense Controversy,* 27–28.
25. The early Nike system (Nike-Ajax) was a two-stage missile, with a range of twenty-five to thirty miles (see Tom Gervasi, *Arsenal of Democracy II* [New York: Grove Press, 1981], 224). Fully operational in 1953, Nike-Ajax batteries eventually were deployed around some thirty potential target areas in the United States (Guerrier and Thompson, *Perspectives,* 11). The second generation Nike system (Nike-Hercules) was created in response to reports in the early 1950s that showed a Soviet buildup of strategic bombers. Operational in 1958, Nike-Hercules provided a new element of defense in that its antiballistic missiles could carry nuclear warheads (Guerrier and Thompson, *Perspectives,* 11). Nike's third generation system (Nike-Zeus) was a three-staged, solid-propellant missile also capable of carrying nuclear warheads (see Ruth Currie-McDaniel, *The U.S. Army Strategic Defense Command: Its History and Role in The Strategic Defense Initiative* [Huntsville, Ala.: U.S. Army Strategic Defense Command, 1987], 1–2; John G. Zierdt, "Nike-Zeus: Our Developing Missile Killer," *Army Information Digest* 15 [December 1960]: 5–6). The Nike-Zeus system was unique in that it included advanced radar equipment and communications links to "tie the subsystems together" (Baucom, *Origins of SDI,* 7); it became "the forerunner to virtually all U.S. BMD systems" (David N. Schwartz, "Past and Present: The Historical Legacy," in *Ballistic Missile Defense,* ed. Ashton B. Carter and David N. Schwartz [Washington, D.C.: Brookings, 1984], 332).
26. Following the USSR's first successful test explosion in the fall of 1949, General Hoyt S. Vandenberg, Air Force chief of staff, warned that "almost any number of Soviet bombers could cross our borders and fly to most targets in the United States without a shot being fired at them" (quoted in Guerrier and Thompson, *Perspectives,* 13). Shortly thereafter, the United States "began to improve air defenses as a hedge against one-way missions by Soviet medium-range bombers" (United States Office of Technology Assessment, *Strategic Defenses: Ballistic Missile Defense Technologies, Anti-Satellite Weapons, Countermeasures, and Arms Control* [Princeton, N.J.: Princeton University Press, 1986], 27).

27. NSC-68 was promulgated in April 1950. This new policy, which emphasized nuclear war as a legitimate military option, led to dramatic expansion of conventional and nuclear forces, including increased expenditures on strategic defense (see Paul Y. Hammond, "NSC-68: Prologue to Rearmament," in *Strategy, Politics, and Defense Budgets*, ed. Warner R. Schilling, Paul Y. Hammond, and Glenn H. Snyder [New York: Columbia University Press, 1962], 267–78; John Lewis Gaddis, *Strategies of Containment: A Critical Appraisal of Postwar American National Security Policy* [New York: Oxford University Press, 1982], 89–126; and Jonathan B. Stein, *From H-Bomb to Star Wars* [Lexington, Mass.: Lexington Books, 1984], 40–42).
28. Five days after the launch of the USSR's first space satellite (*Sputnik I*), Air Force lieutenant general Donald L. Putt submitted a report to the Air Force chief of staff, General Thomas D. White, urging him to make space-based BMD efforts a high priority (see Thomas A. Sturm, *The USAF Scientific Advisory Board: Its First Twenty Years, 1944–1964*, reprint of 1967 edition [Washington, D.C.: GPO, 1986], 81–84). "The launch of Sputnik prompted the U.S. government to undertake a thorough, and often frantic review of the state of 'high technology' in America" (York, "Nuclear Deterrence," 19–20; see also Schwartz, "Past and Present," 332; and Paul B. Stares, *Space and National Security* [Washington, D.C.: Brookings, 1987], 132).
29. These various stimuli for political support of missile defense received reinforcement from startling commentary coming from nuclear strategists, who laid bare many of the frightening dilemmas of nuclear deterrence. For example, Thomas Schelling raised the possibility that deterrence might reward irrational thinking on nuclear strategy (see Thomas Schelling, *The Strategy of Conflict* [Cambridge, Mass.: Harvard University Press, 1960]). In a similar line of reasoning, Herman Kahn argued paradoxically in his classic book *On Thermonuclear War* that in order to bolster the credibility of deterrence and decrease the possibility of a nuclear exchange, nuclear warfare should be treated as a "thinkable" option. As a way out of these unsavory strategic dilemmas, Kahn suggested that advances in computer software technology might yield a radar system that could detect and track incoming enemy missiles (Herman Kahn, *On Thermonuclear War* [Princeton, N.J.: Princeton University Press, 1960], 495). Kahn's speculations hinted that such a development might "revolutionize the offense/defense balance [in nuclear strategy] by pushing us towards a situation in which defensive technology might have the edge" (John Baylis and John Garnett, *Makers of Nuclear Strategy* [London: Pinter Publishers, 1991], 79). Invoking logic that would become a common line of argument in missile defense advocacy, Albert Wohlstetter argued that antiballistic missile's favorable "cost-exchange ratio" was enough to warrant support for missile defense, since U.S. pursuit of such a system would force the U.S.S.R. to expend precious resources in response (see Albert Wohlstetter, "The Delicate Balance of Terror," *Foreign Affairs* 37 [1959]: 211–34).
30. Neil H. McElroy, *Program for Defense against the Intercontinental Ballistic Missile*, memorandum to the secretary of the Air Force, in *Investigation of National Defense Missiles: Hearings Before the Committee Pursuant to H. Res. 67*, 85th Cong., 2d sess.,

House Hearings, Committee on Armed Services [Washington, D.C.: GPO, 1958], 4196–97. Although the McElroy memorandum assimilated most of the air force's Wizard program under the aegis of the Army's Nike-Zeus project, there were several aspects of Wizard (e.g., radar and command and control electronics) that were permitted to continue in a support role for Nike-Zeus (Baucom, *Origins of SDI,* 11).

31. The Department of Defense's Reentry Body Identification Group released findings in the spring of 1958 that Nike-Zeus's radar was vulnerable to enemy jamming, that it could not discriminate effectively between real targets and decoys, and that the system could easily be overwhelmed by the newly created MIRV technology (see Fred Kaplan, *Wizards of Armageddon* [New York: Simon and Shuster, 1983], 343–45; Alexander Flax, "Ballistic Missile Defense: Concepts and History," *Daedalus* 114 [1985], 34–37). Similar conclusions were reached in early 1959 by a panel of the President's Science Advisory Committee, which "included some of the nation's top scientific experts" (Baucom, *Origins of SDI,* 21).
32. See Guerrier and Thompson, *Perspectives,* 19–20.
33. See Yanarella, *Missile Defense Controversy,* 60–63. Yanarella's discussion on this point frames the army's Nike-Zeus promotion campaign as a typical example of the military-industrial complex's cultivation and orchestration of synergistic connections among politicians, journalists, and military officials to win increased budget appropriations for defense. For further specific evidence of the army's strategy to curry media favor by granting "nonpolitical" guided tours of the Kwajalein Island missile test center to some twenty-nine journalists in March 1959, see Arthur Trudeau, Statement, *Department of Defense Appropriations for Fiscal Year 1962, Hearings Before a Subcommittee of the Committee on Appropriations,* 83rd Cong., 1st sess., House Hearings, Committee on Appropriations (Washington, D.C.: U.S. GPO, 1961), 205; James Wakelin, Statement, *Department of Defense Appropriations for Fiscal Year 1962, Hearings Before a Subcommittee of the Committee on Appropriations,* 83rd Cong., 1st sess., House Hearings, Committee on Appropriations (Washington, D.C.: U.S. GPO, 1961), 397–99.
34. Although McNamara entered office with deep concerns about the threatening nature of the Soviet ICBM buildup, he felt that aggressive pursuit of a U.S. BMD system would not shield Americans from this threat, would overlap unnecessarily with "passive defense" measures such as civil defense, and would undermine superpower stability by triggering an "action-reaction" arms race (see Baucom, *Origins of SDI,* 17–19; Schwartz, "Past and Present," 333–35). For a discussion of McNamara's unique approach to defense planning and its major impact on the course of BMD development, see Yanarella, *Missile Defense Controversy,* 43–59.
35. Specifically, McNamara supported Project Defender, a $200 million per year initiative that explored "in the broadest front the principles and techniques that might prove useful in the attempt to solve the antimissile problem" (Baucom, *Origins of SDI,* 15; see also U.S. Congress, *Department of Defense Appropriations for Fiscal 1961,* 86th Cong., 2d sess., House Hearings, Committee on Appropriations [Washington, D.C.: GPO, 1960], 139). Basic research conducted under the umbrella of Project Defender

included investigation of laser beams, particle beams, and "tailored" nuclear explosions as potential BMD technologies (see Guerrier and Thompson, *Perspectives,* 20; Jerome B. Wiesner and Herbert F. York, "National Security and the Nuclear-Test Ban," *Scientific American* 211 [1964]: 33–34). McNamara's distaste for missile defense was exhibited when he abandoned a "no cities" proposal for nuclear targeting after it was pointed out that such a policy would build the strength of justifications for BMD (see U.S. Department of Defense, Biography of Robert McNamara, Past Secretary of Defense, Internet, DefenseLink Website, Posted 24 November 1998, Online at http://www.defenselink.mil/specials/secdef_histories/bios/mcnamara.htm).

36. Schwartz, "Past and Present," 336.
37. The Soviet Galosh system was similar to the U.S. Nike-Zeus system, "composed of a network of radars and a two- or three-stage, solid-fueled interceptor missile designed for long-range, exoatmospheric interception of incoming ICBMs" (Yanarella, *Missile Defense Controversy,* 118). An overview of the Soviet BMD program is provided in David Holloway, "The Strategic Defense Initiative and the Soviet Union," *Daedalus* 114 [1985]: 257–78. See also Michael J. Deane, *The Role of Strategic Defense in Soviet Strategy* (Miami: Advanced International Studies Institute, 1980). For a comparative assessment of the Soviet BMD program relative to U.S. efforts, see Stares, *Space and National Security.*
38. Guerrier and Thompson, *Perspectives,* 20–22; see also Yanarella, *Missile Defense Controversy,* 117–18.
39. For discussion of the Joint Chiefs' pro-BMD position at this time, see Guerrier and Thompson, *Perspectives,* 23–24; Schwartz, "Past and Present," 337–38.
40. See Herbert F. York, "Deterrence by Means of Mass Destruction," *Bulletin of the Atomic Scientists* 30 (1974): 4–9, idem, "Controlling the Qualitative Arms Race," *Bulletin of the Atomic Scientists* 29 (1973): 4–8.
41. Yanarella, *Missile Defense Controversy,* 200. Momentum for this "internal arms race" was created by American scientists set against each other in an interlocking offense-defense cycle of research breakthroughs. Seen in this light, the Soviet Union was less an instigator of the arms race and more the straggling follower of a massive unilateral American military buildup: "Our unilateral decisions set the rate and scale for most of the individual steps in the strategic arms race," noted York (Herbert F. York, *Race to Oblivion: A Participant's View of the Arms Race* [New York: Oxford University Press, 1970], 230). Even John Foster, former director of Lawrence Livermore National Laboratory (LLNL), acknowledged that in each case of strategic weapons development, "it seems to me that the Soviet Union is following the United States lead and that the United States is not reacting to Soviet actions" (quoted in James Dick, "The Strategic Arms Race, 1957–1961," *Journal of Politics* 34 [1972]: 1063).
42. See Yanarella, *Missile Defense Controversy,* 194. McNamara and other BMD skeptics in the Johnson administration began to realize that a legally binding arms control agreement might be the only way to derail the push for quick deployment of Nike-X. However, during a six-month period of intense diplomatic activity in 1967, American negotiators were unable to interest Soviet premier Aleksei Kosygin in pursuing talks

on defensive arms (see Schwartz, "Past and Present," 338). In retrospect, McNamara thus had no choice but to move ahead with limited deployment of a BMD system. In the words of one McNamara aide, the secretary's choice "was not a small ABM versus none at all, but rather a small ABM versus a big one" (quoted in Guerrier and Thompson, *Perspectives,* 24).

43. Yanarella, *Missile Defense Controversy,* 120. The text of the speech is published in Robert S. McNamara, "Text of Address to Press Club of San Francisco," *New York Times,* 19 September 1967, 18. For additional background commentary, see William J. Durch, *The ABM and Western Security* (Cambridge, Mass.: Ballinger, 1988); Kaplan, *Wizards of Armageddon,* 347–48; Robert S. McNamara, *The Essence of Security: Reflections in Office* (New York: Harper and Row, 1968); and Yanarella, *Missile Defense Controversy,* 120–23.
44. Schwartz, "Past and Present," 339.
45. Guerrier and Thompson, *Perspectives,* 25.
46. Yanarella, *Missile Defense Controversy,* 146.
47. Yanarella, *Missile Defense Controversy,* 147; see also Durch, *ABM and Western Security,* 8; Guerrier and Thompson, *Perspectives,* 25.
48. John W. Finney, "A Historical Perspective," in *The ABM Treaty: To Defend or Not to Defend?,* ed. Walther Stützle, Bhupendra Jasani, and Regina Cowen (Oxford: Stockholm International Peace Research Institute/Oxford University Press 1987), 34–35.
49. Although Safeguard utilized many of the same technical components and concepts as its predecessors, Nixon's decision to deploy Safeguard "represented a clear shift in the emphasis of the ABM program. Unlike its predecessors, Safeguard was designed to provide only minimal defense of the American population, though there was still some talk of protection against Chinese or accidental Soviet launches. The real objective of Safeguard was to increase the security of the American strategic nuclear force by defending the nation's ICBMs—and some bombers based nearby—against any Soviet attempt at a disarming first strike" (Guerrier and Thompson, *Perspectives,* 27; see also Durch, *ABM and Western Security,* 8; Alain C. Enthoven and K. Wayne Smith, *How Much Is Enough? Shaping the Defense Program, 1961–1969* [New York: Harper Colophon Books, 1971], 191; Kaplan, *Wizards of Armageddon,* 350; Schwartz, "Past and Present," 341; and Yanarella, *Missile Defense Controversy,* 152).
50. Schwartz, "Past and Present," 341. In the Senate, anti-ABM stalwarts Albert Gore Sr., J. William Fulbright, and Edward Kennedy lined up "noted foes of BMD in the scientific establishment like Herbert York, George Kistiakowsky, Jerome Wiesner, and George Rathjens to appear before the Senate Armed Services Committee, as well as before the Gore Disarmament Subcommittee. No longer were the decisions and recommendations of the old conservative guard uncontested" (Yanarella, *Missile Defense Controversy,* 156; see also Anne Hessing Cahn, "American Scientists and the ABM: A Case Study in Controversy," in *Scientists and Public Affairs,* ed. Albert H. Teich [Cambridge: MIT Press, 1974], 41–120; U.S. Congress, *The Acquisition of Weapons Systems,* 91st Cong., 2d sess., Joint Hearings, Subcommittee on the Economy in Government of the Joint Economic Committee [Washington, D.C.: GPO, 1970],

Changing National Priorities, 91st Cong., 2d sess., Hearings, Joint Economic Committee [Washington, D.C.: GPO, 1970], *Strategic and Foreign Policy Implications of ABM Systems,* 91st Cong., 1st sess., Senate, Committee on Foreign Relations [Washington, D.C.: GPO, 1969]).

For a well-documented summary of the "eight basic propositions" advanced by anti-ABM scientists in this debate, see Yanarella, *Missile Defense Controversy,* 162–65. This broadening of the "great ABM debates" in the U.S. Congress and the souring situation in Vietnam turned public sentiment against Safeguard and Sentinel in 1969–70 (Durch, *ABM and Western Security,* 8). As Yanarella explains, "[b]y mid-July 1969, the anti-ABM movement had gained such strength and momentum that it counted support from forty-seven or forty-eight senators, with three or four uncommitted senators holding the balance of power" (*Missile Defense Controversy,* 157). Against this rising tide of skepticism, however, President Nixon continued to push for his Safeguard system, "but he did so increasingly for reasons unrelated to its stated military value. . . . Nixon came to view Safeguard primarily as an important bargaining chip in future arms control negotiations and fought for the program with this purpose in mind" (Guerrier and Thompson, *Perspectives,* 28; see also Yanarella, *Missile Defense Controversy,* 178–79, 183–84). For partisan commentary on the pro-missile defense side of this debate, see Donald G. Brennan, "The Case for the ABM," in Center for the Study of Democratic Institutions, *Anti-Ballistic Missile: Yes or No? A Special Report from the Center for the Study of Democratic Institutions* (New York: Hill and Wang, 1968), 27–35.

51. Gerard Smith, preface to *Foundation for the Future: The ABM Treaty and National Security* (Washington, D.C.: Arms Control Association, 1990), vii.
52. Douglas P. Lackey, *Moral Principles and Nuclear Weapons* (Totowa, N.J.: Rowman & Allanheld, 1984), 116.
53. One such exception "allowed each side to deploy a limited ABM system of up to one hundred launchers—with one missile each—and six radar complexes within a radius of 150 kilometers around its national capital. The other exception permitted each nation a similar deployment of launchers and missiles, though with more radars (including two large phased-array radars), in defense of an area in which ICBM silos were located" (Guerrier and Thompson, *Perspectives,* 29).
54. Guerrier and Thompson, *Perspectives,* 33.
55. Edward Reiss, *The Strategic Defense Initiative* (Cambridge: Cambridge University Press, 1992), 45.
56. See John E. Pike, Bruce G. Blair, and Stephen I. Schwartz, "Defending Against the Bomb," in *Atomic Audit: The Costs and Consequences of U.S. Nuclear Weapons Since 1940,* ed. Stephen I. Schwartz (Washington, D.C.: Brookings Institute Press, 1998), 288–89.
57. Baucom, *Origins of SDI,* 97.
58. For example, between 1972 and 1975, funding for all aspects of the BMD program fell from $1.4 billion to $440 million per year (Baucom, *Origins of SDI,* 97–98).
59. Gary L. Guertner and Donald M. Snow, *The Last Frontier: An Analysis of The Strategic Defense Initiative* (Lexington, Mass.: Lexington Books, 1986), 2. By the late 1970s,

technological advances enabled BMD advocates to see glimpses of this "better day." As the decade came to a close, Pratt explains that "the Army's BMD insurance policy had evolved into a menu of near-term options for U.S. policymakers. . . . Clearly, the interest of the service managers and the major BMD contractors was to push for near-term deployment" (Erik K. Pratt, *Selling Strategic Defense* [Boulder, Colo.: Lynne Rienner, 1990], 78); see also Hans A. Bethe, Jeffrey Boutwell, and Richard L. Garwin, "BMD Technologies and Concepts in the 1980s," *Daedalus* 114 [1985]: 60–71; Abram Chayes, Antonia Handler Chayes, and Eliot Spitzer, "Space Weapons: The Legal Context," *Daedalus* 114 [1985]: 193–218).

By the early 1980s, signs were emerging that pro-BMD forces were preparing for a political offensive. For example, the 1980 Republican Party platform contained a plank that called for "vigorous research and development of an effective anti-ballistic missile system, such as is already at hand in the Soviet Union, as well as more modern ABM technologies" (quoted in Pratt, *Selling Strategic Defense,* 88). This enthusiasm for missile defense was carried over from the 1980 electoral campaign into the early days of Ronald Reagan's presidency, when Reagan appointed stalwart SDI advocates such as Martin Anderson, Edwin Meese, Richard Allen, and George Keyworth to senior advisory posts in the White House. As domestic policy advisor Martin Anderson recalled in his memoirs, "we did something that, by the book, we should not have done. Without ever formally acknowledging it, even to ourselves, a small, informal group on strategic defense was formed within the White House" (Martin Anderson, *Revolution* [San Diego: Harcourt Brace Jovanovich, 1988], 91; see also Reiss, *Strategic Defense Initiative,* 40–44; Frances FitzGerald, *Way Out There in the Blue: Reagan, Star Wars and the End of the Cold War* [New York: Simon and Schuster, 2000], 196–202).

60. Reagan titled his missile defense program the "Strategic Defense Initiative" (commonly known as SDI) in a pamphlet released by the White House on 28 December 1984 (see "The President's Strategic Defense Initiative," *Department of State Bulletin* 85 [1985]: 65–72).
61. Jesse Helms, "The President Turns a New Page in Defense," *Congressional Record,* 24 March 1983, 7155–57.
62. G. A. Greb, review of *Teller's War,* by William Broad, *256 Science,* 15 May 1992, 1043.
63. Pratt, *Selling Strategic Defense,* 112.
64. FitzGerald, *Way Out There,* 275.
65. Handed down by President Reagan in May 1985, NSDD 172 established a framework for the Pentagon's Star Wars information campaign. It required that all official statements on SDI (including those made in "scientific channels") be cleared by the assistant to the president for national security affairs (see Christopher Simpson, "Commentary on NSDD 172: Publicizing The Strategic Defense Initiative," in *National Security Decision Directives of the Reagan and Bush Administrations,* ed. Christopher Simpson [Boulder, Colo.: Westview, 1995], 449). When statements were made that did not square with the official script set forth in NSDD 172, members of the Reagan administration intervened to "correct the record." A more thorough description and assessment of this practice is provided in chapter two.

66. Titled "Special Plans Guidance-Strategic Defense," this memorandum directed the Pentagon to "improve and update deception plans covering the missile defense program's cost and abilities" (quoted in Tim Weiner, "General Details Altered 'Star Wars' Test," *New York Times*, 27 August 1993, A19).
67. Kevin O'Neill, "Building the Bomb," in *Atomic Audit: The Costs and Consequences of U.S. Nuclear Weapons Since 1940*, ed. Stephen I. Schwartz (Washington, D.C.: Brookings Institute Press, 1998), 82.
68. John Tirman,"The Politics of Star Wars," in *Empty Promise: The Growing Case Against Star Wars*, ed. John Tirman (Boston: Beacon Press, 1986), 1.
69. Hughes, *SDI: A View*, 199.
70. Since the Star Wars address is a rhetorically rich text, it is able to support insightful criticism from several different theoretical perspectives. However, a steadfast focus on this speech as the locus of SDI controversy obscures other important dimensions of the dispute also deserving of analysis, such as public announcements of scientific experiments, and advocacy of SDI within the administration's secret channels of communication. The public address orientation of previous work (with the exception of G. Thomas Goodnight, "Ronald Reagan's Reformulation of the Rhetoric of War: Analysis of the 'Zero Option,' 'Evil Empire,' and 'Star Wars' Addresses," *Quarterly Journal of Speech* 72 [1986]: 390–414) tends to screen out these important factors by telescoping the SDI controversy into a one-time transaction between Reagan and the public. The public address orientation hides the fact that public support for SDI gained its solidity over time, in the face of a long series of events subsequent to and following Reagan's address. Previous rhetorical analyses of missile defense advocacy also overlook the prominent role of institutionalized deception in the official management of public controversy.

    For example, Kenneth Zagacki and Andrew King proceed under the naive assumption that Reagan's "techno-romantic synthesis" rested on a legitimate scientific foundation (see Kenneth Zagacki and Andrew King, "Reagan, Romance and Technology: A Critique of 'Star Wars,'" *Communication Studies* 40 [1989]: 1–12). Zagacki and King's admission that "we know of no empirical evidence indicating that the American public has been duped by Reagan's speech" ("Reagan, Romance, and Technology," 8) is perplexing when assessed in light of the surfeit of evidence that has emerged documenting the existence of institutionalized deception schemes employed by the Reagan administration to exaggerate SDI feasibility. In a similar vein, while Rebecca Bjork's treatment of BMD in the post–cold war environment is groundbreaking, it nevertheless suffers from a significant thinness, in that it provides an inadequate account of Patriot missile performance during the Persian Gulf War (Rebecca Bjork, *The Strategic Defense Initiative: Symbolic Containment of the Nuclear Threat* [New York: SUNY Press, 1992]). Writing with the assumption that the Patriot was largely successful in intercepting Scuds, Bjork uses the discussion of the Patriot merely as a segue into a larger consideration of Global Protection Against Limited Strikes (GPALS). In the process, she overlooks a corpus of technical literature featuring a robust controversy over the issue of Patriot missile accuracy during the Gulf War.

71. Michael Handel, *War, Strategy, and Intelligence* (London: Frank Cass, 1989), 310.
72. Sun Tzu, *The Art of War*, trans. Samuel B. Griffith (New York: Oxford University Press, 1973), 66.
73. Handel, *War, Strategy, and Intelligence*, 380.
74. As historian Barton Whaley describes, "[i]n the 1920s surprise and deception were preserved in German military doctrine and training by General van Seeckt, and by the mid-1930s the Abwehr had restored the specific organizational machinery for that deception. This machinery was the D Group (Gruppe III-D), one of the half dozen main divisions in Colonel von Bentivegni's military security section (Abteilung III)" (Barton Whaley, *Codeword Barbarossa* [Cambridge: MIT Press, 1973], 171). In Soviet military doctrine, deception was codified as a standard element of military art under the title *maskirovka* (often translated as "camouflage"). According to former CIA officer Richards Heuer, "as a standard element of military art, there is considerable writing about maskirovka in the open Soviet military press. Their doctrine identifies four broad categories of maskirovka: 1) Camouflage (skrytiye) . . . 2) Simulation (imitatsiya) . . . 3) Feints and demonstrations (demonstrativnye deystviye or manevry) . . . 4) Disinformation (dezinformatsiya)" (Richards J. Heuer Jr., "Soviet Organization and Doctrine for Strategic Deception," in *Soviet Strategic Deception*, ed. Brian D. Dailey and Patrick J. Parker [Lexington, Mass.: D.C. Heath, 1987], 42–43).
75. See Michael I. Handel, "Intelligence and the Problem of Strategic Surprise," *Journal of Strategic Studies* 7 (1984): 229–81.
76. A secret document signed by General Wilhelm Keitel on 15 February 1941 laid out the "guidelines for deception of the enemy" in Operation Barbarossa, "the greatest deception operation in the history of war" (quoted in Whaley, *Codeword Barbarossa*, 248). On Operation Barbarossa generally, see Russel H. S. Stolfi, "Barbarossa: German Grand Deception and the Achievement of Strategic and Tactical Surprise Against the Soviet Union, 1940–1941," in *Strategic Military Deception*, ed. Donald C. Daniel and Katherine L. Herbig (New York: Pergamon Press, 1982), 195–223; and Whaley, *Codeword Barbarossa*. Allied forces also pursued active deception campaigns during World War II. For example, following Britain's victory in North Africa, allied troops hoped to invade Europe through Sicily, but such a route was blocked by heavy German fortifications on the southern coasts of Italy. "To trick them into moving their forces elsewhere, Lieutenant Commander Ewen Montagu, an officer with the Intelligence Division of the British Admirality, came up with the idea of planting bogus papers on a body that could reasonably be expected to fall into German hands. . . . As planned, the body washed ashore on the coast of Spain carrying bogus papers that suggested a two-pronged allied attack against Sardinia in the Western Mediterranean and Kalamata, on the Peloponnesian coast of Greece. . . . Shortly thereafter, German forces . . . diverted or redeployed other resources from Sicily to Greece and Sardinia, including boats, coastal batteries, and troops" (Dorothy E. Denning, *Information Warfare and Security* [Reading, Mass.: Addison-Wesley, 1999], 106; see also Ronald Seth, *The Truth Benders: Psychological Warfare in the Second World War* [London: Leslie Frewin, 1969], 76–77). In another context, Trevor Barnes,

producer of BBC Television's *Newsnight,* discusses American use of Radio Free Europe to spread misperceptions about Soviet military capabilities during the cold war (Trevor Barnes, "Democratic Deception: American Covert Operations in Post-War Europe," in *Deception Operations,* ed. David A. Charters and Maurice A. J. Tugwell [London: Brassey's, 1990], 297–324).

77. See Amrom H. Katz, "The Fabric of Verification: The Warp and the Woof," in *Verification and SALT: The Challenge of Strategic Deception,* ed. William C. Potter (Boulder, Colo.: Westview Press, 1980), 193–220. Making a similar point, Richards Heuer comments that "[t]he known history of Soviet military deception in World War II is one of passive deception, with the possible exception of the Manchurian Campaign in 1945" (Heuer, "Soviet Organization and Doctrine," 46).
78. Nikolai Lenin, quoted in Joseph D. Douglass Jr., "Soviet Disinformation," *Strategic Review* 9 (1981): 18.
79. John Lenczowski, "Themes of Soviet Strategic Deception and Disinformation," in *Soviet Strategic Deception,* ed. Brian D. Dailey and Patrick J. Parker (Lexington, Mass.: D.C. Heath, 1987), 56.
80. Ladislav Bittman, *The Deception Game: Czechoslovak Intelligence in Soviet Political Warfare* (New York: Syracuse University Research Corporation, 1972), 21.
81. Heuer, "Soviet Organization and Doctrine," 45. These assessments eventually served as the basis for U.S. perception of a "missile gap" during the cold war. "At that time, U.S. intelligence was so dependent on such information that this led to an upward revision of national intelligence estimates of Soviet heavy bomber strength and to claims of a 'bomber gap' that preceded a so-called 'missile gap'" (Heuer, "Soviet Organization and Doctrine," 45). For additional discussion on the specifics of Soviet attempts to distort American intelligence assessments, see Ladislav Bittman, *The KGB and Soviet Disinformation* (Washington, D.C.: Pergamon-Brassey's, 1985); Richard H. Shultz and Roy Godson, *Dezinformatsia: Active Measures in Soviet Strategy* (Washington, D.C.: Pergamon-Brassey's, 1984).
82. See Ronald G. Sherwin and Barton Whaley, "Understanding Strategic Deception: An Analysis of 93 Cases," in *Strategic Military Deception,* ed. Donald C. Daniel and Katherine L. Herbig (New York: Pergamon Press, 1982), 177.
83. See Handel, *War, Strategy, and Intelligence,* 403.
84. See Michael I. Handel, "A 'Runaway Deception': Soviet Disinformation and the Six-Day War, 1967," in *Deception Operations,* ed. David A. Charters and Maurice A. J. Tugwell (London: Brassey's, 1990), 167.
85. For example, consider the absolutist position staked out by historian Ronald Seth on this point: "As for the ethical considerations, these do not exist; the concept of total war is accepted as the basis for waging war. Personally, I am quite unable to understand the attitude of those who oppose black propaganda on ethical grounds" (Seth, *Truth Benders,* 198).
86. Handel, *War, Strategy, and Intelligence,* 35.
87. Barnes, "Democratic Deception," 320.
88. Ibid.

89. Trevor Melia, "Review of Peter Dear's *The Literary Structure of Scientific Argument: Historical Studies;* Alan G. Gross's *The Rhetoric of Science;* Greg Myers's *Writing Biology: Texts in the Social Construction of Scientific Knowledge,* and Lawrence J. Prelli's *A Rhetoric of Science: Inventing Scientific Discourse,*" *ISIS* 83 (1992): 100.
90. Aristotle, *On Rhetoric,* trans. George A. Kennedy (New York: Oxford University Press, 1991), 35.
91. Several rhetoric of science scholars have approached the task of interpreting scientific discourse by employing rhetoric as a "globalizing hermeneutic metadiscourse" that works as a master key of interpretation, said to be capable of unlocking the meaning of the full range of scientific phenomena (for commentary, see Dilip Gaonkar, "The Idea of Rhetoric in The Rhetoric of Science," *Southern Communication Journal* 58 [1993]: 258–95). In the "globalized" account of rhetoric, since "all knowledge is rhetorical," science is reduced to "rhetoric without remainder," that is, an activity composed completely of rhetorical constructions (see Walter Weimer, "Why All Knowing Is Rhetorical," *Journal of the American Forensic Association* 20 [1984]: 63–71; idem, "Science as a Rhetorical Transaction: Toward a Nonjustificational Conception of Rhetoric," *Philosophy and Rhetoric* 10 [1977]: 1–29; and Alan G. Gross, *The Rhetoric of Science* [Cambridge, Mass.: Harvard University Press, 1990]).
92. Gaonkar says rhetoric of science criticism is theoretically "promiscuous." He argues that because rhetoricians of science are motivated by a hankering to see rhetoric in everything, they routinely commit the sin of "deferring the text." When rhetoric is difficult to detect on the textual outer surface, Gaonkar suggests that rhetoric of science critics invent phantom objects of inquiry by conjuring up suppressed or disguised rhetorical forms. This globalizing tendency is a particular concern, says Gaonkar, because it threatens to marginalize the entire discipline of rhetoric (see Gaonkar, "The Idea").
93. Aristotle, *On Rhetoric,* 52–78. Aristotle identifies deliberation as the first of three genres of rhetoric deserving consideration. The others are display oratory ("epideictic" speeches of praise and blame) and litigation ("forensic" arguments in the law courts).
94. Aristotle, *On Rhetoric,* 34.
95. Bruno Latour, *Science in Action: How to Follow Scientists and Engineers through Society* (Cambridge, Mass.: Harvard University Press, 1987), 174–75.
96. Thomas B. Farrell, "An Elliptical Postscript," in *Rhetorical Hermeneutics: Invention and Interpretation in the Age of Science,* ed. Alan G. Gross and William M. Keith (New York: SUNY Press, 1997), 324.
97. John Lyne, "Quantum Mechanics, Consistency, and the Art of Rhetoric: Response to H. Krips," *Cultural Studies* 10 (1996): 128.
98. Philip Wander, "The Rhetoric of Science," *Western Speech Communication Journal* 40 (1976): 226.
99. Steve Fuller, "Who Hid the Body? Rouse, Roth, and Woolgar on Social Epistemology," *Inquiry* 34 (1991): 398.
100. See Aristotle, *On Rhetoric,* 51–77.

101. See K. P. Addelson, "The Man of Professional Wisdom," in *Discovering Reality,* ed. S. Harding and M. Hintikka (Dordrecht, Netherlands: Reidel, 1983); Theodore Porter, *Trust in Numbers: The Pursuit of Objectivity in Science and Public Life* (Princeton, N.J.: Princeton University Press, 1995).
102. Thomas Nagel, *The View from Nowhere* (New York: Oxford Univrsity Press, 1986).
103. See Carl Hempel, *Philosophy of Natural Science* (Englewood Cliffs, N.J.: Prentice Hall, 1966); Nagel, *The View from Nowhere.* Because the "view from nowhere" featured in this traditional empiricist framework is said to be able to produce ahistorical knowledge fundamentally different in kind from the contingent, ephemeral insight generated by investigations not performed under the control of the scientific method, it lends advocates a substantial persuasive edge in public disputes over knowledge claims (see Joseph Rouse, *Engaging Science* [Ithaca, N.Y.: Cornell University Press, 1996]; idem, *Knowledge and Power: Toward a Political Philosophy of Science* [Ithaca, N.Y.: Cornell University Press, 1987]; and idem, "Policing Knowledge: Disembodied Policy for Embodied Knowledge," *Inquiry* 34 [1991]: 353–64).
104. Michael Calvin McGee and John R. Lyne, "What Are Nice Folks Like You Doing in a Place Like This?" in *The Rhetoric of the Human Sciences,* ed. John S. Nelson, Allan Megill, and Donald N. McCloskey (Madison: University of Wisconsin Press, 1987), 393. "The historical tension between rhetorical and Platonic models of authority is frequently represented as one between two mutually exclusive approaches to the manufacture of belief. Yet there is a sense in which antirhetoric's appeal to objective knowledge and its accompanying denunciation of rhetoric is one of the most effective rhetorical strategies available" (McGee and Lyne, "Nice Folks," 393).
105. Chandra Mukerji, *A Fragile Power: Scientists and the State* (Princeton, N.J.: Princeton University Press, 1990), 191. In Richard Harvey Brown's view, the idiom of science works "projectively," that is, rhetors project the objectivity of scientific truth claims "through poetic and political practices of standardization that invite intersubjective confirmations" (Richard Harvey Brown, "Modern Science: Institutionalization of Knowledge and Rationalization of Power," *Sociological Quarterly* 34 [1993]: 157). Although the early steps in the process of scientific discovery often feature local influences that would commonly be seen as inconsistent with objective research methods, Brown explains that as scientists attempt to persuade larger publics, references to these "social influences" are often dropped from descriptions. "In their studies of *Laboratory Life,* [Mulkay and Gilbert] and other researchers found that scientists try to get their claims to knowledge accepted by telling stories of how these claims emerged. But as they become successful with larger publics, such social explanations are soon dropped, and the knowledge claim is described technically as a taken for granted object in nature. The style becomes impersonal, focusing on objective agencies rather than human agents, almost as though the physical world were acting and speaking for itself" (Brown, "Modern Science," 156). See Latour's treatment of "black boxes" for a variation on this theme (Latour, *Science in Action,* 63–144).

    The "rhetoric of realism" (see Steve Woolgar, "The Very Idea of Social Epistemology: What Prospects for a Truly Radical 'Radically Naturalized

Epistemology'?" *Inquiry* 34 [1991]: 381; Charles Bazerman, *Shaping Written Knowledge: The Genre and Activity of the Experimental Article in Science* [Madison: University of Wisconsin Press, 1988]; and Wander, "Rhetoric of Science") often receives its fullest expression in the "final form statements" (Dudley Shapere, "Talking and Thinking about Nature: Roots, Evolution, and Future Prospects," *Dialectica* 46 [1992]: 289) found in articles, research papers, conference proceedings, congressional testimony, and other "official" outlets.

106. Porter, *Trust in Numbers,* 217. Many expect that the elaborate system of checks and balances that make up the process of scientific inquiry will filter out (or at least control) such local factors, leaving unvarnished knowledge in its place. This presumption of the neutrality of scientific practice provides leverage to rhetorical appeals that invoke the rhetoric of science as a trump card against competing knowledge claims that can be shown to be founded in personal opinion or individual preference. "Confidence in the transparency and neutrality of scientific language," explains Keller, "is wondrously effective in supporting [scientists'] special claims to truth" (Evelyn Fox Keller, *Secrets of Life: Essays on Language, Gender, and Science* [New York: Routledge, 1992], 28).

107. Dorothy Nelkin, *Selling Science* (New York: Freeman, 1995). Media reporting of scientific events tends to reinforce the prevailing view that the scientific enterprise has a duty toward progress and has an intrinsic propensity to perform technical miracles. "Historically, the mass media has regarded science more as a cornucopia, as a producer of spectacular blessings and occasional harms, than an ordinary part of society. This image sets science apart, and affords it great power and potential" (Michael LaFollette, "Mass Media News Coverage of Scientific and Technological Controversy," *Science, Technology, and Human Values* 6 [1981]: 25). One consequence of this image of science is that a great deal of weight is often shifted to arguments for technical projects that are couched in the rhetoric of science. The prevailing view that science has been successful in the past affords advocates the rhetorical resources to disarm skeptics questioning the utility of scientific solutions.

108. The "whiggish" view of scientific history as it is presented in scientific textbooks is often cited as a main cause of this widely held belief (see Thomas Kuhn, *The Structure of Scientific Revolutions* [Chicago: University of Chicago Press, 1970]). For a discussion on the origins and dynamics of historical whiggism in the history of natural science, see John G. McEvoy, "Positivism, Whiggism, and the Chemical Revolution: A Study in the Historiography of Chemistry," *History of Science* 35 (1997): 1–33.

109. Drawing upon this presumption, advocates deploying the rhetoric of science can gain space to argue that although research may be progressing slowly, the gradual accumulation of knowledge still justifies confidence in the scientific enterprise. The notion that the quantity and quality of scientific knowledge increases proportionally as research continues has been called the "linear model." Sarewitz explains that according to this model, "the path from fundamental scientific research to useful products is an orderly progression, starting with the creation of new knowledge in the basic research laboratory and moving sequentially through the search for applications, the

development of specific products, and the introduction of these products into society through standard commercial channels or through government programs such as national defense" (Daniel Sarewitz, *Frontiers of Illusion: Science, Technology, and the Politics of Progress* [Philadelphia: Temple University Press, 1996], 97).

110. "The endless frontier is the mythical territory where nature and its laws await their gradual and progressive discovery. Scientists are the explorers venturing into this unknown terrain, and scientific theories and data are the products of their exploration" (Sarewitz, *Frontiers of Illusion*, 99). Widespread allegiance to this "endless frontier" myth permits advocates utilizing the rhetoric of science to cast their claims in an air of excitement and optimism that primes audiences to expect that scientific work will yield a steady stream of fresh discoveries. On the historical origins of the "endless frontier" myth, see David M. Hart, *Forged Consensus: Science, Technology, and Economic Policy in the United States, 1921–1953* (Princeton, N.J.: Princeton University Press, 1998); Timothy M. O'Donnell, "The Rhetoric of American Science Policy" (Ph.D. diss, University of Pittsburgh, forthcoming).
111. Sheila Jasanoff, "Beyond Epistemology: Relativism and Engagement in the Politics of Science," *Social Studies of Science* 26 (1996): 400.
112. See Stewart L. Udall, *The Myths of August* (New York: Pantheon, 1994), 24–33.
113. Although recent erosion of support for civilian nuclear power has dissipated the contemporary currency of some of these attitudes, it is clear that substantial pro-atomic science sentiments persist. For example, as in 1945, today there is widespread support for the idea that the military has sufficient and unmatched scientific expertise to manage the nation's stockpile of nuclear weapons (see Arjun Makhijani, Howard Hu, and Katherine Yih, eds., *Nuclear Wastelands* [Cambridge, Mass.: MIT Press, 1995]). Additionally, support persists for the elaborate classification system that protected military science from outside scrutiny during the cold war (see Herbert N. Foerstel, *Secret Science* [Westport, Conn.: Praeger, 1993]).
114. Several scholars have called attention to the prominence of argumentation as a constitutive dimension of science, highlighting the argumentative aspects of such processes as theory formulation and comparison (see Weimer, "Science as a Rhetorical Transaction," 5), fact construction (see Latour, *Science in Action*, 25–26), and negotiation of the meaning of scientific inquiry itself. On this latter topic, Shapere observes that "[a]ny assessment of science, critical or supportive, presupposes first, a clear and detailed understanding of what science is and does, and second an assessment of the body of arguments that have been put forth as reasons for skepticism with regard to anything and everything that science says and does" (Shapere, "Talking and Thinking," 295–96).
115. See Thomas B. Farrell, *Norms of Rhetorical Culture* (New Haven, Conn.: Yale University Press, 1993); idem, "Knowledge, Consensus, and Rhetorical Theory," *Quarterly Journal of Speech* 62 (1976): 1–14.
116. For discussion of the process through which argumentation in public controversies sifts and reshapes social knowledge, see G. Thomas Goodnight, "Controversy," in *Argument in Controversy: Proceedings of the Seventh SCA/AFA Conference on*

Argumentation, ed. Donn Parson (Annandale, Va.: Speech Communication Association, 1992), 1–13.

117. H. Tristam Engelhardt Jr. and Arthur L. Caplan, *Scientific Controversies: Case Studies in the Resolution and Closure of Disputes in Science and Technology* (Cambridge: Cambridge University Press, 1987), vii.
118. Celeste Condit, "How Bad Science Stays That Way: Brain Sex, Demarcation, and the Status of Truth in the Rhetoric of Science," *Rhetoric Society Quarterly* 26 (1996): 83–109; see also Henry Howe and John Lyne, "Gene Talk in Sociobiology," *Social Epistemology* 6 (1992): 109–64.
119. Latour, *Science in Action*, 63–102.
120. Paul R. Gross and Norman Levitt, *Higher Superstitions: The Academic Left and Its Quarrels with Science* (Baltimore: Johns Hopkins University Press, 1994), 235.
121. See, e.g., Howe and Lyne, "Gene Talk"; Bruno Latour and Steve Woolgar, *Laboratory Life* (London: Sage, 1979); John Lyne and Henry F. Howe, "Punctuated Equilibria," *Quarterly Journal of Speech* 72 (1986): 132–47; Sharon Traweek, "Border Crossings: Narrative Strategies in Science Studies and among Physicists in Tsukuba Science City, Japan," in *Science as Practice and Culture*, ed. Andrew Pickering (Chicago: University of Chicago Press, 1992), 429–65; and idem, *Beamtimes and Lifetimes* (Cambridge, Mass.: Harvard University Press, 1988).
122. In classical rhetorical theory, stasis is a concept that refers to the center of discussion, the locus of dispute, or the key point around which argumentation pivots. For an exhaustive attempt to develop a theoretical approach to rhetoric of science anchored in classical theory, see Lawrence Prelli, *A Rhetoric of Science: Inventing Scientific Discourse* (Columbia: University of South Carolina Press, 1989).
123. Charles A. Taylor, "Defining the Scientific Community: A Rhetorical Perspective on Demarcation," *Communication Monographs* 58 (1991): 403.
124. See, e.g., Condit, "How Bad Science," 95–97; Aant Elzinga, "Science as Continuation of Politics by Other Means," in *Controversial Science: From Content to Contention*, ed. Thomas Brante, Steve Fuller, and William Lynch (New York: SUNY Press, 1993), 127–52; and Brian Martin, "Sticking a Needle into Science: The Case of Polio Vaccines and the Origin of AIDS," *Social Studies of Science* 26 (1996): 265–66.
125. Trevor Pinch, "Cold Fusion and the Sociology of Scientific Knowledge," *Technical Communication Quarterly* 3 (1994): 88.
126. Thomas B. Farrell and G. Thomas Goodnight, "Accidental Rhetoric: The Root Metaphors of Three Mile Island," *Communication Monographs* 48 (1981): 270–300.
127. On the importance of blending synergistically productivist and interpretive aspects of the rhetorical tradition in criticism, see Thomas B. Farrell, "On the Disappearance of the Rhetorical Aura," *Western Journal of Communication* 57 (1993): 147–58; Thomas A. Hollihan, "Evidencing Moral Claims: The Activist Rhetorical Critic's First Task," *Western Journal of Speech Communication* 58 (1994): 229–34; Thomas A. Hollihan, Patricia Riley, and James F. Klumpp, "Greed versus Hope, Self-Interest versus Community: Reinventing Argumentative Practice in Post-Free Marketplace America," in *Argument and the Postmodern Challenge*, ed. Ray McKerrow (Annandale, Va.:

Speech Communication Association, 1993), 332–39; Andrew King, "The Rhetorical Critic and the Invisible Polis," in *Rhetorical Hermeneutics: Invention and Interpretation in the Age of Science*, ed. Alan G. Gross and William M. Keith (New York: SUNY Press, 1997), 299–316; and Michael Leff, "The Idea of Rhetoric as Interpretive Practice: A Humanist's Response to Gaonkar," in *Rhetorical Hermeneutics: Invention and Interpretation in the Age of Science*, ed. Alan G. Gross and William M. Keith (New York: SUNY Press, 1997), 89–100. For a similar discussion in the context of critical theory, see Robert C. Holub, *Jurgen Habermas: Critic in the Public Sphere* (London: Routledge, 1992).

128. John Lyne, "Bio-rhetorics: Moralizing the Life Sciences," in *The Rhetorical Turn: Invention and Persuasion in the Conduct of Inquiry*, ed. Herbert W. Simons (Chicago: University of Chicago Press, 1990), 37.
129. Condit, "How Bad Science," 86–87.
130. Thomas Brante, "Reasons for Studying Scientific and Science-Based Controversies," in *Controversial Science: From Content to Contention*, ed. Thomas Brante, Steve Fuller, and William Lynch (New York: SUNY Press, 1993), 187.
131. Brian Martin, *Scientific Knowledge in Controversy: The Social Dynamics of the Fluoridation Debate* (New York: SUNY Press, 1991), 164.
132. For commentary on the iterative process through which these parallel tasks played off each other, see Gordon R. Mitchell, "Placebo Defense: The Legacy of Strategic Deception in Missile Defense Advocacy" (Ph.D. diss, Northwestern University, December 1997), 265–79.
133. I drew from my analysis of the epistemic costs of strategic deception in the Star Wars era to suggest that excessive compartmentalization and secrecy of the research project would likely threaten the long-term integrity of Ballistic Missile Defense Organization's (BMDO's) scientific mission (see Gordon R. Mitchell, "Another Strategic Deception Initiative," *Bulletin of the Atomic Scientists* 53 [1997]: 22–23). To highlight the deceptive dynamics and arms control stakes implicated by the dispute over THAAD's legality under the ABM Treaty, I published two Space Policy E-Prints on the Federation of American Scientists Website (see Gordon R. Mitchell, "Two-Way Deception Traffic on the Road to Missile Defense," Space Policy Project E-print, Internet, Federation of American Scientists website, December 1997, online at http://www.fas.org/spp/eprint/gm3.htm; and idem, "The TMD Footprint Controversy," Space Policy Project E-print, Internet, Federation of American Scientists Website, 12 September 1996, online at: http://www.fas.org/spp/eprint/gm1.htm). To convey my research findings to members of Congress, I filed a brief with the Senate Committee on Government Oversight, and consulted with minority staff member Rick Goodman concerning BMDO's compliance with the Federal Advisory Committee Act (FACA) rules (see Gordon R. Mitchell, "Sparta's TMD Footprint Advice to BMDO: Overstepping the Boundaries of FACA?" Issue brief for Senator David Pryor and Senator Carl Levin, 24 July 1996, subsequently published as Space Policy Project E-print, Internet, Federation of American Scientists Website, 8 August 1996, online at http://www.fas.org/spp/eprint/gm2.htm).

134. These papers were designed to explain the interweaving dynamics of rhetoric, science, and politics in U.S. missile defense advocacy to European audiences considering whether to support plans for new U.S. missile defense systems. The London office of ISIS-U.K. serves the U.K. policy community, while ISIS-Europe (based in Brussels) is dedicated to providing policy analysis to European Union member states, nongovernmental organizations, and interested citizens and scholars. Amplification of my arguments by ISIS-U.K. enabled me to reach British journalists and MPs (see Gordon R. Mitchell, "Missile Defence Policy: Strident Voices and Perilous Choices," ISIS-U.K. Briefing Series on Ballistic Missile Defence, no. 1, April 2000). ISIS-Europe published an adapted form of this brief that was expanded to reach a wider NATO audience (see Gordon R. Mitchell, "U.S. National Missile Defence: Technical Challenges, Political Pitfalls, and Disarmament Opportunities," ISIS-Europe Briefing Paper, no. 23, May 2000, Internet, ISIS-Europe Website, online at http://www.fhit.org/isis). This brief was distributed to the European Parliament's Foreign Affairs Committee, members of the European Council's Policy Unit, members of the Commission's Directorate General for External Affairs, and the office of Javier Solana, European Union security chief. Portions of the brief were also translated into French and German and posted on the ISIS-Europe website. ISIS-Europe staff sent email advertisements notifying some 500 members of the EU-NGO Common Foreign and Security Policy Contact Group of the brief's electronic publication.
135. These articles condensed and amplified the arguments contained in chapters three and four, carrying my research findings to communication scholars and interdisciplinary audiences in the so-called science studies field (see Gordon R. Mitchell, "Placebo Defense: The Rhetoric of Patriot Missile Accuracy in the 1991 Persian Gulf War," *Quarterly Journal of Speech* 86 [2000]: 121–45; idem, "Whose Shoe Fits Best? Dubious Physics and Power Politics in the TMD Footprint Controversy," *Science, Technology, and Human Values* 25 [2000]: 52–86).
136. See Gordon R. Mitchell and Anthony Todero, "On Viewing Arguments for Japanese Missile Defense as Delicious but Deceptive American Exports," paper presented at the First Tokyo Argumentation Conference, Tokyo Olympic Center, Tokyo, Japan, August 2000, copy on file with the author.
137. This collaboration took place at the National High School Institute, a summer debate workshop hosted by Northwestern University in 1998. At this workshop, a group of high school students (the Cherub Study Group) researched ballistic missile defense as part of their yearlong project to investigate U.S. foreign policy toward Russia. After a week of intensive library and Internet research, I worked with the group to fashion an accessible op-ed piece summarizing our research findings on missile defense. This piece was subsequently published as a Space Policy E-Print on the Federation of American Scientists website, and I discussed the project in an article focusing on the promises and pitfalls of scholarly intervention into public controversies (see Cherub Study Group, "Return of the Death Star?" Space Policy Project E-print, Internet, Federation of American Scientists website, 17 July 1998, online at: http://www.fas.org/spp/eprint/980731-ds.htm; Gordon R. Mitchell, "Pedagogical Possibilities for Argumentative Agency in

Academic Debate," *Argumentation and Advocacy* 35 [Fall 1998]: 55–57). See also Gordon R. Mitchell, "The National Missile Defense Fallacy," *Pittsburgh Post-Gazette*, 29 April 2000, A17; idem, "About Those 'Normal Accidents,'" *Pittsburgh Post-Gazette*, 16 July 2000, E4

138. Martin, "Sticking," 264.
139. Woolgar, "The Very Idea," 382.
140. Irving Horowitz, "Comment," in *Peace and the War Industry*, ed. Kenneth Boulding (New Brunswick, N.J.: Transaction Books, 1973), 55–60.
141. Yanarella, *Missile Defense Controversy*, 195–96.

Cover of Newsweek *magazine, 4 April 1983. Reprinted permission of Newsweek, Inc.*

CHAPTER TWO

Star Wars and the Quest for an Elusive "Leakproof" Missile Shield

Secrecy is not compatible with science, but it is even less compatible with democratic procedure. Two hundred years ago James Madison said, "A popular government without popular information is but a prologue to a farce or a tragedy, or perhaps both." The term credibility gap is a modest description of our monstrous current problem. . . . Adopting a policy of openness as the first and only unilateral step toward disarmament would strengthen our relationships with our allies as well as illustrate the advantages of freedom as compared with the practices of our Soviet colleagues.[1]

—Edward Teller, *Better a Shield than a Sword*

As the valiant hero of a 1940 B movie titled *Murder in the Air,* actor Ronald Reagan safeguarded a futuristic laser weapon, the inertia projector, from a band of Communist spies. "Well, it seems the spy ring has designs on the greatest war weapon ever invented, which, by the way is the exclusive property of Uncle Sam," declared Reagan on the silver screen. At the climax of the film, Reagan pointed the inertia projector at the plane carrying the escaping spies and shot them out of the sky.[2] Forty-three years later, Reagan championed another futuristic weapon system, the X-ray laser missile defense system, as part of his "Star Wars" Strategic Defense Initiative (SDI).[3] While Reagan trumpeted the inertia projector as a Hollywood performer, and Star Wars as president of the United States, there is a growing body of evidence to suggest that he was following a fictional script in both roles.[4] Congressional investigations, whistleblower testimony, and a long streak of technical failure have raised serious questions about the legitimate nonfiction status of scientific research conducted to support Reagan's SDI proposal.

The Star Wars project was an example of politicized science par excellence. Reagan's visionary claims about SDI put enormous rhetorical pressure on the American military research establishment to produce scientific evidence showing the feasibility of a "leakproof" missile defense system. Since official missile defense advocates backed a version of SDI that featured a space-based X-ray laser, the potential brightness and focus of this laser took on considerable political significance as indicators of SDI realism in the public debates over missile defense in the 1980s.

With physics and politics hitched tightly together in the secretive cold war environment, the lines separating legitimate from partisan scientific research blurred. Director Edward Teller transformed the Lawrence Livermore National Laboratory (LLNL) into a pro-missile defense evidence factory able to harness the persuasive clout of science by maintaining its facade as an objective research institution. Fed a steady stream of cooked and garnished Livermore data, Reagan and other missile defense advocates cashed in rhetorically, then kept the cycle churning by earmarking billions of dollars for more research. "As enthusiasts rallied to the cause, politics overshadowed science. Strongly pushed by the President, SDI advanced with the cheerleading of the executive branch science advisors, the evasion of critical reviews, and the stifling of internal dissent," wrote MIT physicist Joseph Romm.[5]

The Star Wars episode raises several questions of appreciable interest to those pursuing understanding of the rhetorical dynamics of strategic deception in missile defense advocacy. What was the rhetorical motivation for Reagan to champion the X-ray laser? How did political factors shape the course of scientific research? How did reported scientific results influence political arguments about SDI? What accounted for the eventual downfall of the X-ray laser program? These questions steer attention to the nexus between laboratory science and public argument, a translation zone where scientific findings are sluiced through political channels to form public arguments, and where, conversely, political imperatives filter down to influence experimental activity in scientific laboratories and test ranges.[6] This chapter is designed to yield insight on the interplay between laboratory science and public argument in the context of the mid-1980s Star Wars debates. The chapter moves first through criticism of Reagan's Star Wars address, then examines the "nuts and bolts" behind LLNL experiments, and finally considers the political stakes involved in SDI strategic deception.

In the mid-1980s, political debates on SDI were colored vividly by the lines of argument sketched in Reagan's famous "Star Wars" address. Accordingly, part one of this chapter embarks on a textual analysis of the Star Wars address to explore the dramatic role Reagan's missile defense vision played in setting the tone for political and military affairs during the late stages of the cold war. To illuminate the larger context of political factors that coalesced to produce Reagan's SDI proposal, official correspondence between Reagan and his science advisors, retrospective historical accounts, and congressional reports are consulted to shed light on the process through which the Star Wars idea took hold in official circles. Since Reagan's Star Wars address ended with a challenge to the American scientific community to produce a workable missile defense system, discussion of the president's speech will flow smoothly into the next section of the chapter, where aspects of strategic deception in scientific research are explored in detail.

On its own terms, the subject of strategic deception in Reagan-era SDI laboratories qualifies as a fascinating case study in aberrant science. The charges of heavy-handed intimidation, excessive secrecy, doctored experimental results, and rigged test designs levied against Star Wars operatives such as Teller and his assistant Lowell Wood deserve investigation, simply because the prospect of systematic fraud in such a massive government research program counts as an alarming public policy concern.[7] Part two of this chapter pursues such a line of investigation by examining internal Department of Energy (DOE) memoranda, correspondence, public reports, congressional hearings, and popular accounts of the ground-level science involved in the X-ray laser program, the technical centerpiece of Reagan's Star Wars vision.

By focusing on the connections between public arguments advanced by SDI advocates and scientific research conducted at Livermore, the third part of this chapter elucidates the ways in which allegedly doctored data served rhetorical purposes in public spheres of deliberation, as well as how broad political currents fed back into Star Wars research and development strategies. The critical perspective afforded by this dialectical emphasis will provide unique insight into the various retrospective judgments that have been advanced by commentators in this case. Was strategic deception on SDI justified? The chapter will close with extended reflection on this controversial question, a question whose answers speak volumes about the general appropriateness of scientific trickery as a tool of missile defense advocacy in contemporary times.

The X-ray Laser and Reagan's MAD Escape

In 1977, Livermore physicist George F. Chapline conceived of the X-ray laser as a tool that could be used for civilian medical and engineering applications.[8] This discovery caught the attention of the man then in charge of the nation's military space program, Air Force lieutenant colonel Dr. Robert M. Bowman (veteran of 101 combat missions in Vietnam and holder of doctoral degrees in aeronautics and nuclear engineering from Cal Tech). Soon after Chapline's discovery, Bowman penned a classified report to the Pentagon in 1977 that probably contained the first-ever reference to missile defense as "Star Wars."

> One of the things that I pointed out in there with regard to the capabilities of laser battlestations, was that these new capabilities come right out of Star Wars. They probably will never have any defensive capability, but they have awesome offensive capabilities, such that if one side successfully deployed a system, they would have absolute military superiority. And if both sides had such systems, it would probably mean inevitable nuclear war. What I recommended was a freeze on nuclear technology and directed energy technologies.[9]

A different story regarding the promise of space weapons emerged from Senator Malcolm Wallop (R-WY), who published an article in *Strategic Review*, claiming that technology for chemically powered lasers was being developed that could render the nuclear balance of terror obsolete.[10] In August 1979, Wallop sent a copy of his article to Reagan, who was then preparing for what would become a successful presidential campaign.[11] Wallop followed up shortly thereafter, lobbying Reagan at a campaign barbecue in the Sierra Nevadas, hosted by Senator Paul Laxalt (R-NV).[12] Reagan received similar visits on the campaign trail. In February 1980, Lt. Gen. Daniel O. Graham, a staunch advocate for space weapons, was called in to brief Reagan in Nashua, New Hampshire, prior to a debate between Reagan and primary opponent George Bush.[13]

Meanwhile, Livermore director Edward Teller and his assistant Lowell Wood recognized the potential of the laser as a mechanism for space-based missile defense. Teller and Wood speculated that when fired from space, the laser could intercept and disable enemy ICBMs during their initial "boost phase" of flight.[14] Motivated by the promise of this potential application, they ignored Bowman's recommendation for a freeze on research, and instead urged Livermore researchers to pursue further X-ray experimentation aggressively.[15]

Teller took the 1980 presidential election of Ronald Reagan as a grand opportunity to secure support for the X-ray laser project at the highest levels of government. Building on a strong personal rapport with Reagan that was first cultivated when Reagan was governor of California,[16] Teller's crusade started with a successful bid to install understudy George Keyworth as science advisor in May 1981.[17] With firebrand X-ray laser proponent Keyworth inside the White House, Teller's influence surged, and it was not long before the first formal White House meeting on missile defense took place, in September 1981. At this meeting, Teller urged several high-ranking Reagan administration officials to accept the imperative of missile defense and develop a nuclear doctrine based on "assured survival," instead of "assured destruction."[18]

On 8 January 1982 another high-level meeting on missile defense took place in the Roosevelt Room of the White House. With President Reagan in attendance, key members of the "High Frontier" SDI group, including staunch missile defense advocate Graham,[19] concluded that the "time was ripe for the president to make a speech announcing the start of a crash program of antimissile research similar to the Manhattan Project."[20] The president responded to the meeting enthusiastically, following up with a letter stating that "you can be sure that we will be moving ahead rapidly with the next phase of the effort."[21]

With momentum behind the SDI project building at the White House, Teller jockeyed to get a private audience with Reagan by making a plea on the 15 June 1982 edition of the television show *Firing Line.* Having seen the program, Reagan responded with an invitation for Teller to come to the White House.[22] Teller accepted the invitation and met Reagan in the Oval Office on 14 September 1982. At this meeting, Teller stressed that space-based missile defense represented a breakthrough type of "third generation" nuclear weapon that promised to completely revamp nuclear strategy. Reagan asked if such a system could actually be designed, and Teller responded, "we have good evidence that it would."[23] Years later, Representative George Brown (D-CA) reflected on this pattern of correspondence: "It's no wonder President Reagan bought the Star Wars concept hook, line, and sinker, when he was getting information such as this from Dr. Teller."[24] Historian Frances FitzGerald notes that it is easy to overestimate the degree of influence that space enthusiasts Wallop, Graham, and Teller probably had on Reagan, because a big part of their missile defense promotion campaign involved nurturing the "Star Wars legend" that Reagan's exuberance for a defensive shield was the direct result of their lobbying.[25] In reality, the influence of space weapon proponents was just one of several factors motivating the president to give his famous 23 March 1983 "Star Wars address."

Inside the White House, there were also strong political currents driving Reagan's decision. Early in 1983, strident critiques of the administration's military buildup picked up, as the movement to freeze deployment of nuclear weapons gathered steam. When Reagan called the Soviet Union an "evil empire" in a fiery speech to the National Association of Evangelicals on 8 March 1983, pressure from the freeze movement increased. With sarcastic editorial cartoons portraying the president as Darth Vader circulating in the media, White House staff members then set about preparing for another address scheduled for 23 March 1983.[26] Initial drafts of this speech, provided by the Department of Defense (DOD), contained shopworn themes about the dangers of Soviet imperialism and the irresponsibility of nuclear freeze proponents.[27] Since Reagan balked at the idea of giving the speech as prepared by DOD, he apparently asked National Security Advisor Robert McFarlane to write up an "insert" on defensive weapons.[28] Reagan requested this revision, in part because he wanted to "break something new" in the speech, but also because he wanted to signal to voters that his massive military buildup against the "evil empire" was not the blunder it was made out to be by freeze movement activists.[29] Deputy Chief of Staff Michael Deaver saw the inclusion of this section on space weapons "as a campaign issue because it held out hope to the American voters that the nuclear threat would be neutralized, thus blunting Democratic attacks on Reagan as a warmonger."[30]

In the days leading up to the Star Wars address, McFarlane, Deaver, and a select few advisors in Reagan's inner coterie revised and refined the new section of the speech. Part of the single entry in Reagan's diary on 22 March 1983 reads: "Another day that shouldn't happen. On my desk was a draft of the speech to be delivered tomorrow night on TV. This one was hassled over by NSC, State and Defense. Finally I had a crack at it. I did a lot of rewriting. Much of it was to change bureaucratese into people talk."[31]

Reagan's dramatic appeal caught many by surprise. Members of the defense establishment and scientific community had not been informed that such a significant shift in strategic doctrine was in the offing.[32] After reading a draft of Reagan's speech hours before it was given, Secretary of State George Schultz responded with skepticism, remarking "[w]e don't have the technology to do this."[33] According to Reagan biographer Dinesh D'Souza, Reagan anticipated such reactions, and "had the concept of the Strategic Defense Initiative (SDI) developed only by the internal White House staff. . . . By the time other cabinet departments and agencies were shown the policy, it was a fait accompli."[34] As the president's proposal circulated after the speech, it "ignited a controversy over what was held by the public and experts alike to be a settled matter: the inability to defend a country from attack by nuclear missiles," explains G. Thomas Goodnight.[35] Indeed, the Star Wars address thrust missile defense, which had been a dormant public issue ever since the signing of the 1972 ABM Treaty, back to the center of national public attention. As Howard Ris Jr. of the Union of Concerned Scientists observed, "[s]eldom has an arms issue moved as quickly from the periphery to the center of the policy arena as the space weapons issue."[36] Soon after the Star Wars address, robust discussion of BMD issues broke out in Congress,[37] in the pages of academic journals,[38] and in the popular press.[39] According to Michael Mazarr, analyst with the Center for Strategic and International Studies, SDI "caused more controversy than probably any other public policy issue of the Reagan years."[40] This policy discussion unfolded against a backdrop of cultural fascination with the fantastic Hollywood images depicted in George Lucas's dramatic *Star Wars* films.[41]

Such a burst of public deliberation on national defense could be seen as an outgrowth of Reagan's speech, where he enacted a rare, participatory moment in the usually sterile and alien expert dialogue on nuclear deterrence. By appearing to draw the solution to the cold war's ominous nuclear dangers from the audience itself, Reagan opened up a space for citizens to see themselves as relevant and potentially efficacious actors in the drama of superpower politics and diplomacy. This maneuver was set up by a vivid exposition on the insidious Soviet threat, and executed through a series of suggestive rhetorical questions. These questions coaxed viewers and listeners to follow their own trains of logic on the way to the seemingly inescapable conclusion that missile defense promised a real way out of MAD's unsavory nuclear dilemmas.

Before Reagan enacted this participatory moment in the closing stages of the address, he set the stage by cataloguing multiple dangers posed by reckless Soviet imperialism. "They're spreading their military influence,"[42] Reagan warned as he showed aerial photographs documenting Communist military excursions into America's "backyard" neighborhood in Central America and the Caribbean Basin. Priming the audience with a sense of cold war claustrophobia, Reagan described a Soviet intelligence facility in Lourdes, Cuba, where, "1,500 Soviet technicians" attended to "acres and acres of antennae fields and intelligence monitors" pointed at the United States (439). The president's narrative situated the U.S.S.R. as an "offensive" military power bent on world domination at any cost (440). "These pictures only tell a small part of the story," Reagan explained. "[T]he Soviet Union is also supporting Cuban military forces in Angola and Ethiopia. . . . They've taken over the port that we built at Cam Ranh Bay in Vietnam," he warned (440). In Reagan's view, this surge in Soviet expansionism was particularly troubling in light of recent declines in U.S. military readiness. "When I took office in January 1981," Reagan recounted, "I was appalled by what I found: American planes that couldn't fly and American ships that couldn't sail" (440). Indeed, for much of the Star Wars address, Reagan painted a gloomy picture of the cold war, where surging Soviet strength combined with increasing American vulnerabilities to cast dark clouds of doubt on the future.

Near the end of the address, however, Reagan struck a lofty tone. Speaking about recent budget victories that had resulted in hefty defense spending hikes, the president said "[t]hanks to your strong support . . . we began to turn things around" (440). With high-tech projects such as the B-1 bomber and conventional force modernization under way, Reagan hinted that American technical know-how and ingenuity might be able to turn the tide of the cold war. Reflecting on the prospects of achieving a more robust defense through buildup of offensive arms, Reagan said, "[t]his approach to stability through offensive threat has worked" (442). However, because such a strategy was based on "deterrence of aggression through the promise of retaliation" (442), Reagan questioned its wisdom because it relied upon the implied threat of savage nuclear attack. "I've become more and more deeply convinced," Reagan said, "that the human spirit must be capable of rising above dealing with other nations and human beings by threatening their existence" (442). There must be some way, Reagan mused aloud, "to break out of a future that relies solely on offensive retaliation for our security" (442).

With a sense of urgency heightened by dark discussion of the encroaching Soviet threat, Reagan then offered "a vision of the future which offers hope" (442). This vision involved a crash program to pursue development of defensive arms that could render nuclear weapons "impotent and obsolete" (443). In the earlier portions of the speech, Reagan had built a case for cold war vigilance by dwelling exhaustively on details of fresh Soviet aggression. When it came time for Reagan to share his vision of defensive arms, however, he eschewed such an expositive strategy, turning instead to a series of rhetorical questions to press his case. "What if free people could live secure in the knowledge that their security did not depend on the threat of instant U.S. retaliation to deter a Soviet attack, that we could intercept and destroy strategic ballistic missiles before they reached our own soil or that of our allies?" Reagan asked (442).

Reagan's approach relied on the enthymeme, a rhetorical strategy that Aristotle singled out as one of the most effective tools of persuasion.[43] The enthymeme is an argument "with one part unspoken or suppressed," explain Kathryn Olson and G. Thomas Goodnight. In enthymematic argumentation, "the arguer depends on the auditor's ability and willingness to supply the missing portion from shared knowledge, experiences, or assumptions and so complete the argument."[44] The enthymeme is said to be particularly persuasive because it involves audience members directly in the creation of proofs, thereby establishing a sense of identification between speaker and audience. Reagan's Star Wars address led audience members to complete his rhetorical questions with answers that pointed to the necessity of building a missile defense. One question boiled the issue down to an obvious level: "Wouldn't it be better to save lives than to avenge them?" (442). Another question dealt with the issue of expense: "Isn't it worth every investment necessary to free the world from the threat of nuclear war?" (443). Yet another question concerned technical feasibility: "Are we not capable of demonstrating our peaceful intentions by applying all our abilities and our ingenuity to achieving a truly lasting stability?" (442).

With these questions, Reagan prompted listeners and viewers to construct their own cases for missile defense, as they listened to the president's address. Using an enthymematic style of argument, he invited audience members to complete his case, using their own logic. Of course it would be better to save lives than to avenge them. Yes, we should spare no expense in pursuit of peace. Absolutely, our ingenuity can surmount any technical challenge. In the coming years, these main points would become bread-and-butter moves in missile defense advocates' rhetorical repertoires. Reagan's speech was designed to allow audience members to walk away with these conclusions in mind, after reaching them on their own accords.

The fulcrum leveraging all these conclusions was Reagan's claim that missile defense could transcend nuclear deterrence and emancipate the

United States from an immoral MAD framework. According to the president, Star Wars was "a long-term research and development program to begin to achieve our ultimate goal of eliminating the threat posed by strategic nuclear missiles" (442). "More than a visionary's dream," writes Goodnight, "'Star Wars' purported to set in motion the means by which the world could be liberated from the terrors of the nuclear age."[45]

During question-and-answer sessions with reporters in the days following the Star Wars address, Reagan invoked the image of adversaries holding cocked guns, pointed at each other, to dramatize what he saw as the madness of nuclear deterrence.[46] There were two possible ways out of this dire situation, Reagan reasoned. One way was arms control. "There is another way, and that is if we could, the same scientists who gave us this destructive power, if they could turn their talent to the job of, perhaps, coming up with something that would render these weapons obsolete," the president told reporters.[47]

The integrity of Reagan's appeal to science depended on a major technical assumption, that it would be feasible to develop a virtually perfect "leakproof" or "Astrodome" defense, capable of neutralizing all incoming enemy missiles. Only such a complete system could transcend deterrence and emancipate the American people from the immoral threats of MAD. A partial system, or a system unable to completely knock out all enemy missiles, would serve only to supplement deterrence by complicating the decision making of adversaries contemplating nuclear strikes on the United States.[48] As Secretary of Defense Caspar Weinberger explained, "[t]he defensive systems the President is talking about are not designed to be partial. What we want to try to get is a system . . . that is thoroughly reliable and total."[49]

By defending a leakproof version of SDI, missile defense advocates were able simultaneously to argue for a continued military buildup and also appeal to popular sentiment against nuclear weapons. "The promise of 'perfect defense' enabled the President to appeal for the SDI program over the heads of Congress, direct to public opinion," writes historian Edward Reiss. "Key Democrats were probably 'swung' this way, especially as the first SDI budget was debated in the shadow of elections for the House of Representatives, as well as for the presidency, in November 1984."[50] Reagan's tone on this point hearkened back to the president's series of meetings with Teller in 1982, where Teller sold Reagan on the notion that the X-ray laser was a "third-generation" weapon, with potential to fundamentally change the nature of warfare. As Janice Hocker Rushing explains, the vision of perfect defense also jibed well with Reagan's use of mythos to cast a government science project (SDI) into a restorative call for science to transcend the "original sin" of the atomic bomb.

> Star warriors and Jedi knights aside, Reagan's vision of the future is not just another sci-fi movie, for its promise is restorative. His challenge to science evokes sacred connotations. . . . 'Those who give us nuclear weapons' now are chosen to atone for the Original Sin by accepting not punishment, but a new challenge. . . . The future is the past, only this time, the same 'snake in the grass' that tempted us out of Paradise in the first place—science—is also the agency of our return.[51]

Instead of committing to a specific technology or approach to missile defense in the Star Wars address, Reagan left it up to missile defense scientists to concoct an optimal mix of systems that would achieve the objectives laid out in the speech.[52] As a first step in this direction, immediately following the Star Wars address, Reagan convened two expert panels (the "Hoffman" and "Fletcher" panels) to study issues of BMD feasibility and political workability. Since many of the members of these panels were drawn from the ranks of private BMD contractors who had already been working on missile defense R & D for years,[53] it was not surprising that the advisory commissions produced rosy assessments of exotic BMD technologies, such as chemical, X-ray and free electron lasers, neutral particle beams, and railguns.[54] BMD advocates utilized this array of technical options to maximize their maneuvering room in public arguments. Even though each specific technology might not be perfect in isolation, advocates asserted that a combination of imperfect technologies working harmoniously in a "layered defense" could compose a virtually impenetrable shield.[55]

While SDI advocates relied heavily on the "layered" defense argument following Reagan's Star Wars address, the White House eventually began citing the X-ray laser as the missile defense technology that had the best chance of fulfilling the Star Wars vision. "The integrity of the X-Ray laser studies is considered a major point by both critics and supporters of the 'Star Wars' research program. By many accounts, the promise of the bomb-powered laser helped influence Reagan to begin the $3 billion-a-year space weaponry program in 1983," writes journalist Deborah Blum.[56] Could Star Wars have been launched successfully without the X-ray laser? As *New York Times* science writer William Broad reflects, "[i]t seems unlikely. No other weapon had the allure. It and it alone suggested the field had reached a new level of maturity.

Whether the silver bullet would actually save America was immaterial. What was important was what it symbolized."[57]

The proposed space-based X-ray laser platforms were each theoretically capable of generating dozens of simultaneous nuclear-pumped laser bursts that could intercept Soviet ICBMs early in the boost phase. This sort of capability attracted the interest of advocates committed to the project of leakproof defense. In its final report, the Fletcher panel recommended more funding for the X-ray laser than for any other SDI technology.[58] The popular media also played a role in highlighting the X-ray laser. For example, the cover of the 4 April 1983 *Newsweek* issue devoted to Reagan's Star Wars address featured artist Jeffrey Mangiat's rendering of an exotic-looking X-ray laser platform orbiting in space.[59]

With the prospects for success of Reagan's rhetorical strategy tied to the possibility of developing a leakproof missile defense, the technical issue of X-ray laser feasibility quickly took on enormous political significance. Reagan's tack fused together the scientific question of whether American researchers could assemble sharply focused, superbright X-ray lasers with the political question of whether the country could escape the balance of nuclear terror inherent in the MAD framework.

Seen in this light, rather than constituting a completely self-sufficient argument, the Star Wars address was really the skeleton of an argument that awaited subsequent validation in the form of scientific proof of X-ray laser feasibility. To make the leap from science fiction fantasy to legitimate military option in credible fashion, Reagan's SDI proposal thus depended on scientific support from Livermore. As Reagan said in the Star Wars address, "I know this is a formidable, technical task, one that may not be accomplished before the end of the century. Yet, current technology has attained a level of sophistication where it's reasonable for us to begin this effort."[60] It was as if Reagan was giving a Jedi-like charge to the missile defense scientists preparing to embark on this monumental project: "May the Force be with you."[61]

Indeed, Reagan's new commitment to defensive arms hitched his political wagons to the laser beams generated by Star Wars scientists in experiments under desert test ranges. According to Rebecca Bjork, "Reagan's moralistic stance would not have been effective without the support of the technological community. He needed at least some testimony that his version of SDI would be technically feasible. Without this grounding in technical 'fact,' his vision of a post-SDI world would be dismissed as mere fantasy."[62]

Desert Tests and Teller's Zest

Scientific support for Reagan's Star Wars vision eventually came in a string of experiments that were hailed as proof that controlled nuclear explosions could generate X-ray lasers of sufficient focus and brightness to intercept enemy missiles from space. Missile defense advocates pointed to the announcement of each successful test as evidence that the Star Wars project was not only realistic but progressing steadily toward the eventual goal of an operational space-based system. According to FitzGerald, "In congressional hearings senior Defense Department officials, distinguished scientists and strategic policy analysts argued about laser weapons and boost-phase defenses as if these weapons were about to jump off the assembly line."[63] Skeptics' arguments that labeled the SDI vision as science fiction fantasy were rebutted with "objective" scientific evidence from Livermore.

The parade of public announcements proclaiming experimental success functioned rhetorically to build momentum for SDI, and by creating the illusion of scientific progress, invite sympathetic projections of future success. This rhetorical approach was strategically codified in official Strategic Defense Initiative Organization (SDIO) documents. In explaining to Congress the role of scientific feasibility experiments, SDIO stated that "[t]hese activities will also provide a timely, visible, and understandable set of milestones with which to measure the program's progress and accomplishments."[64] Guertner notes that through a kind of appetizing effect, this rhetorical approach created significant financial momentum behind the snowballing SDI research program.

> If Congress could clearly see the final price tag, strategic defense would have little chance of surviving the scrutiny of deficit-minded legislators. But incremental funding, technological optimism, ambiguous standards of proof—for example, validation of components rather than systems reliability—and predictably shrill Soviet reactions may combine to propel it through the appropriations process for many years. This process would resemble a recipe the late Illinois Senator Everett Dirksen once read during a Senate filibuster. His recipe (read "tactic") for cooking frogs cautioned against plopping them directly into boiling water because they would jump right out and mess up the kitchen: "better to put them in a pot of cool water, turn the heat on low, cover the pot, and bring the poor critter to a slow boil." Dirksen would have known how to get appropriations for the SDI.[65]

The feasibility evidence marshaled by Star Wars advocates was drawn largely from five experiments conducted beneath desolate stretches of desert in the western United States from 1980 to 1985. Code-named Dauphin, Cabra, Romano, Cottage, and Goldstone, these experiments were cited by missile defense advocates as proof of X-ray laser workability in a variety of forums. However, it now appears that the argumentative proof derived from these experiments was based on counterfeit data. A careful retrospective assessment of the manner in which these experiments were designed and executed, and the way they were interpreted by Livermore's top spokespersons, raises serious questions about the integrity of the scientific process in the X-ray laser program. These questions become apparent when one considers four aspects of management surrounding X-ray laser experimentation: the design and timetable of the experiments themselves, the official interpretation and reporting of experimental data, the containment of local dissent at Livermore, and the global muzzling of controversy over X-ray laser research in a variety of quarters outside the laboratory. The following discussion presents a more complete behind-the-scenes picture of the science practiced in the X-ray laser project, by working through each of these aspects in turn.

Test Design and Timing

In Star Wars research, the design and timing of the X-ray tests were driven largely by political factors, often at the expense of prudent scientific planning. As *Los Angeles Times* reporter Robert Scheer reported, "the [X-ray] program has fallen victim to politics, and the search for 'spectacular results,' in the words of one such critic, 'has overridden careful physics.'"[66] The intense demand for data confirming the viability of the X-ray project sped up research to rates that compromised the ability of scientists to control and understand the experiments. "'Pressure to go faster,' one federal scientist said, 'means making mistakes like relying on a calibration system they didn't fully understand which gave a false large signal.'"[67]

The most vivid specific example of this tradeoff between speed and scientific integrity was the Goldstone experiment. There were serious doubts raised about the reliability of the experimental design of the precursors to Goldstone, the Romano and Cottage tests in 1985. Los Alamos scientists Gottfried Schappert and Donald Casperson found that the beryllium reflector used in these tests inappropriately magnified the nuclear-pumped X-ray laser, producing exaggerated brightness readings.[68] These findings were verified by an independent internal Livermore review later that month,[69] and further supported by a review conducted by the prestigious "JASON" group, an independent committee of senior scientists.[70]

Federal scientists urged that the Goldstone experiment be delayed so that the detector problems could be ironed out.[71] On 4 December 1985 Representative Edward Markey (D-Mass.) took the floor of the House and stated, "the Administration should delay the planned Goldstone test until it has rectified these technical problems. It should not be rushing to continue expensive X-Ray laser tests that may yield little useful data."[72] Thirty of Markey's colleagues echoed his concerns in a letter to Secretary of Defense Caspar Weinberger that urged the Goldstone timetable to be pushed back.[73]

Weinberger ignored the congressional outcry and Livermore proceeded with the Goldstone experiment as planned. "Lab officials accepted the arguments of test proponents that a delay would have unfavorable political repercussions for the program," reported Robert Scheer.[74] Thus, on 28 December 1985 the Goldstone experiment went off as scheduled. However, the results were very discouraging for Livermore researchers; the brightness of the X-ray laser by the nuclear blast was far dimmer than expected. As Broad explained, it was "like the owner of a new car suddenly discovering that his engine produced 10 horsepower instead of the 100 advertised by the dealer."[75]

The Goldstone experiment was not an isolated case of design manipulation. In a separate line of tests, Star Wars researchers outfitted test mechanisms to maximize chances for favorable experimental outcomes. This test series, part of the "Homing Overlay Experiment" (HOE), was conducted to assess the effectiveness of kinetic kill interceptors, small ground-based rockets aimed at incoming missiles. HOE interceptors were designed to fly to an intercept point and destroy target missiles. Although the kinetic kill technology featured in these experiments had only an indirect bearing on the technical prospects of the X-ray laser project, missile defense advocates nevertheless trumpeted HOE success emphatically as general proof of Star Wars feasibility.

HOE involved four series of experiments conducted between February 1983 and June 1984, at Vandenburg Air Force Base in California and Mech Island in the Kwajalein Atoll. In the first three tests of the series, sensor cooling anomalies, software problems, and random failures in guidance electronics prevented HOE interceptors from homing in on and destroying incoming target missiles.[76] In the fourth and final series of tests, conducted on 10 June 1984, the air force reported dramatic success; officials indicated that the HOE interceptor had successfully homed in on, tracked, and destroyed the incoming target missile (see Fig. 1).

Following the tests, missile defense advocates claimed a technical breakthrough; according to commentator David Osborne, "[t]he interception was instantly held up by Washington as the first proof of the feasibility of President Reagan's elaborate proposal."[77] Advocates such as William F. Buckley claimed the successful experiment showed that the "axiomatic geometry of missile defense" had been validated, that it was possible to "hit a bullet with a bullet."[78] A Defense Department official said, "[t]his experiment that resulted in the destruction of a Minuteman reentry vehicle shows that the Strategic Defense Initiative is not the 'star wars' concept that it has been tabbed."[79] According to Reiss, Star Wars advocates broadcast the "bullet hit a bullet" HOE test sequence around the globe, as part of a promotion campaign that "encouraged widespread belief in 'Star Wars' systems."[80]

Speaking about the HOE test, SDIO director James Abrahamson stated that "[t]he test achievements vividly demonstrate the undergirding technical genius resident in our society and confirm the readiness to proceed with the Strategic Defense Initiative."[81] Maj. Gen. Elvin Heiberg, commander of the Army Ballistic Missile Defense Systems Command, trumpeted the fourth HOE experiment as "an absolutely tremendous success," and was quoted on the front page of the *Washington Post* as saying, "[w]e do know we can pick 'em up and we can hit 'em."[82] The official emblem of the HOE project team projected a similar air of invincibility—it showed a war-hardened duck shooting a laser beam out of its forehead to intercept incoming missiles (see Fig. 2). For years, the apparent success of the fourth HOE test was memorialized in an exhibit at the Smithsonian Institution and then the U.S. Space and Rocket Center Museum in Huntsville, Alabama. The exhibit featured a model HOE kill vehicle hanging from the ceiling, with a photograph of the test in the background, accompanied by the caption "HOE 1984 intercept" (see Fig. 3).

Astonishingly, in 1993, nine years after the key HOE test, air force scientists came forward to disclose that they had "rigged" the experiment. After failures in the first three HOE tests, one scientist said, "we would lose hundreds of millions of dollars in Congress if we didn't perform [the fourth test] successfully."[83] To ensure success, scientists set up the experiment in a way that would virtually guarantee that the interceptor missile would home in on and destroy the target Minuteman missile. First, a homing beacon was placed on the target missile and a receiver was fastened to the interceptor, enabling the two to communicate. In effect, the target missile was saying "here I am, come and get me."[84] Second, the target missile was artificially heated before launch; this increased its visibility to the heat-seeking HOE interceptor missile.[85] Third, "to hedge against the possibility of failure . . . test managers secretly installed explosives on the target warheads that would detonate automatically in the case of a near-miss."[86]

The Defense Department's muted reaction to these 1993 disclosures stopped well short of flatly denying such rigging, and Pentagon officials even acknowledged that the HOE deception strategy was more substantial than initially reported. Asked about the disclosures, former secretary of defense Caspar Weinberger told the *New York Times,* "[y]ou're always trying to practice deception. You are obviously trying to mislead your opponents and make sure they don't know the actual facts."[87] Citing an internal Pentagon review, Secretary of Defense Les Aspin said that the deception components were in place for only the first three HOE experiments, in which the interceptor missile did not even come close enough to the target for the tracking and detonation aids to be relevant. The fourth experiment, of "bullet hit a bullet" fame, Aspin explained, took place after the deception program had already been canceled.[88] Aspin thus intimated that the fourth HOE was a legitimate scientific exercise; but his murky explanation failed to establish this fact forthrightly.[89] In a similar manner, retired major general Eugene Fox (in charge of HOE at the time of the tests) denied having knowledge of the fact that the test was rigged, but agreed that it would have been possible for scientists in the field to manipulate the experimental apparatus without oversight.[90]

One further example of politicized test design was a September 1985 experiment in which researchers fired a ground-based directed energy weapon, the Mid-Infrared Advanced Chemical Laser (MIRACL), at a model of a Titan missile, strapped to the ground. While the Titan missile eventually exploded after being subjected to the heat of the laser at close range, the experiment demonstrated nothing; even a "high-powered rifle would have had a similar success against the thin-skinned missile," wrote MIT's Romm.[91] Nevertheless, SDIO director James Abrahamson called the ho-hum event a "world-class breakthrough."[92] Reacting to the MIRACL test, a dissenting scientist at Sandia National Laboratory wrote of "strap-down chicken tests, where you strap the chicken down, blow it apart with a shotgun, and say shotguns kill chickens. But that's quite different from trying to kill a chicken in a dense forest while it's running away from you."[93]

Embellishment of Test Results

The translation zone that connects the scientific laboratory to the political realm presents fertile ground for advocates to craft and spin experimental data so that such data jibes with their political objectives. In the case of Star Wars experimentation, even when scientific tests clearly failed (e.g., Cottage, Goldstone), such failures did little to temper Teller and Wood's overzealous plugging of the X-ray project to external audiences. Their practice of embellishing test results in briefings and statements to policy makers and the media represented a second major strategy employed to promote the X-ray project. "To keep the Excalibur [X-ray] goose laying its golden eggs, a certain level of hype about progress toward laser action was necessary," said physicist Antoine Churg.[94] "Briefings from Teller and Wood may have led President Reagan and other high officials to believe that a wonder weapon—the X-Ray laser—capable of obliterating hundreds of missiles in a single burst was close to becoming a reality," wrote defense analyst Charles Bennett.[95] As Markey commented, "[t]hese letters show Edward Teller making wildly optimistic claims that we were on the verge of developing a silver bullet that could literally end the threat of nuclear war. The motivation behind these claims was Teller's desire to prevent the negotiation of any Star Wars limits."[96]

Closer examination of key Livermore memoranda elucidates the extent to which the technical promise of the X-ray project was oversold. Missile defense advocates began deliberate exaggeration of technical details well before Reagan's Star Wars address. In a 21 October 1981 letter to Attorney General Edwin Meese, High Frontier panel member Karl Bendetsen stated that the High Frontier group had reviewed the X-ray laser project and found that the X-ray device had the capacity to channel the energy from a hydrogen bomb explosion "into one or more tightly focused and independently aimable beams. Each of these beams has about a million times the brightness of the bomb's undirected energy, so that the lethal range of a sub-megaton bomb can readily be extended to distances of thousands of miles."[97] Further, the letter claimed that Livermore had promised to deliver "a fully weaponized [X-ray laser] for ballistic missile defense on a five year time scale."[98]

These extravagant claims of brightness and quick deployment potential diverged wildly from the capabilities of the laser demonstrated at that time,[99] but because Teller had effectively co-opted the High Frontier group, Bendetsen's letter reached the White House without being reviewed by other scientists at Livermore. Teller's unchallenged reportage of events at Livermore cultivated a belief among top White House officials that experiments had established the boost-phase intercept capability of the X-ray laser (its range extended to thousands of miles), and that such a laser could be produced within five years.

Teller further deceived the White House with his announcement of the results of the Romano experiment in a 22 December 1983 letter to George Keyworth, the president's science advisor. In the letter, Teller labeled the experiment a "clear-cut scientific breakthrough,"[100] wrote that the results of the test were so dramatic that "we are now entering the engineering phase of X-Ray lasers," and concluded that "we have developed the diagnostics by which to judge every step of the engineering process."[101] This invocation of engineering terminology by Teller implied that all basic doubts regarding X-ray lasing had been resolved, and that Livermore was ready to make the transition from the research stage to the weapons production stage.[102] However, as Broad noted, the power of the laser "wasn't 1/100th or even 1/1000th of what Teller advertised to the White House."[103] Teller's interpretation of the Romano test results was sharply at odds with views held by senior Livermore researchers at the time, foremost among them Roy Woodruff, leader of Livermore's key "A" Research Division. In response to Teller's wildly optimistic reports of X-ray laser brightness, Woodruff drafted a letter of clarification designed to dispel Teller's exuberance, although lab officials eventually blocked the letter from reaching the White House.[104]

A similar pattern of embellishment could be found in Teller's 28 December 1984 letter to Paul Nitze, the senior arms control advisor to the State Department.[105] In the wake of the 1983 Romano experiment, Teller's letter boasted to Nitze that the lab had made a breakthrough that indicated "a real prospect of increasing the brightness" of the X-ray laser:

> This is an exceedingly large gain, and even if it cannot be fully realized, this approach seems likely to make X-Ray lasers a really telling strategic defense technology. For instance, a single X-Ray laser module the size of an executive desk which applied this technology could potentially shoot down the entire Soviet land-based missile force, if it were to be salvo-launched. . . . It might be possible to generate as many as 100,000 independently aimable beams from a single X-Ray laser module.[106]

These claims were repeated by Teller in a shorter note to National Security Advisor Robert McFarlane sent that same day. But in the note to McFarlane,

Teller added that the dramatic enhancement of the power of the X-ray laser now had "some solid experimental foundation," and that a trillionfold enhancement in laser brightness "might be accomplished in principle in as little as three years."[107] This forecast diverged boldly from the official Livermore projections, which had not made any concrete timetable projection, given the preliminary state of the "Super Excalibur" X-ray project. Further, Teller's promise of dramatic advances in brightness and independent aiming capability were clearly unwarranted extrapolations from the quality of data available at the time (gleaned from the recent Romano test).[108] Nevertheless, Teller's claims were permitted to propagate freely throughout the military and political decision-making establishment. His outrageous assertion that an X-ray laser space platform the size of a desk could "shoot down the entire Soviet land-based missile force" strengthened the hand of missile defense advocates echoing Reagan's line that SDI could render nuclear weapons impotent and obsolete.

Another instance of systematically embellished test results occurred following the 1985 Cottage test, when Lowell Wood delivered a briefing to Central Intelligence Agency director William Casey on 23 April 1985. Following up on Teller's previous claims made to Nitze, Wood told Casey that the X-ray laser had "as many as 100,000 independently aimable beams." Wood also discussed the results of the Cottage experiment, stating that the test produced evidence of dramatic focusing gains and also "weapon-level" brightness.[109] These claims prompted another internal Livermore memo, authored by "R" Program scientist George H. Miller. Miller's memo stated that the sentiment of the "R" Group researchers was that Wood's overselling had created a "serious" problem.

> It is [not] as easy as Lowell states, our current capabilities are not as significant as he states, and the time scales are not nearly as short as he states. This conflict of views is in large part due to Lowell's verbal style in which he contends that anything which he believes is possible is likewise easy. He does not ever mention the inventions which are required, the new knowledge which must be gained or the real difficulties which must be overcome. . . . The problem is that his style is absolutely monolithic, uncompromising and frequently abusive personally to other scientists in the program.[110]

Local Controversy Containment

While memos by skeptical missile defense scientists such as Woodruff and Miller demonstrated the existence of internal doubt in LLNL regarding X-ray laser feasibility, these doubts were never amplified extensively in official circles or public spheres during the Reagan Star Wars era. One factor accounting for the effectiveness of missile defense advocates' strategy of embellishing Livermore's experimental test results was the local containment of controversy at the laboratory. While there were in-house dissenters, Livermore's top leadership made sure that these oppositional voices did not challenge the official public narrative or rock the proverbial missile defense boat.

A typical example of this process of local controversy containment took place in the aftermath of Teller's 1983 letter to the president's science advisor, George Keyworth, in which Teller announced the results of the recently conducted Romano experiment. Although Teller did not circulate that letter to other Livermore scientists prior to sending it,[111] a copy of the letter eventually found its way back to Roy Woodruff, research director of the "A" Division at Livermore, and supervisor of the Romano test. When Woodruff realized Teller had claimed that the Romano results showed the X-ray laser project was now ready for the "engineering phase," he was aghast. Woodruff felt that the Romano test results were far too tentative to warrant such a claim, and he went to Teller in protest, demanding that Teller draft a letter of clarification.

Although Teller agreed that parts of his letter were wrong, he refused to draft such a follow-up letter, reportedly saying, "I cannot. . . . My reputation would be ruined."[112] Rebuffed, Woodruff drafted a letter of his own to Keyworth which corrected Teller's account on several key points. As the leader of X-ray laser research at Livermore, Woodruff explained that the program had no "solid predictive ability" for laser results, and that "the X-Ray laser is nowhere near the engineering phase at this time."[113] However, when Woodruff went to see laboratory director Roger Batzel for permission to send the letter, he was stonewalled, and Teller's original claims were allowed to stand as the official LLNL position.

A similar course of events took place in the case of Teller's 1984 letters to Nitze and McFarlane. As with the earlier Romano episode, Woodruff learned of the existence of the letters only when George Budwine, Teller's aide, leaked them to Woodruff after they had been sent. Again Woodruff drafted letters of clarification, stating that Teller and Wood were being "overly optimistic" about the prospects of the X-ray laser, and that the military utility of the device was still a "matter of speculation."[114] Specifically, Woodruff disputed Teller's claims that fundamental advances could be made "in this decade," and that the X-ray device could generate as many as "100 independently aimable beams" capable of simultaneously demolishing as many targets.[115] In response to Teller's claim that the X-ray beams, with a trillionfold multiplication in brightness, could "shoot down the entire force of Soviet land-based missiles," Woodruff responded with a rhetorical question: "Will we ever develop a weapon close to the characteristics described? Not impossible, but very unlikely."[116] Once again, Woodruff's skepticism was squelched; Nitze and McFarlane never saw the clarifications, because Livermore director Batzel blocked the letters from ever being sent. Batzel added insult to injury by penalizing Woodruff with a demotion.

Woodruff subsequently filed a grievance with the University of California, LLNL's parent institution, charging that he was being blackballed for his candid criticism of the X-ray laser project. On 17 September 1987, a grievance panel ruled unanimously in Woodruff's favor. In the panel's decision, it noted that what had emerged at Livermore was a case of systematically distorted communication, an instance "of a senior employee unable to communicate with the Director despite considerable evidence of effort, and who consequently becomes isolated with no remedy available (e.g., relocation within the Laboratory, etc.), because of a lack of communication. The creation of such an 'unperson' status is unacceptable in an organization that endorses the principle of accountability of actions."[117]

Global Controversy Containment

The previous examples of controversy containment within the Livermore research community do not exhaust the full extent of efforts undertaken by missile defense advocates to frustrate the full development of frank, critical dialogue about the X-ray laser project in public spheres of discussion. There were attempts to dampen criticism in quarters outside the laboratory as well. These instances of what might be termed global controversy containment constitute a fourth form of strategic deception that contributed significantly to the SDI promotion campaign.[118]

One dramatic example of global controversy containment involved executive reaction to Woodruff's efforts to rein in Teller and Wood. After the University of California review board issued its finding that the LLNL leadership had erred in muzzling Woodruff's efforts to blow the whistle, Secretary of Energy John Herrington held a remarkable press conference at Livermore where he chided Woodruff for publicly airing his viewpoints. "I think there should be freedom of expression within the laboratory," Herrington said, "but I don't favor having scientists going public on opposite sides of the issue if it's going to be damaging to the laboratory. I think all this needs to be fought out inside the lab."[119] Herrington further castigated Woodruff for becoming "personally involved" in the dispute,[120] labeled the disagreement a "little squabble,"[121] and clearly expressed his preference for Teller's brand of science: "My personal feeling as Secretary of Energy is that Dr. Edward Teller is a national asset."[122]

Another instance of global controversy containment involved sanitization of the so-called Fletcher report. Following his Star Wars address, President Reagan commissioned a large-scale Defense Department study on SDI feasibility. Chaired by former NASA head James C. Fletcher, the panel delivered what appeared to be a solid endorsement of the X-ray laser system in October 1983, recommending that the government spend $1 billion on that form of missile defense. However, among the panel's conclusions was the finding that Soviet countermeasures could seriously complicate the capability of the X-ray laser to function as an effective mechanism for antimissile defense. Such doubts, coming so closely on the heels of Reagan's Star Wars address, and issued by a research board appointed by Reagan himself, would have surely given SDI opponents considerable argumentative fodder. As William Broad described, however, "political pressures mounted to keep the bad news from the public. . . . The seventh volume of the secret report—possible Soviet countermeasures—was left out of the overall antimissile assessment."[123] This omission colored dramatically the overall picture presented by the report. "If you read volume seven first, you wouldn't bother reading the rest of the report. It presents an overwhelming case against the possibility of a hope of mounting something useful. It quite unambiguously indicates the problem was insoluble unless certain things were solved that no one even knew how to address," said Theodore A. Postol, then a science advisor for the navy.[124]

With this substantial reservation shielded from public view, media accounts of the Fletcher report were unduly optimistic. *Aviation Week and Space Technology* quoted a highly placed Defense Department official as saying the Fletcher group "provided a report far more positive and enthusiastic on directed-energy weapons than any of us expected."[125] The *New York Times* reported that "the Fletcher panel report placed great emphasis on X-

Ray lasers as perhaps the most promising future technology to block hostile missiles."[126]

Reagan's subordinates also pursued global controversy containment by simply bribing foreign leaders into expressing support for SDI by dangling the prospect of scientific pork barrel contracts. After the British prime minister expressed doubts about SDI's feasibility to Reagan, National Security Advisor McFarlane explains that Reagan dispatched him on an international spin control mission to dampen growing scientific criticism across the Atlantic.

> [The president] called me into the Oval Office the next morning and he said, "Bud, Margaret and I are just not getting along on this SDI issue. I wish you'd go to London and see if you can't at least lower the level of criticism publicly. We're going to have a rough time getting appropriations if this keeps up." She gave me the same lecture she had given two weeks before, and, seeing I was getting nowhere, I interjected during a pause, "Prime Minister, President Reagan believes that there is at least $300 million a year that ought to be subcontracted to British companies that would support SDI." And there was a long pause. She finally said, "There may be something to this after all."[127]

Scrutiny of the Star Wars X-ray laser program reveals a pattern of strategic communicative deception practiced on four levels: experimental design and timing, reportage and spin of test results, local controversy containment, and global controversy containment. Together, these forms of manipulation enabled missile defense advocates to bolster rhetorical arguments for Reagan's Star Wars vision. Such arguments were supported by appeals based on space-based missile defense using the X-ray laser.

Brilliant Bluff or Dangerous Delusion?

Each of the modes of strategic deception practiced by the Pentagon during the Star Wars era was an outgrowth of an overall government policy of strategic disinformation, codified at the highest official levels, designed to inflate perceptions of U.S. military capabilities and needs during the cold war arms buildup. At the CIA, the policy took the form of a "perception management" program, outlined in a defense guidance document that prescribed: "We should seek to open up new areas of military competition and obsolesce previous Soviet investment or employ sophisticated strategic deception options to achieve this end."[128] At the Department of Defense, deceptive practices received their official sanction in a wide-ranging "coordinated disinformation effort" covering fifteen to twenty programs, one of them SDI.[129] In describing the program, a Defense Department official said, "disinformation can be injected at every stage of a weapon program from documentation and the test envelope, to the actual operation of the system."[130]

On 30 May 1985 Ronald Reagan issued National Security Decision Directive 172, "Presenting the Strategic Defense Initiative." The stated purpose of NSDD 172 was to "provide guidance in the manner in which I want the Strategic Defense Initiative and the SDI research program presented."[131] In the final subsection of the Decision Directive, "Managing the Presentation of the U.S. SDI Program," Reagan fashioned the institutional framework for enforcement of the deception program by creating a presumption of review for all public comment on SDI.[132]

> To ensure that we are indeed clear in presenting U.S. policy in this crucial area, all of our presentations on the SDI program must be coordinated effectively. Therefore, all major public statements, briefings, reports, speeches, articles, op ed pieces, etc., which are generated by officials of this Administration and which involve the U.S. SDI program will be cleared in advance by the Assistant to the President for National Security Affairs. No statements by officials of this Administration will be made publicly, or in diplomatic, military, or scientific channels,[133] that have not been so cleared.[134]

The network of secrecy shrouding the Star Wars program grew like a web around this document. NSDD 172, which served as the framework for the strategic deception program, was itself secret for many years until aggressive post–cold war Freedom of Information Act campaigns forced its release. As Christopher Simpson points out, "[t]he principal reason these orders are kept secret cannot be traced to a concern that they might reveal compromising technical details of new weapons, because such data simply are not found in them. Instead, the continued classification seems to be an effort to preserve the 'plausible deniability' that permits the administration in power to tailor its publicity on controversial topics to whatever audience is at hand."[135]

An April 1987 Joint Chiefs of Staff memorandum entitled "Special Plans Guidance—Strategic Defense" declared that "the Pentagon should improve and update deception plans covering the missile defense program's cost and abilities."[136] Program scientists said that application of the strategic deception policy to the area of missile defense was approved by Secretary of Defense Caspar Weinberger.[137] Such official backing could have been expected, especially given Weinberger's penchant for disinformation as a routine tool of foreign policy. "You are always trying to practice deception," Weinberger said in an interview years later. "You are obviously trying to mislead your opponents and to make sure that they don't know the actual facts," he said.[138]

The numerous deceptive maneuvers executed by missile defense officials and scientists during the Star Wars era appear to have been authorized by several legal instruments that codified governmental prerogatives to censor speech, manipulate scientific results, and deceive external audiences. Were such official government policies justified? The following discussion addresses this question. First, strategic deception is investigated from a purely scientific perspective, where Teller and Wood's X-ray laser promotion campaign is evaluated on the basis of its qualitative effects on scientific research at Livermore. Second, the advantage of historical hindsight is utilized to explore the status of Reagan's Star Wars address as a democratic, participatory moment in U.S. history. Third, adherents of the so-called Reagan victory school are brought into the conversation, with critical attention converging on their argument that, although fake, SDI was nevertheless justified because its symbolic force sped up collapse of the Soviet state. This discussion promises to yield valuable insights that not only tie back into points made by Reagan in his 1983 Star Wars address but also bear directly on concerns associated with contemporary missile defense debates.

Scientific Costs of Strategic Deception

In the short term, systematic distortions in communication engineered by X-ray laser proponents boosted Star Wars' prospects by permitting missile defense advocates to advance their arguments under the protective cover of scientific realism.[139] In the long run, however, this strategy proved to be unsustainable. The manipulation of experimental design, embellishment of test results, and muzzling of internal and external controversy were all factors that contributed to the eventual failure of the X-ray laser project. On a most basic level, this failure can be explained as an unraveling of the scientific research process itself.

There have been projects where military scientists have produced sound technological devices while working under strict constraints of secrecy. In the case of Star Wars, however, excessive secrecy and intolerance for dissent at the highest levels of management seriously compromised the capability of government scientists to produce reliable scientific results. For example, the inordinate resources and attention devoted to fraudulent experiments traded off with progress in other, more promising, lines of research. "Scientists from across the political spectrum have pointed out that such early technology demonstrations do more to slow progress than to enhance it," wrote physicist Joseph Romm, "freezing-in technology prematurely and draining funds from more promising approaches."[140] Politicization of the SDI research agenda also compromised the technical integrity of the program. "I also agree with Professor Makowski of the Technion who wrote, 'Overfunded research is like heroin, it leads to addiction, weakens the mind, and leads to prostitution,'" stated computer scientist and SDI defector David Parnas.[141]

Perhaps the most vivid evidence of this failure was the parade of established scientists who left the project under protest. Determining that the scientific community at LLNL had lost its ability to produce objective work, these whistle-blowing researchers chose to resign rather than allow their expertise and reputation to continue supporting fantastic scientific claims. An examination of the public rationales provided for such departures provides a revealing picture of the scientific costs of Star Wars strategic deception.

In October 1985 Roy Woodruff resigned as associate director for defense systems at LLNL, citing as the reason in his letter of departure "[nearly two years] of potentially misleading" X-ray laser reports passed on to "the leaders and policy makers of this Administration."[142] Assessing the scientific integrity of LLNL following his resignation, Woodruff said, "I think the laboratory is losing its way. It once stood for technical excellence and technical integrity. But I think it's become politicized during the Reagan Administration."[143]

Senior researcher Lowell Morgan also defected from LLNL, voicing similar concerns. "I quit the program for two reasons," he said. "The first is that I felt that the program was going nowhere. The experiments were very difficult, yielding minuscule returns of poor quality data for a tremendous investment of effort and, in my opinion, an obscene investment of money. The second reason is that I felt that the few scientific results that we had were being grossly misrepresented in their support of the fantastic claims about the X-Ray laser."[144]

In 1992, physicist Aldric Saucier quit the army's Strategic Defense Command (set up in 1985 as the army's component of the SDI program), stating in an affidavit, "Star Wars has been a high-risk, space age, national security pork barrel for contractors and top government managers."[145] Saucier went public after attempting unsuccessfully for years to voice through internal channels criticism of the SDI research program. In 1986, he submitted a paper titled "Lessons Learned" to SDIO director James Abrahamson, warning that "contractors working on Star Wars architectures or systems designs had overwhelmed SDI management of the program to the point where officials could not understand what the contractors were doing." In 1989, Saucier raised these same issues at an Army Science Board briefing.[146] Saucier's arguments were ignored, and eventually he was issued a "proposed removal" notice for "unsatisfactory research."[147] As Louis Clark, a lawyer with the Government Accountability Project, commented on the Saucier case, "[t]he Pentagon tends to move quickly against people perceived to be threats and makes examples of them."[148]

This culture of secrecy and intimidation eventually compromised the scientific process itself. As whistle-blowing SDI physicist Leonardo Mascheroni stated in his departing comments, "[t]here's a lot of fear here. Management may have good intentions, but too many of its decisions are based on politics, not technical merit. We need a dialogue with the best brains in the country in order to better judge the merit of projects."[149] Cut off from intersubjective criticism, the Livermore Star Wars researchers even drifted into isolated patterns of language that reflected their cloistered existence: "The fabric of friendship extended even to the language they spoke. Classified projects led to classified jokes. After a while, the young scientists began to be cut off from the spontaneity of the outside world. A visitor could engage them in polite conversation, but so much of their world revolved around secret research that free-ranging discussions could take place only with those who had the proper security clearances," Broad describes.[150]

The long list of respected scientists who defected from the Star Wars research project, and the ultimate failure of the project to produce anything close to a weapons-grade X-ray laser were indicators that the effort never actually reached the status of a legitimate scientific research enterprise. "An Orwellian atmosphere where truthfulness is scorned and deception rewarded is sure to send any technical organization into decline," wrote Broad, "much less an elite one such as Livermore that studies some of the most subtle questions of nature and seeks to hire the most talented and creative scientists in the nation."[151]

Prior to the ultimate collapse of the Star Wars project, missile defense advocates still succeeded in convincing Congress and other key actors that the SDI research effort was legitimate, objective, and supported by a consensus formed through a process of robust peer review and healthy intersubjective criticism.[152] When LLNL director Roger Batzel went to Washington on 23 October 1987 to brief Congress in light of Woodruff's claim that research was being systematically distorted at the laboratory, he issued the following statement: "My policy is that when individual scientists wish to speak out with differing points of view on technical issues, the Laboratory does not intervene."[153] Officials from the Department of Energy expressed a similar public commitment to the norm of open debate in a written reply to questions from GAO investigators in 1988. When asked whether LLNL officials "took [internal X-ray] criticism fully into account in their planning for future research and testing of this device,"[154] DOE officials responded with the assertion that research decisions were based on a robust consensus, forged out of freewheeling and independent peer review.

> In all the scientific and program reviews, the LLNL staff have used the most current and most accurate information available. Most of the scientific reviews have, in fact, been requested by LLNL in order to provide independent peer review of the results and progress. In all cases, we have accurately conveyed the current status of the x-ray laser program to all levels of government and the scientific community. No major disagreements with LLNL's presentation have been expressed. The outcome of the reviews have, in general, been enthusiastic support for the program as laid out by LLNL.[155]

Director Batzel and the Department of Energy advanced these statements to convince Congress that the LLNL research effort on Star Wars was conforming to the highest standards of scientific peer review,[156] yet during this same period, LLNL and other components of the government SDI apparatus were engaged in a systematic deception program designed to stifle internal dissent, distort research, and release exaggerated public reports.[157] It was exactly this tension which convinced a stream of well-qualified researchers such as Mascheroni, Morgan, Parnas, Saucier, and Woodruff to desert the research program in protest. Their exits were symptomatic of a larger erosion of objectivity within the SDI research program, an erosion that brought on the dry up of congressional support for X-ray laser research.

Over time, declassification, leaks, and whistle-blower accounts have shown that Star Wars was more an exercise in strategic deception than a scientifically legitimate research program. From our post–cold war vantage point, it has become clear that much of the optimism and fanfare surrounding Star Wars in the early 1980s was stoked by an illusion of objectivity, and the attribution of scientific validity to rigged experiments that were designed, timed, and reported with maximum rhetorical spin. "There has been a long history of misinterpretation, distortion or fraud throughout the life of this project," said John Pike.[158]

By the mid-1980s, congressional doubts about the X-ray laser program began to mount as whistle-blower leaks and sharp rebuttals from civilian scientists increased in frequency. In response, President Reagan re-asserted that SDI was feasible. As Stephen Guerrier and Wayne Thompson explain: "Reagan's statement backfired: some 3,700 science and engineering faculty members—including a majority of the top 59 physics departments—and 2,800 graduate students signed a pledge not to accept SDI funding."[159]

This massive demonstration of skepticism began to erode SDI's political base. "All this negative publicity combined with the boycott made it easier for critics in Congress to make their case," explained physicist Lisbeth Gronlund.[160] One Senate aide involved in the missile defense issue said "[y]ou have to look at the sum total of scientific statements. Some said, 'We won't work on it.' Others said, 'We work on it, but we're very concerned.' That leads to a sense in Congress that all is not well, that the program has been oversold, overhyped."[161]

Although SDI appropriations continued to grow gradually through the middle of the decade, by 1988, significant doubts about the X-ray laser program led to a shift in priorities for SDIO, as LLNL pursued other projects based on kinetic-kill technology.[162] Troy E. Wade II, the Energy Department's acting assistant secretary for defense programs, provided a casual postmortem for the X-ray laser project during 1988 congressional testimony: "To get from the understanding of physics to a weapon is perhaps a much bigger job than Dr. Teller thought," Wade stated dryly.[163]

Late into President Reagan's second term in office, growing congressional doubts about the feasibility of SDI led to budget cuts, and deployment schedules began to slip.[164] Defense Secretary Frank Carlucci admitted that SDI could not provide "an impenetrable shield."[165] With the presidential election of Vice President George Bush in 1988, the rising tide of fanfare for missile defense that had swelled earlier in the decade receded even further. "After Reagan's retirement," explained Reiss, "SDI was no longer sheltered from technical, political and fiscal realities."[166] Members of the new Bush administration sought to carve out the basis for a more pragmatic missile defense program by distancing themselves from Reagan's heady Star Wars visions. Secretary of Defense Richard Cheney described SDI as "an extremely remote proposition."[167] Vice President Dan Quayle said that Reagan "talked about this impenetrable shield that was going to be completely leakproof. . . . I believe that, in the semantics of let's say political jargon, that was acceptable. But it clearly was stretching the capability of a strategic defense system."[168] Finally, on 21 July 1992, Energy Secretary James Watson "quietly canceled the system's last test."[169]

The Cold War's Dirty Nuclear Secrets

In a 1989 article, Kenneth Zagacki and Andrew King celebrated Reagan's Star Wars vision as an archetypal example of a powerful rhetorical trope they called the "techno-romantic synthesis."[170] Zagacki and King applauded Reagan's ability to weave together technical claims with political arguments in an overall narrative that enacted an inclusive moment of participatory democracy. A techno-romantic synthesis, "works within the idiom of the people, therefore preserving public sentiment and creating a sense—*not necessarily a false sense*—of participation in public debate."[171] When Zagacki and King published these words in 1989, although there were widespread doubts about the technical credibility of the Star Wars program, there were also still plenty of diehard adherents who insisted that Reagan's "Astrodome" defense might still be achievable, even after six years of technical futility. As Zagacki and King commented, "we know of no empirical evidence indicating that the American public has been duped by Reagan's speech. In contrast, we argue that Reagan's rhetoric calls for greater public inclusion in the arms debate, and deflates, but does not deny the importance of the contributions of the techno-scientific community."[172]

Writing in a cold war environment, Zagacki and King were disadvantaged by walls of superpower secrecy that blocked them from grasping the full picture of Star Wars' scientific flaws, since their research efforts were complicated by information restrictions mandated by government officials. It is understandable that in such a context, Zagacki and King would make such a statement as: "We know of no empirical evidence indicating that the American public has been duped by Reagan's speech."[173] However, this chapter's examination of Livermore's research program has revealed that Reagan's vision of an X-ray laser-based "Astrodome" defense was based on blatantly fraudulent scientific data funneled to the White House by Edward Teller and others. The lofty promises in Reagan's speech turned out to be scientific chimeras. Even in the midst of today's hoopla over the possibility of "limited" missile defense, precious few physicists endorse the idea that a Reagan-style "Astrodome" system could ever be built to neutralize thousands of incoming enemy ICBMs.

These findings do not reflect well on Zagacki and King's favorable judgment that Reagan's Star Wars address enacted a participatory, democratic moment in missile defense history. If such a moment was enacted, it occurred in a brief, flickering instant on the night of 23 March 1983. After Reagan's speech, SDI officials embarked on a systematic deception program designed to manipulate public understanding and subvert open dialogue on missile defense issues. "[T]he legacy of Star Wars," speculated Zagacki and King, "may be that Reagan's rhetoric pointed to the potential for continued democratic control over technology, by advocating a nascent argumentative framework accessible to the public."[174] A decade after Zagacki and King's speculation on this point, evidence has mounted that the cold war military apparatus (including SDIO) used frustration of democratic control over technology as a political modus operandi to trigger cycles of defense-spending jackpots.

The collapse of public deliberation occasioned by this suppression of controversy left a vacuum of accountability resulting in mismanagement, waste, and abuse. Institutionalized strategic deception requires perpetuation of a vast bureaucracy designed to protect secret military operations from public scrutiny. While such a bureaucracy sometimes provides cover that enables military officials to better defend the nation, at other times it prevents citizens from learning about shocking abuses committed during military operations and research. Given a virtual blank check to pursue qualitative scientific advantage with minimal accountability during the cold war, American scientists waged what Carol Gallagher calls a "secret nuclear war."[175] Although this secret war did not involve the spectacular detonation of nuclear bombs over cities, it did involve what the Physicians for Social Responsibility describe as a "creeping Chernobyl": the slow, steady contamination of the lifeworld of unsuspecting civilians and enlisted personnel.[176] "While no nuclear weapons have been detonated in war since Hiroshima and Nagasaki, a kind of secret low-intensity radioactive warfare has been waged against innocent populations," explains Nobel Laureate and Harvard Professor Dr. Bernard Lown.[177]

The casualties of this secret war were the unwitting citizens and vulnerable soldiers exposed to radioactive poison during medical experiments and covert weapon tests sponsored by the DOE, Department of Defense (DOD), Atomic Energy Commission (AEC), and other governmental agencies. "We were in a Cold War that was real," says Newell Stannard, an expert on radiation and health who worked on many AEC projects in the early years.[178] Some of these casualties were sacrificed as medical guinea pigs, like the nineteen teenage boys who were enrolled at a state school for mentally retarded children in Waltham, Massachusetts. These students were fed radioactive iodine in their breakfast cereal to provide data on the health effects of nuclear war.[179] Washington state prisoners had their testicles X-rayed as part of cold war nuclear research carried out by military-funded doctors who conducted the experiments in the name of national security, even though there was no potential therapeutic benefit for the incarcerated research subjects.[180]

Other casualties included workers and residents associated with the network of American weapons plants. At bomb plants like the one in Hanford, Washington, runaway cold war logic translated into direct physical harm in ways that Lown says "would have been hardly imaginable under open and accountable decision-making systems."[181] Historian Michael D'Antonio describes how secrecy combined with Cold War paranoia to sustain corrupt and antidemocratic administrative cultures at the national labaratories.

> Hanford and the other bomb factories were allowed to endanger their neighbors and foul the environment because government scientists and officials didn't trust the people they served. Coming out of World War II, the government maintained its wartime claim to extraordinary secrecy. Elected and nonelected authorities didn't believe that, given all the facts, citizens would make the right choices. So they simply declared the facts secret, and retained the power for themselves. For their part, everyday citizens, motivated by fear of the Soviet menace and of an awesome technology, abdicated their responsibility to determine the national interest in a time of peace.[182]

The sordid legacy of death, illness, birth defects, and environmental pollution left over from this "secret nuclear war" casts serious doubt on the widely held perception that the U.S. victory in the cold war was a "clean win." There were major costs incurred during the process of out-arms-racing the Soviet Union. During the cold war, these costs were largely hidden. Secrecy surrounding the military establishment kept them from public view, and the voracious logic of national security prevented bureaucrats and military personnel from standing up to protect the interests of people who were subjected to nuclear experimentation. "As early as 1949, federal officials knew that radiation could be causing cancer in workers and people living near weapons plants," writes D'Antonio. "However, because virtually everything to do with atomic weapons manufacturing was classified, much of the debate . . . was kept from the public."[183] In the immediate aftermath of the cold war, the euphoria associated with American "victory" provided an excuse not to engage in a more sober assessment of the hidden costs entailed in the American arms race strategy. These days, it is hard to overlook the consequences of nuclear secrecy. Since DOE secretary Hazel O'Leary's "Openness Initiative" in 1993, a steady stream of disclosures have dramatized the appalling cold war excesses of the U.S. nuclear war machine.[184] Already, taxpayers have spent $2.05 billion investigating and settling claims of negligence and misconduct on the part of U.S. nuclear scientists.[185]

Open scientific discussion is necessary to craft internal agreement within the scientific community and to enable what social theorist Jürgen Habermas calls "politically effective discussion"[186] in the public sphere regarding the proper pace and direction of scientific investigation. Where such discussion is thwarted by systematically distorted communication, one can expect to find a short-circuiting of the scientific process of inquiry, as well as aberrant scientific practices motivated by partisan political interests. In the Star Wars era, both of these failures were manifest in the history of the missile defense project. Secrecy hamstrung the ill-fated research effort and supported a manipulative project of deliberate political distortion, a project that flouted basic democratic principles of accountability. While Star Wars research may not have harmed innocent civilians directly, it nevertheless fed the culture of secrecy that gripped the entire military research establishment during the cold war.

The consequences of this legitimating function could be detected in areas far beyond the realm of missile defense research. The experience of former Secretary of Energy James Herrington is instructive in this regard. When it came to Star Wars research, Herrington insisted that research disputes between Los Alamos and Livermore remain hidden behind closed doors,[187] but he found that the same policy of strict information control interfered with his ability (as DOE director) to manage the environmental and health risks associated with the nuclear bomb plants. According to D'Antonio, "[t]he entrenched culture of secrecy affected everyone in the DOE bureaucracy, including Secretary Herrington, who found that the secrecy made it difficult for him to assess the condition of the plants that were his responsibility."[188] The DOE's official policy of misinformation boomeranged on the agency. While instrumental in the missile defense area, systematic communicative distortion compromised DOE initiatives in the area of nuclear cleanup.

More insidiously, the reported successes of Star Wars experimentation may have neutralized public attempts to counter the secretive nuclear culture of the cold war. For example, some argue that the apparent fantastic breakthroughs of Livermore SDI tests worked as distractions that provided media cover for military abuses, working to insulate human radiation experimentation from scrutiny and criticism. "Just as excitement over Star Wars technology in the '80s suppressed disapproval of the radiation tests," write journalists Ron Grossman and Charles Leroux, "war and postwar fear of the 1940s shaped the attitudes which gave birth to those tests."[189] Members of the public were not generally cognizant of cold war human radiation experiments until 1993, when media disclosures triggered widespread alarm and led to a Presidential Advisory Committee investigation of the tests. Curiously, many of these same revelations were made nearly a decade earlier in a report released by Markey.[190] Although this report was "carefully documented" and "cited specific published reports on the studies," George J. Annas notes that it went "virtually unrecognized and unheralded primarily because the administration of Ronald Reagan dismissed it as overblown."[191] Political scientist Howard Ball describes the logic of the cold war as a "controlled hysteria. . . . We were trying to figure out how to win a war against the Soviets and we just used people."[192] When it came to the arms race and technological competition with the Soviet Union, the "lust to be first, whatever the cost or risk, left a path strewn with victims," explains journalist James Goldsborough.[193]

Propagation of the Star Wars illusion helped shield the nuclear establishment from scrutiny not only by projecting a false image of competence and responsibility, but also by co-opting the argumentative telos of groups dedicated to challenging the hegemony of nuclearism. Consider the case of the American peace movement. For much of the cold war, the core of the movement's appeal was its opposition to MAD, the doctrine of nuclear deterrence that held that first use of nuclear weapons by either superpower would be unthinkable, given the swift and devastating nuclear retaliation that would surely follow. Peace movement activists contended that MAD was immoral because it cast innocent civilians in the role of hostages whose lives were wagered in the high-stakes game of nuclear diplomacy. The only way out of the deadly spiral of the nuclear arms race, they contended, was the rejection of MAD, through the outright abolition of nuclear weapons.

While the abolitionist imperative of the peace movement had torque when directed against efforts to bolster deterrence through upgrades in the U.S. offensive nuclear arsenal, it experienced slippage when directed against SDI. Because advocates of ballistic missile defense promised a way to neutralize the nuclear threat without relying upon the allegedly immoral (and unreliable) persuasion of deterrence, SDI helped pro-militarization forces blunt the peace movement's call for worldwide disarmament and the elimination of nuclear weapons. "[Reagan's] Grand Vision speech brought about a striking transformation in conservative rhetoric," writes historian Edward Linenthal. "Sounding like members of the antinuclear movement, for which they previously had voiced nothing but contempt, some SDI enthusiasts began attacking deterrence as an immoral national strategy," he explains.[194] This strategy allowed missile defense advocates to win the deliberative contest of utopian futures, since their plan for technological triumph seemed more realistic than the peace movement's vision of worldwide elimination of nuclear weapons. According to Linenthal, this was possible because "abolition seemed more 'utopian' than missile defense, largely because the shapers of strategic defense ideology were able to locate missile defense within recognizable and appealing cultural traditions."[195]

Missile defense advocates' strategy of rhetorical co-option in this regard was not a fortuitous accident—it flowed directly from argumentative blueprints handed down by the top leadership of BMD promotion campaigns. Commenting on the prospects that such a strategy might turn the tide in nuclear freeze debates, High Frontier consultant Greg Fossedal argued that SDI advocacy could be an opportunity to "fast thaw the nuclear freeze movement."[196] Writing with key SDI insider Lt. Gen. Daniel O. Graham, Fossedal explained the mechanism for co-option in an influential 1983 book, *A Defense That Defends: Blocking Nuclear Attack:* "If strategic defense has remained off the agenda by its implicit threat to both the disarmers and the rearmers, so defense, now on the agenda, offers a hope of strategic consensus. Defense offers the peace marcher one of his most cherished hopes: a world free from the threat of total nuclear annihilation."[197]

Remarkably, it appears that marching orders for this rhetorical strategy of co-option were laid out in a Heritage Foundation "Proposed Plan for Project on BMD and Arms Control" with a cover sheet listing the document as "High Frontier: A New Option in Space."[198] According to former aerospace industry insider Carol Rosin, the objectives of this plan were to "get an early BMD underway and develop enough political support to ensure that it could not be turned off," and to force a drastic reorientation of the U.S. arms control debate in such a way as to "make it politically risky for BMD opponents to invoke alleged 'arms control arguments' against an early, or any other, BMD system."[199] Commenting on the report, Bowman said "The way they framed it, and the way they presented it, they were almost gleeful as to what they were doing to the nuclear freeze movement, steal the moral high ground, all that stuff."[200]

Specifically, the Heritage Foundation document instructed BMD advocates to "unambiguously seek to recapture the term 'arms control' and all of the idealistic images and language attached to this term."[201] In a section on "Tactics Considered for Project Plan," the report recommended a "radical approach that seeks to disarm BMD opponents, either by stealing their language and cause, 'arms control,' or by putting them into a tough political corner through their explicit or de facto advocacy of classical anti-population war crimes."[202] The report continued: "[T]o the extent possible, BMD proponents should begin to stress nuclear disarmament as a new end point, perhaps using such descriptions as arms control through BMD = nuclear disarmament. This description is directly contrary to the mainstream arms control community's belief that BMD is directly antagonistic to 'real' arms control."[203] The sample list of groups and individuals targeted to receive this message included the science fiction community (e.g., "Jerry Pournelle's Advisory Group"), Representative Newt Gingrich (R-GA), veterans groups, Catholic bishops, the Physicians for Social Responsibility, the American Bar Association, the American Medical Association, evangelists, celebrities, representatives of the Jewish community, George Will, William Safire, the RAND Corporation, Freeman Dyson, the Federation of American Scientists, and SANE/Freeze.[204] In concluding remarks, the Heritage Foundation Report outlined the rationale for an SDI "public argumentation" campaign that would target such a diverse array of individuals and groups spanning the political spectrum.

> (a) Depending on choice of action plan ("forceful personality" vs. an umbrella group or "political mobilization group"), there may be considerable variation in required products and in HF's level of involvement.
>
> (b) Given the outside prospect of a Reagan defeat in 1984, *it is essential to broaden the base of support for BMD. It presently risks being labeled permanently as a "right wing cause,"* primarily due to a lack of appreciation for the remarkable arms control payoffs of a BMD program. *However, it*

> *would be very risky for HF to start this new line of public argumentation, given its existing orientation* (i.e. fundraising literature) to "*liberal bashing.*"
>
> (c) Unless a pro-BMD coalition is quickly created and broadened, anti-BMD forces (which are well entrenched through Gary Hart and Mondale campaigns) could either kill off the program immediately if they win in 1984, or could successfully throttle it after 1984. MAD-modeled arms control is the *sine qua non* of Democratic models and approaches to arms control in this campaign so far—and the anti-BMD/pro-MAD reaction to Reagan's and HF's proposals will take a predictably "purist" form as a result, no matter who wins this year. Result: fanatical opposition to BMD in any form.
>
> (d) Consequently, *quick co-opting of "arms control" arguments is very important to the long-term survival of BMD.*
>
> (e) *A near-term BMD program (HF's longtime argument) will be driven by ethical urgency,* rather than by technology availability (HF's argument so far). *It is important to try to capture—and capitalize on—existing "urgent" movements,* of which the freeze is the most prominent example.
>
> (f) Preference should be given to an "umbrella group" if a "forceful personality" approach is impossible.[205]

The notion that this Heritage Foundation secret report was a key touchstone for mid-1980s BMD promotion campaigns is strengthened by uncanny parallels between the report's contents and actual discourse patterns exhibited by missile defense advocates during the era. For example, the report established a rhetorical foundation for strategic deception on the technical aspects of SDI, by suggesting that ethical arguments for BMD should be uncoupled from scientific concerns about technical feasibility: "A near-term BMD program (HF's long-time argument) will be driven by ethical urgency, rather than by technological availability (HF's argument so far)."[206] This rhetorical slant eventually became a key move for missile defense advocates facing criticism on technical grounds; when skeptics would question the scientific feasibility of a "leakproof" SDI system, missile defense advocates would respond with moral arguments about the necessity of transcending MAD.[207] Echoing the Heritage Foundation's "public argumentation" campaign in 1984, *Wall Street Journal* editorialist Gregory Fossedal wrote that arms control advocates "prefer a perpetual balance of nuclear terror," and that "defense could be useful not just once we arrive at disarmament, but as a road to getting there."[208] In another register, the Heritage Foundation report anticipated European advocacy efforts by directing SDI stalwarts to reach abroad in their BMD promotion campaigns, stipulating that "the strategy proposed should be appealing to U.S. allies," and that advocates "should spend a great deal of time trying to get an offshore constituency, particularly the governments of major U.S. allies, the more vocal and outspoken the better."[209] This recommendation presaged efforts by U.S. officials such as National Security Advisor Robert McFarlane to lobby European leaders with brazen cash offers for their support of SDI.[210]

The impact of these "steal the language" rhetorical strategies on the peace movement was significant. "The Strategic Defense Initiative (SDI), which had almost universally seemed a crackpot scheme in its earlier, 'High Frontier' stage, responded so well to the claims of freezers that it quickly subdued their movement," recounts William Chaloupka.[211] SDI "destroyed the base of the nuclear freeze movement," D'Souza observes, "because Reagan showed himself to be more deeply committed than its leadership to reducing the danger to Americans posed by the Soviet nuclear arsenal."[212] The movement struggled to recover politically by aligning with establishment doves, fighting for the mainstream goal of arms control. In making this adjustment, however, the movement mortgaged much of its radicalism, substituting a ringing call for abolishment of nuclear weapons and suspension of nuclear weapons research and testing with a more incremental and reformist program. This more modest agenda acknowledged the continuation of the nuclear establishment as an unavoidable contingency, and could not sustain the utopian aspirations of many movement activists.

The largely successful campaign by pro-BMD forces to co-opt and neutralize the American peace movement was a remarkable rhetorical phenomenon. Using the equation "BMD = disarmament" as a template for public argumentation, Star Wars visionaries were able to "steal the language and cause" of activists calling for nuclear disarmament. As space-weapon enthusiasts Daniel O. Graham and Gregory A. Fossedal suggested, a crash buildup of defensive arms would offer "the peace marcher one of his most cherished hopes."[213] According to Herbert Marcuse, such a "unification of opposites" is "one of the many ways in which discourse and communication make themselves immune against the expression of protest and refusal."[214] Notions such as "the Freedom Academy of cold war specialists," the "clean bomb," and the "cozy bomb shelter" fuse together terrifying concepts with hopeful allusions, and the effect is "a magical and hypnotic one—the projection of images which convey irresistible unity, harmony of contradictions."[215]

Would the curve of history have been different if the peace movement's pressure on the nuclear establishment had not been defused by SDI strategic deception? Would the "secret nuclear war" carried out by the American military research establishment have been exposed sooner if the American public had assumed a more active role in oversight? Would fabricated intelligence reports responsible for generating false momentum behind the arms race have been flushed out at an earlier time if the government had been held to a higher standard of accountability?

Speculative answers to these questions are necessarily bound up with judgments regarding the wisdom and efficacy of the antinuclear movement's agenda and tactics. In *The Nuclear Freeze Campaign,* J. Michael Hogan offers a skeptical commentary on the nuclear freeze movement's rhetorical strategies. Hogan bases his indictment of the freeze movement on rhetorical criticism of advocacy efforts conducted by Jonathan Schell, Helen Caldicott, Randall Forsberg, members of the National Education Association, and other activists fond of "apocalyptic scenarios and celebrity appeals."[216] According to Hogan, these players "focused upon the horrors of nuclear war and spoke in politically innocuous metaphors. Skirting the major controversies of the nuclear debate, they refused to debate technical issues, and they refrained from attacking Ronald Reagan or any other specific 'enemy.'"[217]

Hogan overlooks several important issues in his account of cold war antinuclear activism. First, he does not evaluate the significant effects of transnational movement organizing. Security studies scholar Matthew Evangelista discusses this significant aspect of peace movement protest in his recent book, *Unarmed Forces.* Evangelista notes, for example, that "Forsberg, a founding member of the Nuclear Freeze Movement and a major theorist of alternative security policy, was well respected in the community of Soviet academic specialists on military policy, many of whom she had known since at least the 1980s."[218] Challenging Hogan's account of freeze movement activists who "refused to debate technical issues," Evangelista notes that Forsberg and others associated with her Boston-based Institute for Defense and Disarmament Studies (IDDS), used their transnational contacts to build momentum for Soviet denuclearization, floating specific arms control and confidence-building measures, some of which "had a profound effect" on high-ranking Soviet officials such as Eduard Shevardnaze, who "came into his job as foreign minister in July 1985 with no experience or background in security affairs."[219] In the closing pages of this chapter, the precise impact of these peace movement overtures will be explored in more detail.

A second oversight in Hogan's analysis of antinuclear activism involves lack of attention to a major wing of the peace movement that did debate technical issues and challenge Reagan's policies directly—that is, the large group of scientists belonging to such organizations as the Union of Concerned Scientists, Scientists and Engineers for Social and Political Action, and the Federation of American Scientists. By September 1986, 3,850 faculty members and senior researchers and 2,850 graduate students and junior researchers had signed the National Pledge of Non-Participation in SDI research.[220] Over half of the entire membership of the National Academy of Sciences sponsored a petition calling for a complete ban on all space weaponry in May 1986.[221] Antinuclear activism by scientists extended well beyond the SDI context, as well.[222] Although these scientists were fighting an uphill battle, competing against the utopian rhetoric and deep pockets of missile defense advocates, the historical record shows that they affected the cold war endgame in significant ways.

Inside the "Reagan Victory School"

Given the ultimate failure of SDI scientists to deliver on Reagan's promise of "leakproof" missile defense, some chastise the entire Star Wars project as a colossal waste of resources, arguing that the $70.7 billion spent on research and development since 1983 could have been much better spent addressing other social needs.[223] However, others insist that although the Star Wars project may have rested on an illegitimate scientific foundation, it still paid concrete political dividends. For example, former national security advisor Robert McFarlane claims that Star Wars was a laudable military project because it forced the Soviet Union to the arms control table and sped up the collapse of the Soviet state by five years.[224] McFarlane's argument rests on the logic that although never scientifically sound, Reagan's SDI program was nevertheless valuable as a symbolic decoy, what some defense analysts have termed a "virtual sword."[225] This line of reasoning has a lineage that traces back to Lt. Gen. Daniel O. Graham, chief author of a 1983 High Frontier study that argued for pursuit of Star Wars on the grounds that such a path could "severely tax, perhaps to the point of disruption, the already strained Soviet technological and industrial resources."[226] Other cold war strategists echoed similar arguments. For example, shortly after Reagan's Star Wars address, Teller suggested that even if Star Wars' only benefit would be to force the Soviet Union to increase defense expenditures, "we would have accomplished something."[227]

Missile defense advocates such as McFarlane, who tout Star Wars' utility as a tool of cold war victory, belong to the what political scientists Daniel Deudney and John Ikenberry dub the "Reagan victory school" of historical thought.[228] In general, Reagan victory school historians endorse the "peace through strength" thesis and subscribe to the notion that "the Reagan administration's ideological counteroffensive and military buildup delivered the knock-out punch to a system that was internally bankrupt and on the ropes."[229]

Since the fall of the Berlin Wall, a parade of commentators has lined up to corroborate the Reagan victory school's historical narrative. "The Cold War's end was a baby that arrived unexpectedly," observe Deudney and Ikenberry, "but a long line of those claiming paternity has quickly formed."[230] Writing a brand of "vindicationist" history,[231] scholars such as Jeane J. Kirkpatrick, Adam Meyerson, Peter Schweizer, and Jay Winik tell the story of a glorious American cold war victory wrought from the mighty steel of U.S. military strength.[232]

Reagan victory school narratives draw heavily on Star Wars as a key example to prove that American military prowess was a major causal factor responsible for the fall of the Berlin Wall in 1989. For example, former British prime minister Margaret Thatcher is reported to have said, "I firmly believe that it was the determination to embark upon the SDI program and to continue it that eventually convinced the Soviet Union that they could never, never, never achieve their aim by military might."[233] Commentator James Fallows says that SDI "seems to have played a real part in ending the Cold War."[234] Former national security advisor Zbigniew Brzezinski writes that "SDI both shocked the Soviets and then strained their resources. Its scale, momentum and technological daring had been totally unexpected . . . [and] it dawned on Soviet leaders that they could neither match nor keep up with the American efforts."[235]

Although these comments seem to validate McFarlane's hypothesis that the mere illusion of Star Wars effectiveness hastened the Soviet Union's collapse, there is a chorus of dissenting voices that questions the link between inflated perceptions of Star Wars feasibility and Soviet willingness to opt out of the cold war arms race. "The first, most obvious point to make in regard to these notions is that none of the conjectured Soviet responses to Star Wars had occurred before Gorbachev came into office," notes Evangelista, "even though Reagan's SDI speech was delivered two years earlier."[236] Vladimir

Chernyshev, former Soviet General Staff officer, raises similar doubts about the timeline of events in McFarlane's hypothesis: "[Chernyshev] was more dubious about the impact of SDI on the Soviet economy. He said much of the damage was done earlier, when the Soviet Union failed to dismantle its massive defense industries at the end of World War II."[237] This perspective receives support from former Soviet Red Army colonel Anatoly Kravtsov, who scoffed at the suggestion that SDI was simply a ploy to hasten the demise of the Soviet Union: "It was a good idea, but really we were quite capable of bankrupting ourselves. It wasn't necessary."[238]

As early as 1983, the CIA anticipated in a secret study that "[i]t is highly unlikely that the Soviets will undertake a 'crash' program in reaction to U.S. BMD developments, but rather will seek to counter them by steadily placed efforts over the decades the United States will need to develop and deploy its overall defenses."[239] Noting that this is in fact the way things turned out, defense commentator Joseph Camilleri concludes that "there is little to suggest that the emphasis on technological competition, or even the Strategic Defense Initiative, did much to tilt the military balance in favor of the West."[240] Roald Sagdeev, former head of the Soviet Space Research Institute, maintains that SDI had "absolutely zero influence" on the origins or course of Soviet reforms.[241] As the Federation of American Scientists' John Pike argues, "there is no evidence whatsoever that the Soviet Union materially altered any of its military plans or budgets in response to Star Wars. It is ludicrous to even suggest that a system that survived the onslaught of Hitler's legions would implode in the face of a pile of viewgraphs and a few special effects tricks."[242]

Other Russian sources suggest that SDI may have achieved the opposite of McFarlane's hypothesis, interfering with efforts to bring the cold war to a close. "Star Wars (SDI) was exploited by hard-liners to complicate Gorbachev's attempt to end the Cold War," states official policy advisor Aleksandr Yakovlev.[243] Security studies scholars Bruce Blair and Stephen Schwartz corroborate Yakovlev's argument that SDI's most likely effect was to prolong, rather than shorten, the cold war: "If anything, the program was responsible for exacerbating tensions and delaying agreements to reduce nuclear arsenals (McFarlane's stated motivation for backing the program). It thus delayed the eventual end of the cold war and increased U.S. costs."[244] Evaluating McFarlane's spend them into the ground hypothesis, Bowman says, "I think it's totally false. Reagan and his belligerents, including Star Wars, actually prolonged the fall of the Berlin Wall by two years."[245]

These dissenting views suggest that the causal accounts of cold war events provided by Reagan victory school adherents might be "misleading and incomplete,"[246] particularly in the case of Star Wars. There were many other historical currents at work that could be isolated as causal factors accounting for the draw-down in superpower tensions prior to the fall of the Berlin Wall in 1989.[247] Evangelista notes that two such factors were peace movement pressure and transnational scientific dialogue. Even with some of the freeze movement's popular appeal defused by Star Wars, the combination of residual movement pressure and international scientific exchanges combined to create the conditions necessary for the end of cold war hostilities. It is worth quoting Evangelista at some length here, because his findings are based on exhaustive research of Soviet open source materials, and they bear directly on the previous discussion of the relationships between SDI, the peace movement, and the end of the cold war.

> In addition to coordination of specific policy initiatives, the transnational network of U.S. and Soviet disarmament supporters also worked together to create an overall atmosphere conducive to restraint on each side. In order for Gorbachev to succeed in cutting back Soviet military programs and military spending, he had to make a plausible case that the United States did not pose a serious threat to Soviet security. . . . As the Nuclear Freeze movement sought to persuade Ronald Reagan that he had to tone down his harsh rhetoric about the Soviet Union and careless comments about nuclear war, Soviet reformers pushed initiatives that would diminish the 'enemy image' of the USSR in Reagan's eyes. . . . The warming of U.S.-Soviet relations would not have been possible had Reagan not been pushed by the U.S. peace movement to address the threat of nuclear war. . . . American transnational activists, while trying to constrain U.S. military programs, also considered it important to persuade the Soviet government that it did not pose a threat so grave that Soviet unilateral restraint or even negotiated settlements would be dangerous. Much attention in this regard was focused on the Strategic Defense Initiative (SDI). . . . U.S. and Soviet members of the transnational scientists' movement all considered 'Star Wars' a dangerous waste of money, but they did not want it to stand in the way of negotiating deep reductions in nuclear forces. The Americans kept their Soviet colleagues apprised of the fate of SDI in congressional deliberations, the astronomical cost estimates, and the technical critiques. They managed to persuade Gorbachev, sometimes in direct discussion, that the Soviet Union should 'unlink' the signing of a strategic weapons reduction treaty from U.S. pursuit of SDI. Star Wars,

> they argued, would eventually fade away, especially if the Soviet Union continued to pursue its reformist course in defense and disarmament, not to mention internal democratization.[248]

These comments suggest the need for revision of "Reagan victory school" narratives based purely on celebration of brute "peace through strength" logic. Evangelista's careful historical analysis reveals that transnational peace activism was perhaps the primary factor driving official decisions by both superpowers to back away from the nuclear brink. Rather than Star Wars forcing such a retreat (McFarlane's thesis), it appears that SDI actually interfered with efforts of such peace activists to bring the necessary pressure on their governments to move toward more stable and democratic defense postures.

On another level, some historians question the underpinnings of Reagan victory school narratives by casting doubt on the realpolitik threat assessments that backed cold war defense policy and diplomacy. These historians suggest that the apparent U.S. cold war "victory" was a hollow one, because it was achieved against a phantom enemy, constructed by those with vested political and financial interests in perpetuation of the arms race. "[D]uring the Eisenhower years, there never was a missile gap," writes historian Ronald Powaski. Nevertheless, "the pressure generated by the military-industrial complex compelled Eisenhower to accelerate the U.S. missile program."[249] Historian Richard Miller suggests that in this propaganda campaign, defense officials generated misleading and invalid threat data to stoke American fears and build political support for perpetuation of the arms race: "The policy of frightening the American people in order to create support for a massive military machine was based on lies about Soviet intentions. These were not honest evaluations that were disproved by later events, but deliberate falsehoods."[250] During the cold war, "[o]ne president after another has not only accepted worst-case scenarios as valid bases of American deterrence strategy," Powaski writes, "but also has used them to garner congressional support for nuclear weapons programs."[251] It is commonly conceded, writes Noam Chomsky, that during the cold war, the "threat of Soviet aggression was exaggerated. . . . Threats have regularly been concocted on the flimsiest evidence and with marginal credibility at best."[252]

For example, historian Tom Gervasi argues that by excluding bomber weapons from comparative counts of strategic weapons deliverable by each superpower, Reagan officials strategically exaggerated the level of Soviet military danger posed to the United States.[253] This phenomenon of threat inflation was not confined to U.S.-based information sources. As Milton Leitenberg, of the Swedish Institute of International Affairs, observes, biased cold war threat comparisons could be found in publications as respectable as *The Military Balance,* an annual book put out by the reputable London-based International Institute for Strategic Studies (IISS). According to Leitenberg, the selective omissions in *The Military Balance* "have a clear bias. They uniformly understate the capabilities of the United States and of the West, and overstate those of the Soviet Union."[254] As Gary Chapman and Joel Yudkin, of the Council for a Livable World, point out, it now appears that the "bomber gap," "missile gap," and "window of vulnerability," were fearsome cold war slogans connected only loosely to the material realities of actual superpower military deployments.

> In the early 1950s alarmists alleged that the United States lagged behind the Soviet Union in the number of strategic bombers available for attack. The "bomber gap" was completely false; the United States always had more strategic bombers. In 1960 President John F. Kennedy based a good part of his campaign on a growing "missile gap." President Kennedy later admitted that the United States was ahead of the Soviet Union in missiles. In the mid-1960s the U.S. government discovered a disparity in warheads and in anti-ballistic missile capabilities, so the United States started a crash program of deploying multiple warheads on missiles and started to build an expensive nationwide anti-ballistic missile system. In the 1970s military analysts opened an alleged "window of vulnerability" of U.S. silo-based missiles. To close it, the Carter administration proposed a vast underground shell game with missiles being shuffled back and forth between reinforced concrete bunkers. The "window of vulnerability" argument persisted into the 1980s when the Reagan administration proposed an even more expensive and dubious technological fix, the Strategic Defense Initiative, or "Star Wars." Like its predecessors, the "window of vulnerability"never existed, because it was impossible for the Soviet Union to destroy all U.S. land-based missiles in one strike.[255]

From this historical perspective, cold war threats of Soviet world conquest were largely American constructions, woven out of a fabric of fear, secrecy, and paranoia that blanketed elite policy makers and average citizens alike. Officials employed secrecy and deception on a wide spectrum of defense topics, ranging from U.S. military capabilities to Soviet threat assessments. During the cold war, we were in what Miller describes as a "state of siege from within."[256] In this environment of limited information, worst-case assessments became routine, with estimates of enemy strength inflated by cold warriors' worries about their own perceived technological shortcomings. Momentum for military buildup was driven by what physicist Herbert York calls an "internal arms race"[257] that pitted American defense planners and threat assessors against American military contractors in a self-reinforcing spiral.

> [T]he inherent uncertainties involved in weapons assessment give rise to the self-contained logic of perpetual military build-up. The primary effect of competition based on so much uncertainty is to push both sides toward worst-case judgments about the effectiveness of their own and their adversary's defensive systems. . . . The same logic that drives one to doubt one's own capabilities results in inflation of the enemy's. . . . To remedy these perceived shortcomings and reduce uncertainties, both sides are likely to undertake defensive and offensive force improvements that could only prompt similar, redoubled efforts by the other side. These uncertainties combine to create a strategic environment in which crude estimates replace tangible evidence as building blocks of perceived reality.[258]

This historical perspective challenges the validity of Reagan victory school narratives at fundamental levels. These victory narratives rest on the assumption that the cold war was a grave contest between superpowers, narrowly won by the United States. But what if much of the contest itself was illusion, manufactured in order to create the perception that the United States was locked in a monumental struggle for national survival? What good is vindication for the victor who has prevailed in a rigged game? These questions complicate the cut-and-dried "vindicationist" logic undergirding Reagan victory school narratives. The zest of post–cold war victory celebrations was magnified by years and years of cold war threat inflation. However, celebration now may be premature. Chomsky suggests that because the cold war was actually driven by a constellation of interlocking interests tied to unrestrained weapons production, the conflict is not over but "has perhaps half-ended; Washington remains a player."[259] Instead of ending the struggle, the fall of the Berlin Wall has merely forced the United States to cast new foes in enemy roles to replace the departed Soviet Union as the key foil for what physicist Herbert York calls a perpetual "internal arms race"[260] between American defense planners and weapons producers.

Conclusion

Missile defense debates in the United States are as much about the relationships between the military, the public, and science in a democracy as they are about national security. As such, the oft-ignored consequences of institutionalized deception and secrecy in scientific research deserve to be thematized as topics in public debates on missile defense policy. During the cold war, the trump card of national security effectively blocked serious consideration of these factors. Missile defense advocates utilized the topos of scientific objectivity, as well as strategic misinformation and communicative deception, to control the scope and tenor of discussion, frustrating competent public involvement. As a result, Congress was forced to make BMD funding decisions with distorted data in a threat-inflated environment, and robust public input into the debate was foreclosed systematically. There was only one purpose for SDI, former Soviet Army colonel Anatoly Kravtsov said in hindsight: "It was an excuse for military complexes on both sides to milk their governments."[261]

This experience provides valuable insight regarding the nature of institutional argument in the missile defense context. As the X-ray laser episode demonstrates, the missile defense bureaucracy has shown a proclivity to dominate public argumentation, by attaching the weight of scientific objectivity to their claims in one breath, then subverting the norms of scientific objectivity in the next. The tension inherent in this strategy could be effectively rationalized during the height of the cold war with the contention that absolute secrecy was required to prevent sensitive information from being acquired by the United States' nuclear enemy, the Soviet Union.

With the fall of the Berlin Wall, this strategy is becoming increasingly less tenable. The costs associated with secrecy have grown to the point where they now challenge the trump status of national security as an argumentative warrant. It is indeed a sign of the times that veteran nuclear war strategists such as McGeorge Bundy, Admiral William Crowe, and Sidney Drell now argue that understanding and open communication are more important to national security than nuclear secrecy.

> But the general atmosphere of secrecy that has surrounded the whole nuclear-weapon enterprise, in country after country, is unhealthy. It has had a destructive effect on the level of public understanding in every nuclear-weapon state, and it has also increased fear and mistrust between competing

> countries. . . . [U]nderstanding, shared by citizens and their governments, is much more valuable for our survival than most of our nuclear secrets.[262]

Voicing a similar concern from a Russian perspective, Valery Yarynich recently called for more openness and dialogue on nuclear weapons issues. A retired colonel in the Russian Strategic Rocket Forces, Yarynich argued that because secrecy and mistrust tends to complicate safe command and control of nuclear weapons, it is imperative for the United States and Russia to move beyond cold war postures of secrecy and mistrust.

> Contrary to popular belief, the reduction of nuclear inventories has not decreased the danger of their use. The key to lowering the risk of accidental or unauthorized use lies in the systems of command and control. But there can be no meaningful discussion of such issues as long as most information about these systems remains so tightly held within both governments and completely closed to the public. . . . Without an open examination of the issue, it is difficult to judge the merits of giving top commanders this [nuclear] power. . . . The lack of open discussion also perpetuates the dangerous hair trigger on our two command systems.[263]

This realization should prompt citizens interested in nuclear safety and government accountability to adopt a stance of responsible skepticism, insisting that in BMD debates missile defense advocates meet minimum burdens of proof by disclosing the evidentiary support backing their claims. It should also prompt scholars to engage in public critique of institutional argumentation that falls short of meeting such burdens.

While Ronald Reagan has retired to his ranch and recently withdrawn from political life, Edward Teller continues the zealous campaign of missile defense advocacy that he started as director of the Lawrence Livermore National Laboratory during the cold war. "We're in trouble," Teller stated in a 12 January 1997 interview.[264] Asked to estimate, on a scale from one to ten, the possibility of the United States being drawn into a nuclear war, Teller replied:

> Look, if you asked me that question in the last year of the Bush administration, I would have said zero or one, considering the possibility of an accidental chance. If the current administration policies continue for another four years, we will have 10 years of almost no significant weapons development. The chances then would be five or six out of 10.[265]

Clearly, Teller's propensity for hyperbole has not dulled with age. As readers consider arguments such as these advanced by Teller and other missile defense advocates, they would do well to bear in mind the remarkable and startling lessons of the extravagant X-ray laser program's historical legacy. One of the most ironic elements of this legacy is Teller's 1987 proposal of "adopting a policy of openness as the first and only unilateral step toward disarmament."[266] Although Teller's masterful record of using the classification system to advance partisan political objectives makes his 1987 suggestion a curious proposal, perhaps it is an idea whose time has come.

Notes

1. Edward Teller, *Better a Shield than a Sword* (New York: Free Press, 1987), 119–20.
2. See Philip M. Boffey, William J. Broad, Leslie H. Gelb, Charles Mohr, and Holcomb B. Noble, *Claiming the Heavens* (New York: Times Books, 1988), 3–6.
3. For a comprehensive treatment of the key terms and concepts used in the SDI debate, designed to "clarify and simplify SDI's complex terrain, lay out the technology involved, and identify those areas under debate," see Harry Waldman, *The Dictionary of SDI* (Wilmington, Del.: Scholarly Resources, 1988).
4. There may have been another film that inspired Reagan to entertain visions of arming the heavens. According to historian Frances FitzGerald, "Alfred Hitchcock's *Torn Curtain,* a 1966 film . . . revolves around an attempt to develop to develop an anti-missile missile. In it an American agent played by Paul Newman declares, 'We will produce a defensive weapon that will make all nuclear weapons obsolete, and thereby abolish the terror of nuclear warfare'" (Frances FitzGerald, *Way Out There in the Blue: Reagan, Star Wars and the End of the Cold War* [New York: Simon and Schuster, 2000], 23). Apparently, Reagan aides noticed the striking parallels between Newman's words and the president's own rhetorical themes advanced during official speeches and appearances (see Strobe Talbott, *Master of the Game: Paul Nitze and the Nuclear Peace* [New York: Alfred A. Knopf, 1988], 188). Perhaps these moments of uncanny historical parallelism are confirmation of science fiction author Jerry Pournelle's observation that science fiction "is intended to be realistic—that is, the author and reader believe the story might happen" (Jerry Pournelle, "The Construction of Believable Societies," in *The Craft of Science Fiction,* ed. Reginald Bretnor [New York: Harper and Row, 1976], 105). During the 1970s, science fiction writing was characterized by "a spirit of outwardness" (Poul Anderson, "Star-flights and Fantasies: Sagas Still to Come," in *The Craft of Science Fiction,* ed. Reginald Bretnor [New York: Harper and Row, 1976], 34). This trend resulted in a spate of works designed to show what "may become if new and unanticipated doors into knowledge or into other worlds suddenly are opened by men using the scientific method and its technologies" (Reginald Bretnor, "SF: The Challenge to the Writer," in *The Craft of Science Fiction,* ed. Reginald Bretnor [New York: Harper and Row, 1976], 4; see also Thomas N. Scortia, "Science Fiction as the

Imaginary Experiment," in *Science Fiction, Today and Tomorrow,* ed. Reginald Bretnor [New York: Harper and Row, 1974], 147).

The significance of this development was not lost on actual policy makers, who drew on science fiction writers for formal political advice. "A modern list of some of those who've done such consulting reads like a contemporary Who's Who of Science Fiction, including Gregory Benford, Ben Bova, Pat Cadigan, William Gibson, Larry Niven, Jerry Pournelle, Neal Stephenson, Bruce Sterling and Vernor Vinge," explains journalist Elizabeth Weise ("Stranger than Fiction," *USA Today,* 22 July 1998, 5D). According to Weise, "[s]cience fiction writers also played a role in helping end the Cold War, as members of Ronald Reagan's Citizens' Advisory Council on National Space Policy, which created The Strategic Defense Initiative (SDI), known as Star Wars. They produced a set of papers outlining the notion that space was the key to the Cold War and strategic defense was the key to space" ("Stranger than Fiction," 5D). Advisory Council member Pournelle even anticipated what would become one of the major defenses of SDI, years after it was proven to be a hoax: "If you applied the right kind of research in the right places, you could force the Russians to go bankrupt," (quoted in Weise, "Stranger than Fiction," 5D).

5. Joseph Romm, "Pseudo-Science and SDI," *Arms Control Today* 19 (1989): 15.
6. Jürgen Habermas identifies this interface between government agencies and private research facilities as a crucial translation zone connecting the "scientific" and "practical" realms of discourse: "Communication between politically authorized contracting agencies and objectively knowledgeable and competent scientists at major research and consulting organizations marks the critical zone of the translation of practical questions into the scientifically formulated questions and the translation of scientific information back into answers to practical questions" (Jurgen Habermas, *Toward a Rational Society: Student Protest, Science, and Politics,* trans. Jeremy J. Shapiro [Boston: Beacon Press, 1970], 70).
7. To get at the root of SDI's rhetorical appeal, it may be even more appropriate to focus on the communications between Teller and Reagan than it is to train critical attention on Reagan's rhetorical dynamic vis-à-vis the American public. It is quite possible that the basis for Reagan's rhetorical flourishes was rooted in the exaggerated scientific claims that Teller fed to the White House through official and informal channels. For a description of Teller's influence on Reagan during Reagan's tenure as governor of California, see Boffey et al., *Claiming the Heavens,* 7; James Chace and Caleb Carr, *America Invulnerable* (New York: Summit, 1988), 294. For description of the magnitude of Teller's influence over Reagan during Reagan's presidency, see William Broad, *Teller's War* (New York: Simon and Schuster, 1992).
8. See William Broad, *Star Warriors* (New York: Simon and Schuster, 1985), 107. This was the same year that missile defense firebrand General George J. Keegan stoked fears of Soviet BMD breakout, speculating that by 1980 the USSR would "have technically and scientifically solved the problem of the ballistic missile threat," and that by 1980 it would have tested successfully such a ballistic missile defense (see "Charged Particle Beam: Anatomy of a Defense Scare," *F.A.S. Public Interest Report* 30 [1977]: 1–2).

9. Robert Bowman, telephone interview with the author, 21 July 1999.
10. Malcolm Wallop, "Opportunities and Imperatives of Ballistic Missile Defense," *Strategic Review* 7 (1979): 13-21. According to FitzGerald, "Wallop had come to this idea thanks to the work of Angelo Codevilla, a young staff assistant on the Select Committee on Intelligence. . . . Though entirely self-taught, he [Codevilla] could have been said to belong to the Curtis Le May - Phyllis Schlafly school of strategy, for he maintained that there was no essential difference between nuclear and conventional weapons and that with enough counter-force weapons and civil-defense measures the U.S. could protect 90 percent of its population in a full-scale nuclear war" (FitzGerald, *Way Out There*, 122).
11. FitzGerald, *Way Out There*, 121.
12. FitzGerald, *Way Out There*, 124.
13. See Angelo Codevilla, *While Others Build* (New York: Free Press, 1988), 73-75.
14. The X-ray option was particularly desirable as a technology for missile defense because of this boost-phase targeting capability. Tracked and targeted in their boost phase, ICBMs are rendered much less threatening because they cannot yet MIRV, or separate into a multitude of separate warheads. Interception at the pre-MIRVed stage greatly reduces the complexity of the missile defense task by removing decoys and countermeasures from the equation.
15. Romm, "Pseudo-Science and SDI," 18. This campaign trumpeting technological wizardry as the basis of military strength hearkened back to General William Westmoreland's "electronic battlefield" initiatives. Testifying before Congress in 1970, General Westmoreland waxed eloquent in support of technological advance as a source of security: "On the battlefield of the future, enemy forces will be located, tracked, and targeted almost instantaneously through the use of data links, computer assisted intelligence evaluation, and automated fire control. . . . I am confident that the American people expect this country to take full advantage of its technology—to welcome and applaud the developments that will replace wherever possible the man with the machine" (William Westmoreland, statement, *Congressional Record*, 13 July 1970, 23823–25). Dr. Malcolm Currie, director of defense research and engineering, echoed similar thoughts: "[A] remarkable series of technical developments have brought us to the threshold of what will become a true revolution in conventional warfare." (Quoted in Phil Stanford, "The Automated Battlefield," *New York Times Magazine*, 23 February 1975, 12; see also Richard Burt, "New Weapons Technologies: Debate and Directions," Adelphi Paper 126 [London: Brassey's/International Institute for Strategic Studies, 1976]).

    On the Soviet side, defense analysts V. V. Druzhinin and D. S. Kontorov made similar claims in their 1972 book, *Concept, Algorithm, and Decision*, "The scientific technical revolution has embraced all fields of human endeavor, especially military. Human creative capabilities will increase and expand as man is liberated from technological functions and the results of solved problems are turned over to computers" (V. V. Druzhinin and and D. S. Kontorov, *Concept, Algorithm, Decision* (*A Soviet View*), trans. U.S. Air Force [Washington, D.C.: GPO, 1972], 294, 296).

16. Teller first met Reagan in Sacramento in the fall of 1966. Reagan returned the visit with a trip to Livermore in the winter of 1967. "It was Ronald Reagan's first chance to hear about defense weapons," Teller later recalled (Edward Teller, "SDI: The Last, Best Hope," *Washington Times,* 28 October 1985, 75).
17. As Keyworth recounted, "[b]luntly, the reason I was in that office is because Edward first proposed it to me, and the president very much admires Edward" (quoted in Gregg Herken, *Counsels of War* [New York: Oxford University Press, 1987], 397).
18. See "Teller Said to Urge Development of X-ray Laser," *Aerospace Daily,* 1 December 1982, 158–59.
19. The "High Frontier" group included the key White House players involved in pursuit of the SDI project, such as General Daniel O. Graham, George Keyworth, Attorney General Edwin Meese, and National Security Advisor William Clark. The High Frontier group based much of its pro-BMD arguments on a 1983 Heritage Foundation report which Graham authored (see Daniel O. Graham, *The Non-Nuclear Defense of Cities: The High Frontier Space-Based Defense against ICBM Attack* [Cambridge: Abt Books, 1983]). In this report, Graham argued for a space-based missile defense system by pointing to the recent success of the space shuttle as a harbinger of future potential in this area. He claimed that BMD "appears remarkably inexpensive, and can probably be deployed in a relatively short time," and that such deployment could be key to a "technological end-run which would offset current Soviet strategic nuclear advantages" (Graham, *The Non-Nuclear,* vii). As commentator Paul Weyrich recounted, "Gen. Graham soon found a willing listener in one President Ronald Reagan" (Paul M. Weyrich, "A True Patriot," *Washington Times,* 8 January 1996, 34).
20. Broad, *Teller's War,* 114.
21. Reagan, letter to Karl Bendetsen, 20 January 1982, Bendetsen Collection; see also Dinesh D'Souza, *Ronald Reagan: How an Ordinary Man Became an Extraordinary Leader* (New York: Simon and Shuster, 1997), 175.
22. Stanley A. Blumberg and Louis G. Panos, *Edward Teller: Giant of the Golden Age of Physics* (New York: Scribners, 1990), 7.
23. Edward Teller, statement, *Firing Line,* 19 October 1987, Internet, Online, Lexis/Nexis (an electronic information database with over 28,000 full-text sources online at http://www.lexis-nexis.com/lncc).
24. George Brown, "Representatives Release Teller X-Ray Laser Letters, Demand Release of Documents on DOE Investigation," in press release by Rep. Ed Markey, 1 August 1988, copy on file with the author.
25. FitzGerald, *Way Out There,* 121. Nevertheless, FitzGerald points out that Wallop, Graham and Teller still "played an important role in launching the SDI program. In the years before the speech they campaigned mightily for their own particular strategic-defense concepts, besieging the Congress and the Pentagon as well as the White House, and publicizing their concepts through the press. In the process they created a public stir around the idea of missile defenses and kept the issue alive throughout the days of apostasy in Weinberger's Pentagon. Then, too, partly as a result of their efforts,

the particular technologies they espoused later became major elements of the SDI program" (FitzGerald, *Way Out There*, 121).

26. FitzGerald, *Way Out There*, 203.
27. See Douglas C. Waller, *Congress and the Nuclear Freeze* (Amherst, Mass.: University of Massachusetts Press, 1987), 192.
28. Donald R. Baucom, *The Origins of SDI, 1944-1983* (Lawrence: University of Kansas Press, 1992), 193
29. Baucom, *Origins of SDI*, 193.
30. FitzGerald, *Way Out There*, 208.
31. Ronald Reagan, *An American Life* (New York: Simon and Schuster, 1990), 571.
32. See George Yonas, "The Strategic Defense Initiative," *Daedalus* 114 (1985): 73–90; D'Souza, *Ronald Reagan*, 175.
33. Quoted in D'Souza, *Ronald Reagan*, 175.
34. D'Souza, *Ronald Reagan*, 175.
35. G. Thomas Goodnight, "Ronald Reagan's Reformulation of the Rhetoric of War: Analysis of the 'Zero Option,' 'Evil Empire,' and 'Star Wars' Addresses," *Quarterly Journal of Speech* 72 (1986): 403–4.
36. Howard C. Ris Jr., preface to *The Fallacy of Star Wars: Why Space Weapons Can't Protect Us*, ed. John Tirman (New York: Vintage Books, 1984), vii.
37. Senator Larry Pressler's (R-SD) book provides a useful overview of the arguments advanced in the House and Senate during the three years following Reagan's Star Wars address (see Larry Pressler, *Star Wars: The Strategic Defense Initiative Debates in Congress* [Westport, Conn.: Greenwood Press, 1986]). The massive cost of the Star Wars program dictated that many of the key debates took place during appropriations hearings (see, e.g., U.S. Congress, *Department of Defense Appropriations for Fiscal Year 1985*, 98th Cong., 2d sess., Senate Hearings, Commitee on Appropriations [Washington, D.C.: GPO, 1984]; U.S. Congress, *Department of Defense Authorization for Appropriations for Fiscal Year 1985*, 98th Cong., 2d sess., Senate Hearings, Committee on Armed Services [Washington, D.C.: GPO, 1984]). There were also numerous hearings convened to address BMD issues specifically (see, e.g., U.S. Congress, *Arms Control and the Militarization of Space*, 97th Cong., 2d sess., Senate Hearings, Committee on Foreign Relations [Washington, D.C.: GPO, 1982]; U.S. Congress, *Controlling Space Weapons*, 98th Cong., 1st sess., Senate Hearings, Committee on Foreign Relations [Washington, D.C.: GPO, 1983]; U.S. Congress, *Strategic Defense and Anti-Satellite Weapons*, 98th Cong., 2d sess., Senate Hearing, Committee on Foreign Relations [Washington, D.C.: GPO, 1984]).
38. During the Star Wars era, articles on BMD appeared frequently in journals such as *Daedalus, International Security, Foreign Policy,* and *Foreign Affairs.* Journal editors such as Stephen R. Graubard even viewed the publication of such articles as an important public service. Introducing the first of two special issues of *Daedalus* devoted exclusively to SDI, Graubard explained, "[t]o try to comprehend what is now in prospect, to reflect with sobriety and judgment on the implications of what is being set in motion,

will require new voices, trained in international security matters, to join the 'veteran' analysts of the 1950s, 1960s, and 1970s in developing accurate information, reasonable hypotheses, and persuasive criticism of public policy" (Stephen R. Graubard, preface to special issue, "'Weapons in Space,' vol. 1," *Daedalus* 114 [1985]: v–vii).

39. Steven Anzovin's *The Star Wars Debate* offers a compendium of articles on Star Wars published in popular journals such as *Business Week, Commentary, The New Yorker,* and *The Progressive* (Steven Anzovin, ed., *The Star Wars Debate,* [New York: H.W. Wilson, 1986]). For an excellent presentation and treatment of the role of political cartoons in shaping Star Wars dialogue, see Edward Tabor Linenthal, *Symbolic Defense* (Chicago: University of Illinois Press, 1989).
40. Michael J. Mazarr, *Missile Defenses and Asian-Pacific Security* (London: Macmillan Press, 1989), 3.
41. As journalist Charles Ealy notes, "[t]he Star Wars phenomenon, which began in 1977, was far more significant than first suspected, helping set the tone for the Reagan-era view that the Soviet Union was the 'evil empire' and making arguments for building a 'Star Wars' missile defense system sound more realistic" (Charles Ealy, "Lucas Made American Missile System Sound Realistic: Popular Culture's Phantom Menace," *Toronto Star,* 14 May 1999, 17). Although missile defense detractors attempted to saddle Reagan's initiative with science fiction baggage, these efforts largely failed. "Critics, in an effort to kill the project, quickly labeled it 'Star Wars,' but given the popularity of George Lucas' trilogy, that label only enhanced it" (Paul Weyrich, "A True Patriot," 34). Worried about a possible reverse effect, whereby SDI's negative connotations would tarnish the image of the *Star Wars* films, Lucasfilm, Ltd. filed a remarkable lawsuit in 1985, alleging trademark infringement by the "Coalition for The Strategic Defense Initiative." A U.S. District Court judge dismissed the case when he could find no evidence of trademark infringement (see *Lucasfilm Ltd. v High Frontier, et. al.,* 622 F. Supp. 931).
42. Ronald Reagan, "Address to the Nation on Defense and National Security," *Public Papers of the Presidents of the United States, Ronald Reagan 1983,* vol. 1 (Washington, D.C.: GPO, 1984), 439. Further references to this speech will be made by page number in the text.
43. See Aristotle, *On Rhetoric,* trans. George A. Kennedy (New York: Oxford University Press, 1991), 46–47.
44. Kathyrn M. Olson and G. Thomas Goodnight, "Entanglements of Consumption, Cruelty, Privacy, and Fashion: The Social Controversy over Fur," *Quarterly Journal of Speech* 80 (1994): 250.
45. Goodnight, "Reagan's Reformulation," 405. As Jeffrey Simon of the National Defense University observes, "SDI augured in a 'new' defensive doctrine or way of viewing the world, requiring a radical departure in U.S. force structure with significant consequences for security" (Jeffrey Simon, ed., *Overview to Security Implications of SDI* [Washington, D.C.: National Defense University Press, 1990], 13). For discussion of Soviet reactions to changes in strategic doctrine prompted by SDI, see Stephen Blank, "SDI and Defensive Doctrine: The Evolving Soviet Debate," Kennan Institute for

Advanced Russian Studies Occasional Paper no. 240 (Washington, D.C.: Woodrow Wilson Center, 1990); Mary C. Fitzgerald, "Soviet Views on SDI," Carl Beck Paper in Russian and East European Studies no. 601 (Pittsburgh: University of Pittsburgh Center for Russian and East European Studies, 1987); Kevin N. Lewis, "Possible Soviet Responses to the Strategic Defense Initiative: A Functionally Organized Taxonomy," RAND Note N-2478-AF (Santa Monica, Calif.: RAND Corporation, 1986); and Dmitri Mikheev, *The Soviet Perspective on the Strategic Defense Initiative* (New York: Pergamon-Brassey's, 1987).

46. FitzGerald, *Way Out There,* 208.
47. Ronald Reagan, "Remarks and a Question-and-Answer Session With Reporters on Domestic and Foreign Policy Issues, March 25, 1983," *Public Papers of the Presidents of the United States, Ronald Reagan: 1983,* vol. 1 (Washington, D.C.: GPO, 1984), 448.
48. For discussion of the significance of this distinction, see Pressler, *Star Wars.*
49. Quoted in Edward Reiss, *The Strategic Defense Initiative* (Cambridge: Cambridge University Press, 1992), 53.
50. Reiss, *Strategic Defense Initiative,* 54–55.
51. Janice Hocker Rushing, "Ronald Reagan's 'Star Wars' Address: Mythic Containment of Technical Reasoning," *Quarterly Journal of Speech* 72 (1986): 421–22; see also Rebecca Bjork, *The Strategic Defense Initiative: Symbolic Containment of the Nuclear Threat* (New York: SUNY Press, 1992).
52. This initial strategy led to substantial confusion abroad regarding U.S. intentions. "The Reagan administration sought not to get hoisted with its own petard in discussions about what SDI was and what it was not. At the same time, there were so many 'spokesmen' in and out of government presenting contradictory versions of SDI that the allies [in Europe] rightly did not know which version to deal with" (Robert C. Hughes, *SDI: A View from Europe* [Washington, D.C.: National Defense University Press, 1990], 27).
53. See Erik K. Pratt, *Selling Strategic Defense* (Boulder, Colo.: Lynne Rienner, 1990), 111; Reiss, *Strategic Defense Initiative,* 72–78.
54. Yonas provides a thorough description of the Fletcher and Hoffman panels' assessments of BMD technologies ("Strategic Defense Initiative," 73–90). For technical accounts of the various BMD systems considered by the Hoffman and Fletcher panels, see Harold Brown, "Is SDI Technically Feasible?" in *The Search for Security in Space,* ed. Kenneth Luongo and W. Thomas Wander (Ithaca, N.Y.: Cornell University Press, 1989), 205–22; Hans A. Bethe and Richard L. Garwin, "Appendix A: New BMD Technologies," *Daedalus* 114 (1985): 331–68; Heritage Foundation, "Strategic Defense: The Technology That Makes It Possible," in *Anti-Missile and Anti-Satellite Technologies and Programs,* ed. U.S. Department of Defense, U.S. Office of Technology Assessment, and the Heritage Foundation (Park Ridge, N.J.: Noyes, 1986), 17; and United States Office of Technology Assessment, *Strategic Defenses: Ballistic Missile Defense Technologies, Anti-Satellite Weapons, Countermeasures, and Arms Control* (Princeton, N.J.: Princeton University Press, 1986), 27–28.
55. See Heritage Foundation, "Strategic Defense," 4.
56. Deborah Blum, "'Star Wars' Based on Lie?" *Sacramento Bee,* 25 October 1987, A1.

57. Broad, *Teller's War,* 35.
58. See "Defensive Technologies Study Sets Funding Profile Options," *Aviation Week and Space Technology,* 24 October 1983, 50
59. See "Star Wars," *Newsweek,* 4 April 1983, cover.
60. Reagan, "Address," 443.
61. The validity of such critical speculation appeared to be strengthened by Strategic Defense Initiative Organization (SDIO) director James Abrahamson's statement in a 12 March 1985 interview on the *McNeil-Lehrer NewsHour.* Putting Star Wars' political fortunes in perspective, Abrahamson offered the following comment: "I say that the Force is indeed with us" (James Abrahamson, statement, *McNeil-Lehrer NewsHour,* 12 March 1985, Internet, Online, Lexis/Nexis).
62. Bjork, *Strategic Defense Initiative,* 131.
63. FitzGerald, *Way Out There,* 16.
64. United States, Strategic Defense Initiative Organization, "SDI: Goals and Technical Objectives," reprinted in Kenneth Luongo and W. Thomas Wander, eds., *The Search for Security in Space,* (Ithaca, N.Y.: Cornell University Press, 1989), 156.
65. Gary L. Guertner, "What is Proof?" in *The Search for Security in Space,* ed. Kenneth Luongo and W. Thomas Wander (Ithaca, N.Y.: Cornell University Press, 1989), 191.
66. Robert Scheer, "Scientists Dispute Test of X-Ray Laser Weapon," *Los Angeles Times,* 12 November 1985, 1.
67. Quoted in Scheer, "Scientists Dispute," 1.
68. In a secret review of the Livermore data, Los Alamos scientists discovered systemic problems in the measurement apparatus used for the entire series of tests. "Three months after the reportedly successful Cottage test, scientists at the Los Alamos weapons lab reviewed the highly classified data and warned Livermore officials that the results had been distorted because the device used to measure the laser's intensity cannot provide accurate readings and, therefore, should not be used" (Scheer, "Scientists Dispute," 1).
69. Livermore physicist Joseph Nilsen corroborated the Los Alamos group's findings in a 27 June classified report, circulated at LLNL. See Scheer, "Scientists Dispute," 1.
70. Scheer, "Scientists Dispute," 1.
71. Los Alamos physicists urged Teller's group to put off Goldstone in order to buy time for improvements to be made in the experimental apparatus. "A design error in a key measuring device used in all past tests has caused it to give false readings. Los Alamos scientists urged Livermore to develop a new mechanism to measure the laser, which would have caused an estimated delay of six months to a year in the program" (Scheer, "Scientists Dispute," 1).
72. Edward Markey, "Goldstone X-Ray Laser Test Should Be Delayed," *Congressional Record,* 5 December 1985, E5406.
73. William Broad, "30 Lawmakers Urge Delay in Laser Weapon Test," *New York Times,* 7 December 1985, A7.
74. Scheer, "Scientists Dispute," 1.

75. Broad, *Teller's War,* 205.
76. Clarence A. Robinson, "BMD Homing Interceptor Destroys Reentry Vehicle," *Aviation Week and Space Technology,* 18 June 1984, 19–20.
77. David Osborne, "Reagan's Great Lie in the Sky," *Independent* (London), 29 August 1993, 13.
78. William F. Buckley, "Intercepting the Truth about SDI Testing," *Buffalo News,* 14 October 1993, 3.
79. Quoted in Robinson, "BMD Homing Interceptor," 19.
80. Reiss, *Strategic Defense Initiative,* 56.
81. Quoted in John Pike, "Strategic 'Deception' Initiative," *Arms Control Today* 23 (1993): 4.
82. Quoted in Pike, "Strategic 'Deception,'" 4. For additional examples of claims that HOE success proved SDI feasibility, see Kim Holmes, "Technology Speeds the Strategic Defense Initiative Timetable," Heritage Foundation Backgrounder no. 557, 13 January 1987; Osborne, "Reagan's Great Lie"; Phil Reeves, "Star Wars 'Fake' Fooled the World," *Independent* (London), 19 August 1993, 8; and Jeffrey R. Smith, "3 'Star Wars' Tests Rigged, Aspin Says," *Washington Post,* 10 September 1993, A19.
83. Quoted in Tim Weiner, "Lies and Rigged 'Star Wars' Test Fooled the Kremlin, and Congress," *New York Times,* 18 August 1993, A1+. For a corroborating account, see Ronald E. Powaski, *The Cold War* (New York: Oxford University Press, 1998), 249.
84. Quoted in Weiner, "Lies," A15.
85. Weiner, "Lies," A15. In an approach that gave new meaning to the notion of "cooked" experimental results, "One missile specialist conceded that Army engineers 'put a shed over the missile (before launch) and blew in warm air'" (quoted in Pike, "Strategic 'Deception,'" 6). Once heated, the target missile became easy prey for the heat-seeking HOE interceptor to track and destroy.
86. Smith, "3 'Star Wars' Tests," A19.
87. Quoted in Weiner, "Lies," A15.
88. See Eric Schmitt, "Aspin Disputes Report of 'Star Wars' Rigging," *New York Times,* 10 September 1993, B8. Aspin claimed the deception program was canceled on 3 September 1983. The fourth HOE test took place 10 June 1984. But consider Pike's analysis that the narrowness of Aspin's explanation still left room for the possibility of test rigging: "[T]he purported termination of one deception special access program in no way implies the termination of all deception efforts related to SDI generally, or the HOE in particular" ("Strategic 'Deception,'" 6).
89. The ambiguity of Aspin's account of the HOE experience was captured in the widely divergent headlines in three major newspapers' 10 September 1993 coverage of the story. The headlines ranged from "Aspin Denies 'Star Wars' Tests Were Rigged" (*Los Angeles Times,* A22), to "Aspin Disputes Report of 'Star Wars' Rigging" (*New York Times,* B8), to "3 'Star Wars' Tests Rigged, Aspin Says" (*Washington Post,* A19). Senator David Pryor (D-Arkansas) felt Aspin's explanation failed to clear the air on HOE: "The Pentagon is like the student grading its own exam paper in these operational test situations, and we're going to have to end that" (quoted in Schmitt, "Aspin Disputes,"

B8). Perhaps the most definitive overall assessment of the HOE affair was provided by John Pike of the Federation of American Scientists: "Did the SDI program include an unacknowledged deception effort? Were there unacknowledged technical characteristics of the final HOE test which diminished the apparent significance of the test? And was the initial reporting by the *New York Times* and the *Washington Post* well founded? Despite recent efforts by senior Clinton administration officials to obscure the record, including the explanations by Aspin, September 9, the answer to all three questions is clearly yes" (Pike, "Strategic 'Deception,'" 3; see also John Horgan, "Lying by the Book," *Scientific American* 267 [October 1992]: 20; John Wagner, "Drafting Rules of Deception: How and When," *Washington Post,* 6 August 1992, A23).

90. Eugene Fox, statement, *Crossfire,* Cable News Network, Atlanta, Ga., 18 August 1993, Internet, Online, Lexis/Nexis. In a National Public Radio interview, Weiner discounted Fox's denial, explaining that "it is the nature of deception programs as run by the Pentagon that not every participant in a program knows that he is part of a deception program" (Weiner, interview with Noah Adams, *All Things Considered,* National Public Radio, 18 August 1993, Internet, Online, Lexis/Nexis).
91. Romm, "Pseudo-Science and SDI," 19.
92. Quoted in Romm, "Pseudo-Science and SDI," 19.
93. Quoted in Union of Concerned Scientists, *Empty Promise: The Growing Case Against Star Wars,* ed. John Tirman (Boston: Beacon Press, 1986), 18.
94. Quoted in Delthia Ricks, "Scientists Say Teller Misled Government on Star Wars," *UPI Newswire,* 22 October 1987, Internet, Online, Lexis/Nexis.
95. Charles Bennett, "The Rush to Deploy SDI," *Atlantic Monthly,* April 1988, 56.
96. Ed Markey, "Representatives Release Teller X-Ray Laser Letters, Demand Release of Documents on DOE Investigation," press release by Rep. Ed Markey, 1 August 1988, copy on file with the author.
97. Karl Bendetsen, letter to Edwin Meese, 20 October 1981, copy on file with the author.
98. Ibid.
99. By 1981, the only tangible evidence of X-ray lasing was generated by the 14 November 1980 Dauphin test, a crude preliminary experiment which did not produce a laser even close to weapon-grade intensity.
100. Teller, letter to George Keyworth, 22 December 1983, copy on file with the author.
101. Ibid.
102. See United States General Accounting Office (USGAO), *SDI Program: Evaluation of DOE's Answers to Questions on X-Ray Laser Experiment,* NSIAD-88–181BR, Washington, D.C.: GPO, 1988.
103. William Broad, interview with Noah Adams, *All Things Considered,* National Public Radio, 6 March 1992, Internet, Online, Lexis/Nexis.
104. See Woodruff, letter to George Keyworth, 28 December 1983, copy on file with the author. Woodruff specifically disputed Teller's claim that the X-ray laser project had moved into the engineering phase, stating, "the x-ray laser is nowhere near the engineering phase at this time" (Letter to Keyworth, 4). Woodruff closed his letter with the caution that "it is premature to extrapolate present successes to the conclusion that a

viable weapons system is possible in the near term" (Letter to Keyworth, 5).

105. Edward Teller, letter to Paul Nitze, 28 December 1984, copy on file with the author.
106. Ibid.
107. Edward Teller, letter to Robert McFarlane, 28 December 1984, copy on file with the author.
108. Later, Livermore's George Maenchen would formalize brewing doubts about the Romano experiment in an internal review. See George Maenchen, *Livermore Report* COPD 84–1993, 14 August 1984, copy on file with the author.
109. Lowell Wood, "Soviet and American X-Ray Laser Efforts: A Technological Race for the Prize of a Planet," briefing presented to William Casey, 23 April 1985, copy on file at the Federation of American Scientists Space Policy Project Archive, Washington, D.C.
110. Quoted in U.S. GAO, *SDI Program.*
111. The decision to exclude certain Livermore scientists, particularly "A" Group director Woodruff, from the review process was very unusual. "In all cases Lowell [Wood] assisted with the drafting of these letters and in no instance was I asked to review them for accuracy . . . [even though] responsibility for attempting to meet the programmatic goals called for in these letters would clearly fall to me" (Roy D. Woodruff, letter to Roger Batzel, 19 October 1985, copy on file with the author).
112. Teller's statement was recounted in William Broad's 18 July 1989 interview with Roy Woodruff (quoted in Broad, *Teller's War,* 154).
113. Roy D. Woodruff, letter to George Keyworth, 28 December 1983, copy on file with the author.
114. Roy D. Woodruff, letter to Paul Nitze, 31 January 1985, copy on file with the author.
115. Ibid.
116. Ibid. To emphasize the point, Woodruff added that the odds of success for realizing the advances described by Teller were similar to the likelihood that a group of Soviet missiles would be struck down by a meteor shower (Woodruff, letter to Nitze).
117. Quoted in Broad, *Teller's War,* 218–19. For further commentary, see Kevin O'Neill, "Building the Bomb," in *Atomic Audit: The Costs and Consequences of U.S. Nuclear Weapons Since 1940,* ed. Stephen I. Schwartz (Washington, D.C.: Brookings Institute Press, 1998), 81–82. Woodruff charged that even after promised changes, "the harassment continued unabated, including needless security investigations" (quoted in William Broad, "Dispute Settled at Weapons Lab," *New York Times,* 23 May 1990, A18).
118. In a general sense the strategy of global controversy containment was executed routinely through the strategic use of classification: "[B]oth critics and supporters of SDI say the anti-missile research program also classifies details on its budget and test plans to sidestep political controversy" (Bennett, "The Rush," 56; see also Paul Walton, "'Top Secret' SDI Technologies Ban?" *Jane's Defence Weekly,* 15 February 1986, 246). This strategy also involved the denial of clearance for Pentagon insiders wishing to publish critical articles on SDI in the public press: "Several articles discussing the controversial missile defense system—even some only mildly critical—have been refused clearance for publication in military journals. Insiders suspect the censoring has more to do with politics than with national security" ("Washington Whispers," *U.S. News and*

World Report, 16 June 1986, 12). The Pentagon even classified a 1986 General Accounting Office report that utilized exclusively declassified information. In explaining the rationale for such a measure, SDI spokesman Lt. Col. Lee DeLorme said that pieces of unclassified data could still be damaging when assembled in such a way that they would give adversaries unique insight: "The Russians could put them together, but then they would never be certain of what they have. . . . By compiling it all and putting it all in a single government document, you have made their task much easier" (quoted in "Hill SDI Report Stamped Secret by Pentagon," *Washington Post,* 5 March 1986, A23). In response to the Pentagon's decision to classify the report, one congressional aide scoffed, "If the KGB has anybody with an IQ of 100 working on this, they've already put together a more detailed report than we have" (quoted David J. Lynch, "Pentagon Drawing the Shades on Star Wars, Senators Fret," *Defense Week,* 24 February 1986, 1). Selected classified information favorable to SDI was also leaked to the media as part of this Pentagon strategy, explained Livermore physicist Hugh DeWitt: "Classified information about a new weapons system has often been leaked to the press for the purpose of creating a spectacular image in the public mind that will lead to great increase in funding for the project" (Hugh E. DeWitt, letter to David Saxon, 3 October 1982, copy on file with the author).

119. John Herrington, quoted in Dan Morain, "Energy Secretary Warns Scientists Not to Disagree in Public," *Los Angeles Times,* 23 July 1988, 1.
120. John Herrington, press conference, Lawrence Livermore National Laboratory, 22 July 1988, copy on file at the Federation of American Scientists Space Policy Project Archive, Washington, D.C.
121. John Herrington, quoted in Flora Lewis, "A 'Star Wars' Cover-Up," *New York Times,* 3 December 1985, A16.
122. Herrington, press conference.
123. Broad, *Teller's War,* 148–49.
124. Theodore A. Postol, quoted in Tina Rosenberg, "The Authorized Version," *Atlantic,* February 1986, 26.
125. "Energy Weapons," *Aviation Week and Space Technology,* 5 September 1983, 50.
126. Charles Mohr, "Scientist Quits Antimissile Panel, Saying Task Is Impossible," *New York Times,* 12 July 1985, A6.
127. Robert McFarlane, statement, "The Cold War," *CNN International News,* 21 March 1999, transcript #99032100V16, Internet, Online, Lexis/Nexis.
128. Quoted in Bruce van Voorst, "The Ploy That Fell to Earth: Star Wars Suffers Another Blow with Charges That an Antimissile Test Was Faked," *Time,* 30 August 1993, 26+.
129. Although a 1986 internal review of the disinformation policy found that it "would be more costly than its benefits," the DOD decided to keep the policy intact in order to inflate assessments of weapons systems performance, including SDI: "The Defense Department, however, will retain its established practice of providing disinformation on specific military programs, including strategic defense. . . . Release of deceptive information will continue to primarily involve weapon system performance claims at the program official level, with such disinformation closely monitored by a technical advi-

sory group" ("Disinformation," *Aviation Week and Space Technology,* 28 July 1986, 15).

130. Quoted in David North, "U.S. Using Disinformation Policy to Impede Technical Data Flow," *Aviation Week and Space Technology,* 17 March 1986, 16–17. Interestingly, the quoted official added that "[i]f some of the disinformation activity on a particular program get [*sic*] passed to Congress through hearings or other means, there are channels on the Hill that can be used to get the correct information to the people who need to know" (quoted in North, "U.S. Using Disinformation, 17). In the case of SDI, such a "channel of truth" was not so readily available; consider Roy Woodruff's futile efforts to secure a Washington audience to correct distorted SDI test data sent to Congress (see Broad, *Teller's War,* 220–21).
131. Ronald Reagan, "Presenting the Strategic Defense Initiative," National Security Decision Directive 172, 30 May 1985, reprinted in *National Security Decision Directives of the Reagan and Bush Administrations,* ed. Christopher Simpson (Boulder, Colo.: Westview, 1995), 535–48.
132. NSDD 172 set the tone for official discourse on SDI. As Simpson explains, "the authorized explanation of, and rationale for, SDI laid out here [in NSDD 172] were often repeated verbatim for the remainder of the administration" ("Commentary on NSDD 172: Publicizing the Strategic Defense Initiative," in *National Security Decision Directives of the Reagan and Bush Administrations,* ed. Christopher Simpson [Boulder, Colo.: Westview, 1995], 449).
133. Presumably, inclusion of the phrase "scientific channels" in this sentence of NSDD 172 empowered administration officials to intervene directly into scientific discussions to steer public presentation of SDI on a favorable course. On 22 July 1988 Department of Energy secretary James Herrington appeared to act on this authority, ordering scientists from LLNL and Los Alamos to cease public debate over the validity of key LLNL X-ray laser tests (see Herrington, press conference). A similar argument was advanced in a 1986 letter from the DOE assistant secretary for nuclear weapons research to the director of LLNL, written in the wake of the dispute between Woodruff and Batzel. In a clear attempt to dampen public discussion of the X-ray laser issue, the DOE official wrote, "we do not believe the discussion of nuclear directed energy laser weapons concepts during media interviews is in the best interests of the Department of Energy or the national SDI program" (quoted in Union of Concerned Scientists, *Empty Promise,* 20; see also David J. Lynch, "DOE May Discourage Journalists from Talking to Most Federal Scientists: Inverviews on SDI Could Be Sharply Curtailed," *Defense Week,* 22 September 1986, 1; Keith Rogers, "Lab 'Gag Order' by Energy Dept. Alleged," *Valley Times,* 7 February 1987, 1).

    These attempts to foreclose public discussion on Star Wars even extended to former employees of the SDI bureaucracy: "The Strategic Defense Initiative Organization is taking stringent steps to stop news leaks on the program. It now requires that reporters obtain permission from SDIO before they can interview former SDIO employees working as defense contractors. Both the former employees and contractors have strict orders to comply. Also, SDIO is attempting to stop leaks at the Energy Department. It has ordered scientists involved in SDI programs to decline interviews unless SDIO has granted approval" ("News Leaks," *Aerospace Daily*, 9 June 1986, 386). The initiative to discourage media scrutiny of Star Wars research also extended to discussions that featured exclusively nonclassified material (see Keith Rogers, "DOE Wants Rules Tightened on 'Star Wars' Talk," *Valley Times*, 11 February 1986, 1).

134. Reagan, "Presenting the Strategic Defense Initiative," 14.
135. Christopher Simpson, introduction to *National Security Decision Directives of the Reagan and Bush Administrations*, ed. Christopher Simpson (Boulder, Colo.: Westview, 1995), 1–7.
136. Tim Weiner, "General Details Altered 'Star Wars' Test," *New York Times*, 27 August 1993, A19.
137. Weiner, "Lies," A1.
138. Quoted in Weiner, "Lies," A15.
139. As an editorial in the *St. Petersburg Times* recounted, "Congress was having trouble paying its own bills. But with President Reagan stumping hard for the program, and reports of successful tests seeping into closed briefings with legislators, Congress authorized more than $30 billion for Star Wars research and testing" ("The Big Lie in the Sky," editorial, *St. Petersburg Times*, 20 August 1993, 16A).
140. Romm, "Pseudo-Science and SDI," 19. For example, in 1985 LLNL scientist Roy Woodruff explained that excessive funding for the X-ray laser project, won from Congress by Teller and Wood in their fraudulent promotion campaign, distorted research priorities at the laboratory and compromised the quality of staff hired. "The redirection of funds to nuclear directed energy weapons efforts, which has taken place during the last several years, has already caused an imbalance within the program which is at the bound of tolerance. . . . The long-term consequences of such actions are, in my view, life threatening to this institution. A significant number of our key scientists would no doubt move to other programs, or more likely out of the Laboratory entirely. Once this rare talent and experience is lost to the Nation's weapons program, it cannot be restored for at least a decade" (Woodruff, letter to Kenneth Withers, 6 February 1985, copy on file with the author).

    Four years later, it appeared that Woodruff's concerns had indeed materialized: "Allegations that two physicists in the mid-1980s hyped a strategic-defense weapon that would use a nuclear explosion as its power source have led to serious budget cuts for the nation's entire nuclear-weapons research program, Energy Department officials say. The budget cuts are driving away scientists at federal nuclear weapons laboratories in California and New Mexico, who are vital for maintaining a viable nuclear arsenal, the officials say. 'You're talking about several hundred people gone from the program never to return,' said Dick Hahn, director of the Energy Department's weapons research, development and testing. The technology base is shrinking as a result'" (Vincent Kiernan, "Excess Hype of X-Ray Laser Causes Funding Gap for Weapons Research," *Space News*, 9 October 1989, 8).
141. David Lorge Parnas and Danny Cohen, "SDI: Two Views of Professional Responsibility," Institute on Global Conflict and Cooperation Policy Paper no. 5 (Berkeley, Calif.: Regents of the University of California, 1987), 12.

142. William Broad, "Beyond the Bomb: Turmoil in the Labs," *New York Times,* 9 October 1988, 6–23.
143. Ibid.
144. Lowell W. Morgan, letter to Rep. George E. Brown Jr., 3 November 1987, copy on file with the author.
145. Quoted in George Lardner, "Scientist Says Army Seeks to Fire Him for Criticizing SDI," *Washington Post,* 10 January 1992, A17.
146. Ibid.
147. After six scientists contacted by the Federation of American Scientists upheld Saucier's viewpoints as part of a Government Accountability Project lawsuit, the army reinstated Saucier (see George Lardner, "SDI Scientist Wins Reprieve from Firing," *Washington Post,* 7 March 1992, A8).
148. Another example of a disgruntled scientist who walked away from Star Wars research was Canadian computer scientist David L. Parnas. In resigning from a $1,000 per day post as researcher on a battle management panel, Parnas said, "many regard the program as a 'pot of gold' for research funds or an interesting challenge. . . . I believe that it is our duty, as scientists and engineers, to reply that we have no technological magic that will accomplish [Star Wars]. The President and the public should know that" (quoted in Mohr, "Scientist Quits," A6).
149. Quoted in Broad, "Beyond the Bomb." In a similar vein, LLNL physicist Hugh DeWitt wrote to the president of the University of California in 1982 with a request for a review of the X-ray laser program by independent civilian scientists *outside the official Pentagon hierarchy.* "I believe that the X-ray laser and other such proposals should have a searching and careful scientific and technical review by competent scientists outside the nuclear weapons labs. Scientists such as Teller in the nuclear weapons establishment may be very sincere in their belief in the possibility of a technological fix for U.S. security; they also have a strong vested interest in these large expensive and *secret* projects. I do not believe that it is beneficial to U.S. security that 'experts' in the nuclear establishment be able to promote these kinds of weapons systems in secrecy aided by selective leaks. I think the X-ray laser should be carefully studied by such people as Wolfgang Panofsky and Sidney Drell at Stanford, Marvin Goldberger at Cal Tech, Richard Garwin at IBM, Kosta Tsipis at MIT, Hans Bethe at Cornell, and other scientists who are competent and objective. An adequate review of the X-ray laser as a weapon system might save the country from pouring resources into a huge secret project that could turn out to be ineffective" (DeWitt, letter to Saxon, emphasis in original).
150. Broad, *Star Warriors,* 49.
151. Broad, *Teller's War,* 286.
152. Secretary of Defense Caspar Weinberger utilized a particularly curious strand of this strategy when he defended his research chief, Donald A. Hicks, on "freedom of expression" grounds after Hicks was criticized by Senator William Proxmire for stating that "I am not particularly interested in seeing department money going someplace where

an individual is outspoken in his rejection of department aims, even for basic research" (quoted in Rona Peligal, "Weinberger Stands Up for Hicks' Right to 'Freedom of Expression,'" *Defense News,* 9 June 1986, 4). Proxmire responded that such a partisan pro-SDI funding policy constituted "repressive measures . . . against legitimate dissent within the scientific community." Weinberger dismissed Proxmire's complaints and spoke in defense of Hicks, stating, "I know you feel strongly about freedom of expression and I am sure you would agree that even defense officials are entitled to exercise such freedom" (quoted in Peligal, "Weinberger Stands Up," 1). Hicks subsequently followed up with letters to *Science,* in which he continued his odd double standard. First, Hicks wrote that "[a]ll the internal memos in the world are terrific, but when a guy stands up and gives an interview and goes on television, somehow he's not one of us" (Statement, News and Comment, *Science* 231 [25 April 1986]: 444.) Then, in a subsequent letter of clarification, Hicks wrote, "We also try to foster an environment that encourages controversy and diverse viewpoints. Intellectual ferment breeds scientific and technical progress. We do not apply political 'litmus tests' to individuals or institutions" (letter to the editor, *Science* 232 [6 June 1986]: 1183). For a defense of the importance of public debate in the context of Hicks's comments, see Sidney D. Drell, letter to the editor, *Science* 232 [6 June 1986] 1183.

153. Roger Batzel, written statement, Lawrence Livermore National Laboratory, 23 October 1987, copy on file at the Federation of American Scientists Space Policy Project Archive, Washington, D.C.
154. U.S. GAO, *SDI Program.*
155. Quoted in U.S. GAO, *SDI Program.*
156. In presentations to "senior government officials," Batzel said that the positions advanced by Teller and Wood "reflect a common judgment of the technical status of the program. . . . The X-Ray laser program has been reviewed frequently and extensively by a variety of external and internal scientific groups. These reviews have all been positive and have complimented the scientific content of the program" (quoted in Keith Rogers, "Livermore Lab Director Disputes Teller Allegation," *Valley Times,* 24 October 1987, 3A).
157. In another example of missile defense advocates' failure to follow through on promises of openness and accountability, SDIO director James Abrahamson released a 1986 report to Congress after an official request. In remarks to reporters, Abrahamson called the report "beneficial and useful. . . . I think it's in keeping with the generally open philosophy that we are trying to maintain for the program" (quoted in "Pentagon Blacks out SDI Budget," *Military Space,* 7 July 1986, 1). However, Abrahamson's answer glossed over the ironic fact that eleven pages of the report were intentionally left blank and that the report contained "no breakdowns of SDI research budgets below the program element level" (quoted in "Pentagon Blacks Out," 1).
158. Quoted in Osborne, "Reagan's Great Lie," 13.
159. See "The National Pledge of Non-Participation: A Boycott of SDI Research," in *Perspectives on Strategic Defense,* ed. Steven W. Guerrier and Wayne C. Thompson (Boulder, Colo.: Westview, 1987), 323–25; "Appeal by American Scientists to Ban

Space Weapons," in *Perspectives on Strategic Defense,* 327.

160. Quoted in Steve Nadis, "After the Boycott," *Science for the People* 20 (1988): 23. "Incidents like the Woodruff document and the brightness controversy call into question the integrity of the whole program," said Stephen Aftergood of the Federation of American Scientists (quoted in Blum, "'Star Wars,'" A12). As Hughes documents, this skepticism spilled over into Europe. "[T]he cumulative weight of negative criticism from other parts of the scientific community (apart from the Brilliant Pebbles concept), particularly with significant press coverage of their commentaries, helped feed the mistrust in Europe about what the administration was doing with SDI. The more Machiavellian among the European commentators began to think of SDI as possibly the biggest confidence game the United States had ever tried to put over on Europe, and, more important, on the Soviet Union" (*SDI: A View,* 81).
161. Nadis, "After the Boycott," 23.
162. See Reiss, *Strategic Defense Initiative,* 85–113.
163. Troy E. Wade II, statement, *Energy and Water Appropriations for 1989,* 100th Cong., 1st sess., House Hearings, Committee on Appropriations, Energy and Water Development Subcommittee (Washington, D.C.: GPO, 1989), 787.
164. See "SDI Budget Cutbacks to Delay Near Term Weapons Deployments," *Aviation Week and Space Technology* (22 May 1989): 22; "'Star Wars' Goal Cut, Quayle Says," *Los Angeles Times,* 7 September 1989, 1.
165. Quoted in "Hackers Steal SDI Information in Internet System," *Washington Times,* 24 December 1990, 3.
166. Reiss, *Strategic Defense Initiative,* 180.
167. Quoted in Reiss, *Strategic Defense Initiative,* 180.
168. Quoted in "'Star Wars' Goal Cut," 1.
169. See John Stehr, "Nuclear-Powered X-Ray Laser Meets a Silent Death," *CBS Morning News,* 21 July 1992, Internet, Online, Lexis/Nexis.
170. Kenneth Zagacki and Andrew King, "Reagan, Romance and Technology: A Critique of 'Star Wars,'" *Communication Studies* 40 [1989]: 1–12.
171. Ibid., 9, emphasis added.
172. Ibid., 8.
173. Ibid.
174. Ibid., 10.
175. Carol Gallagher, *American Ground Zero: The Secret Nuclear War* (New York: Random House, 1993), xxxi. Thomas Goodnight first brought the notion of a "secret nuclear war" to my attention in a 1994 discussion that took place after a flurry of disclosures about Pentagon-sponsored human radiation experimentation. These disclosures revealed that during the cold war, military scientists conducted scores of nuclear operations and research on unwitting human subjects. "From the 1940s into the 1970s, more than 23,000 people were subjected to 1,400 different radiation experiments, many without their informed consent. The subjects were diverse in terms of gender, race, and age, but in general they were chosen from vulnerable populations such as the poor and prisoners" (Arjun Makhijani and Stephen I. Schwartz, "Victims of the

Bomb," in *Atomic Audit: The Costs and Consequences of U.S. Nuclear Weapons Since 1940,* ed. Stephen I. Schwarz [Washington, D.C.: Brookings Institution Press, 1998], 395–432). For further details, see Advisory Committee on Human Radiation Experiments, *Final Report* (Washington, D.C.: U.S. GPO, 1995).

176. Physicians for Social Responsibility, quoted in Arjun Makhijani, Howard Hu, and Katherine Yih, eds., *Nuclear Wastelands* (Cambridge: MIT Press, 1995), xiv.
177. Bernard Lown, quoted in Makhijani, Hu, and Yih, *Nuclear Wastelands,* xiii; see also Bernard Lown, "Clearing the Debris," *Technology Review,* August/September 1995, Internet, Technology Review Website, online at http://www.techreview.com/articles/aug95/AtomicLown.html.
178. Quoted in Michael D'Antonio, "Scars and Secrets: The Atomic Trail," *Los Angeles Times,* 20 March 1994, 14.
179. See Advisory Committee on Human Radiation Experiments, *Final Report,* 320–65.
180. Keith Schneider, "Nuclear Scientists Irradiated People in Secret Research," *New York Times,* 17 December 1993, A1. See also Ron Grossman and Charles Leroux, "Radiation Tests: Needed or Horrid?" *Chicago Tribune,* 9 January 1994, A18.
181. Bernard Lown, quoted in *Nuclear Wastelands,* 3.
182. Michael D'Antonio, *Atomic Harvest* (New York: Crown Press, 1993), 6–7.
183. D'Antonio, "Scars and Secrets," 14.
184. See, e.g., Arjun Makhijani and Stephen I. Schwartz, "Victims of the Bomb," 353–94.
185. Ibid., 394.
186. As Habermas explains, "[t]hrough the unplanned sociocultural consequences of technological progress, the human species has challenged itself to learn not merely to affect its social destiny, but to control it. This challenge of technology cannot be met with technology alone. It is rather a question of setting into motion a *politically effective discussion* that rationally brings the social potential constituted by technical knowledge and ability into a defined and controlled relation to our practical knowledge and will. On the one hand, such discussion could enlighten those who act politically about the tradition-bound self-understanding of their interests in relation to what is technically possible and feasible. On the other hand, they would be able to judge practically, in light of their now articulated and newly interpreted needs, the direction and the extent to which they want to develop technical knowledge for the future" (Habermas, *Toward a Rational Society,* 61, emphasis added).
187. See Morain, "Energy Secretary Warns," 1.
188. D'Antonio, "Scars and Secrets," 237.
189. Grossman and Leroux, "Radiation Tests," A18.
190. Staff of the House Subcommittee on Energy Conservation and Power, *American Nuclear Guinea Pigs: Three Decades of Radiation Experiments on U.S. Citizens,* Committee Print, 99th Cong., 2d Sess. (Washington, D.C.: GPO, 1986).
191. George J. Annas, "Questing for Grails: Duplicity, Betrayal, and Self-Deception in Postmodern Medical Research," in *Health and Human Rights: A Reader*, ed. Jonathan M. Mann, Sofia Gruskin, Michael A. Grodin, and George J. Annas (New York: Routledge, 1999), 316.

192. Quoted in Jon Healey, "Congress Ponders Compensation for Radiation Test Subjects," *Congressional Quarterly,* 8 January 1994, 22.
193. James O. Goldsborough, "Victims of a Perverted Policy: Compensate People Seriously Harmed by Nuclear Tests," *San Diego Union-Tribune,* 13 January 1994, 10.
194. Linenthal, *Symbolic Defense,* 65.
195. Ibid.
196. Quoted in Patricia M. Mische, *Star Wars and the State of Our Souls* (Minneapolis: Winston Press, 1985), 7.
197. Daniel O. Graham and Gregory A. Fossedal, *A Defense That Defends: Blocking Nuclear Attack* (Old Greenwich, Conn.: Devin-Adair Publishers), 1983, 11.
198. Heritage Foundation, "A Proposed Plan for Project on BMD and Arms Control," NSR #46, *High Frontier: A New Option in Space,* labeled "Not for release," copy on file with Carol Rosin. As Robert M. Bowman explains, "like the Pentagon Papers, [this report] was smuggled out of the Heritage Foundation by someone who was disenchanted with what they were doing. It got into the hands of Carol Rosin" (Bowman, telephone interview with the author, 21 July 1999, transcript on file with the author). An aerospace industry insider, Rosin first worked with Wernher Von Braun and held the position of corporate manager at Fairchild Industries in Germantown, Maryland (1974–77), then worked as a consultant with the Redondo Beach, California-based TRW, Inc. (1977–79). In 1983, she founded the Institute for Security and Cooperation in Outer Space (ISCOS), a Washington, D.C., educational think tank. According to Rosin, shortly after this, "someone called me and said, 'I need to give you a paper, and it's very important to your work, and I'm going to meet you on Connecticut Avenue and Massachusetts on the corner and I'll be in a trenchcoat.' . . . I took this paper, and he said '[t]his is the Heritage Foundation plan of action for putting weapons into space'" (Carol Rosin, telephone interview with the author, 26 September 1999). Rosin circulated the document, which was subsequently quoted by Bowman in an Institute for Space and Security Studies newsletter (Robert M. Bowman, "BMD & Arms Control," ISSS Issue Paper, January 1985, 3), and by author Patricia Mische (*Star Wars,* 7–8; see also Patricia M. Mische, "Star Wars and the State of Our Souls," in *Securing Our Planet,* ed. Don Carlson and Craig Comstock [Los Angeles: Jeremy P. Tarcher, 1986], 216–17). Mische followed up on the matter in July 1999: "I . . . called the Heritage Foundation to verify the information. They orally verified this as their position, but would not send me the original documents" (Mische, letter to the author, 17 July 1999, copy on file with the author). Bowman also confirmed the existence of the report: "I saw the material. She [Rosin] had the material" (Bowman, telephone interview with the author).

    Although in 1999 the Heritage Foundation apparently confirmed that its overall position on Star Wars was consistent with the contents of this 1983 secret report, it seems that Heritage Foundation officials nevertheless blocked efforts to publish the report's contents into the record of hearings held by the Senate Committee on Foreign Affairs in 1983 (U.S. Congress, *Controlling Space Weapons,* 98th Cong., 1st sess., Senate

Hearings, Committee on Foreign Relations, 14 April and 18 May 1983 [Washington, D.C.: GPO]). After testifying at this hearing, Rosin attempted to enter a copy of the report into the record as a supplement to her testimony. "They told me that it was not going to be printed. I talked with senators on that committee and I got no satisfactory response, only that it was not going to be presented, because the Heritage Foundation said they never wrote it. . . . Senators told me that the Heritage Foundation denied writing it, and the congressional staff ripped off the cover page of the report, took it away, and that was it" (Rosin, telephone interview with the author). While intriguing, the dispute over the report's authenticity has taken on the color of an academic debate of marginal significance in light of the surfeit of evidence confirming that BMD advocates in fact pursued a rhetorical strategy designed to co-opt rhetorically the "high ground" of antinuclear activism in the 1980s, as well as the fact that such a strategy was advocated in several open-source documents published in the public domain, such as Graham and Fossedal, *A Defense that Defends.*

199. Rosin, telephone interview with the author.
200. Bowman, telephone interview with the author.
201. Heritage Foundation, "A Proposed Plan."
202. Ibid.
203. Ibid.
204. Ibid.
205. Heritage Foundation, "A Proposed Plan," emphasis in original. Note that in this quotation, the reference "HF" could mean "Heritage Foundation" or "High Frontier."
206. Ibid.
207. See, e.g., Reagan, "Address," 442.
208. Gregory A. Fossedal, "A Star Wars Caucus in the Freeze Movement," *Wall Street Journal,* 14 February 1985, 1.
209. Heritage Foundation, "A Proposed Plan."
210. For a description of McFarlane's successful efforts to buy Margaret Thatcher's support for SDI with a $300 million guarantee, see McFarlane, statement, "The Cold War."
211. William Chaloupka, *Knowing Nukes: The Politics and Culture of the Atom* (Minneapolis: University of Minnesota Press, 1992), 74.
212. D'Souza, *Ronald Reagan,* 179.
213. Daniel O. Graham and Gregory A. Fossedal, *A Defense that Defends: Blocking Nuclear Attack* (Old Greenwich, Conn.: Devin-Adair Publishers, 1983), 11.
214. Herbert Marcuse, *One-Dimensional Man* (Boston: Beacon Press, 1964), 90.
215. Marcuse, *One-Dimensional Man,* 93.
216. J. Michael Hogan, *The Nuclear Freeze Campaign: Rhetoric and Foreign Policy in the Telepolitical Age* (East Lansing, Mich.: Michigan State University Press, 1994), 5.
217. Ibid., 19.
218. Matthew Evangelista, *Unarmed Forces: The Transnational Movement to End the Cold War* (Ithaca, N.Y.: Cornell University Press, 1999), 314.
219. Ibid., 310, 314.
220. See Guerrier and Thompson, *Perspectives on Strategic Defense,* 323–24.

221. Ibid., 327.

222. For a detailed treatment of the range of such political resistance across a broad spectrum of issues, see the January/February 1988 special issue of *Science for the People* titled "Science and the Military: Who's Pulling the Strings?"

223. The $70.7 billion figure comes from an analysis conducted by the Congressional Research Service (see Steven A. Hildreth, Congressional Research Service memorandum to Rep. David Pryor, 1 August 1995). For commentary on the political fallout of these estimates, see Jonathan S. Landay, "Pentagon Hit for Hiding Spending on 'Star Wars,'" *Christian Science Monitor,* 5 September 1995, 3; John E. Pike, Bruce G. Blair, and Stephen I. Schwartz, "Defending Against the Bomb," in *Atomic Audit: The Costs and Consequences of U.S. Nuclear Weapons Since 1940,* ed. Stephen I. Schwartz (Washington, D.C.: Brookings Institute Press, 1998), 295. Several authors have criticized this exorbitant amount by highlighting the stark opportunity costs entailed in such expenditures. "As the threat of war recedes, the United States and the Soviet Union (as well as other nations) could begin to divert their financial resources, their technology, their brainpower, and their productive capacity to the solving of other problems," explains Bowman. "The world is rapidly approaching military expenditures of a trillion dollars a year. Most of the world's scientists and engineers are engaged in work on means of destruction. Freed of this burden, we could end hunger everywhere in the world, provide better housing and mass transit, improve health care, create sources of renewable energy, turn seawater into fresh to make the world's deserts bloom, tackle air and water pollution, and make Millennium Three an era of hope and fulfillment for all mankind" (Robert M. Bowman, *Star Wars: A Defense Insider's Case Against SDI* [New York: St. Martin's Press, 1986], 114).

   Making a similar point, Paul Joseph, economist with the Labor Research Association, notes that "[t]he arms race and arms trade represent an enormous waste of resources and a diversion of the world economy. To begin with, consider nutrition. Half a billion people throughout the world are severely undernourished. It has been estimated that funds equivalent to one percent of the military budgets of industrialized countries would be sufficient to close the gap of needed development assistance to agriculture. As to health, the eradication of some of the most important communicable diseases would cost trifling amounts compared to the arms race. The World Health Organization (WHO) spent about $83 million over 10 years to eradicate smallpox in the world. That amount would not suffice to buy a single modern strategic bomber. The program to eradicate malaria which affects more people in the world is dragging owing to lack of funds" (Paul Joseph, *Search for Sanity: The Politics of Nuclear Weapons and Disarmament* [Cambridge, Mass.: South End Press, 1984], 259).

224. Robert McFarlane, "Consider What Star Wars Accomplished," *New York Times,* 24 August 1993, A15.

225. In "Long Shadows and Virtual Swords," defense analysts Ted Gold and Rich Wagner develop a "long shadow effect" theory explaining how "R & D activities afford value in and of themselves before (and in many cases independent from) any production or deployment" (Ted Gold and Rich Wagner, "Long Shadows and Virtual Swords:

Managing Defense Resources in the Changing Security Environment," unpublished paper, June 1990, 3, copy on file with the author). Using SDI as a "prime current example" of the "long shadow effect," Gold and Wagner suggest that the mere possibility of future U.S. BMD deployment had substantive effects on Soviet military planning and weapons acquisition strategies during the cold war. Gold and Wagner conclude that in the present security environment, there is an "aggressive role" for military research programs to incorporate the "long shadow" strategy, and thereby "help shape events and influence the behavior of potential adversaries" ("Long Shadows," 12).

226. Quoted in K. Subrahmanyam, "The 'Star Wars' Delusion," *World Press Review,* June 1983, 22.
227. Quoted in Charles Mohr, "Reagan Is Urged to Increase Research on Exotic Defenses Against Missiles," *New York Times,* 5 November 1983, A32.
228. See Daniel C. Deudney and G. John Ikenberry, "Who Won the Cold War?" *Foreign Policy* 87 (1992): 124; Powaski, *Cold War,* 260.
229. Deudney and Ikenberry, "Who Won," 124.
230. Ibid., 125.
231. Allen Hunter, "The Limits of Vindicationist Scholarship," in *Rethinking the Cold War,* ed. Allen Hunter (Philadelphia: Temple University Press, 1998), 1–34.
232. See Jeane J. Kirkpatrick, "Beyond the Cold War," *Foreign Affairs* 69 (1989–90): 1–16; Adam Meyerson, "The Battle for the History Books: Who Won the Cold War?" *Policy Review* 52 (1990): 2–3; Peter Schweizer, *Victory: The Reagan Administration's Secret Strategy That Hastened the Collapse of the Soviet Union* (New York: Atlantic Monthly Press, 1994); Jay Winik, *On the Brink: The Dramatic, Behind-the-Scenes Saga of the Reagan Era and the Men and Women Who Won the Cold War* (New York: Simon and Schuster, 1996).
233. Quoted in Henry Cooper, "Trying to Get SDI Off the Ground," *Washington Times,* 14 September 1990, F3.
234. James Fallows, interview with Neal Conan, *Morning Edition,* National Public Radio, 10 March 1993, Internet, Online, Lexis/Nexis.
235. Zbigniew Brzezinski, "The Cold War and its Aftermath," *Foreign Affairs* 71 (1992): 42; see also Michael E. Salla, "The End of the Cold War: A Political, Historical, and Mythological Event," in *Why the Cold War Ended,* ed. Ralph Summy and Michael E. Salla (Westport, Conn.: Greenwood Press, 1995), 255.
236. Evangelista, *Unarmed Forces,* 240.
237. Paul Quinn-Judge, "Reagan-Era Officials Deny Faking SDI Test," *Boston Globe,* 19 August 1993, 3.
238. Quoted in Bill O'Neill, "Fear and Laughter in the Kremlin," *New Scientist,* 20 March 1993, 37. For discussion of nineteen factors much more significant than SDI in bringing about the fall of the Soviet Union, see Reiss, *Strategic Defense Initiative,* 192–93. For analysis of CIA data on Soviet military spending that disproves McFarlane's "spend them into the ground" hypothesis, see Pike, Blair, and Schwartz, "Defending Against the Bomb," 295–97; Raymond L. Garthoff, *The Great Transition: American-Soviet Relations and the End of the Cold War* (Washington, D.C.: Brookings, 1994),

506–8; Franklyn D. Holtzman, "Politics and Guesswork: CIA and DIA Estimates of Soviet Military Spending," *International Security* 14 (1989): 101–31; and Richard Ned Lebow and Janice Gross Stein, *We All Lost the Cold War* (Princeton, N.J.: Princeton University Press, 1994), 369–76.

239. United States Central Intelligence Agency, "Possible Soviet Responses to the U.S. Strategic Defense Initiative," NIC M 83–10017, copy 458 (12 September 1983), formerly secret, Internet, Federation of American Scientists Website, online at http://www.fas.org/spp/starwars/offdocs/m8310017.htm.

240. Joseph A. Camilleri, "The Cold War . . . And After: A New Period of Upheaval in World Politics," in *Why the Cold War Ended,* ed. Ralph Summy and Michael E. Salla (Westport, Conn.: Greenwood Press, 1995), 241.

241. Roald Sagdeev, quoted in Evangelista, *Unarmed Forces,* 242. Evangelista gives further details about Sagdeev's position: "Sagdeev was particularly critical of the argument that a fraudulent test of an SDI interceptor missile in June 1984 'helped persuade the Soviets to spend tens of billions of dollars to counter the American effort to develop a space-based shield.' The test, which entailed putting a beacon on the target and a receiver on the interceptor, evidently fooled the U.S. Congress into spending billions on the Star Wars program, but to think that it fooled Soviet military experts, argued Sagdeev, is 'naive.' Sagdeev, who closely followed developments in strategic defense from Moscow at the time, said the 1984 test was hardly noticed. He pointed out that the Soviet Union had already achieved the successful interception of a warhead by an ABM defense missile—without fraud—in March 1961. Therefore, for the United States to achieve the same result twenty-three years later was meaningless." (Evangelista, *Unarmed Forces,* 242–43).

242. Pike, "Strategic 'Deception,'" 4. Pike's position was corroborated in statements made by a senior member of the Soviet general staff to the *Washington Post* in 1985: "We are not going to take the path that the U.S. administration is trying to force us into. We have made it clear that we will not ape the United States" (quoted in Jim Hoagland and Dusko Doder, "Moscow Won't 'Ape' SDI, Top Soviet General Says," *Washington Post,* 9 June 1985, A1).

243. Quoted in Richard Ned Lebow and Janice Gross Stein, "Reagan and the Russians," *Atlantic Monthly,* February 1994, 37.

244. Pike, Blair, and Schwartz, "Defending Against the Bomb," 298.

245. Bowman, telephone interview with the author.

246. Deudney and Ikenberry, "Who Won," 124.

247. See ibid.; Reiss, *Strategic Defense Initiative,* 192–93.

248. Evangelista, *Unarmed Forces,* 383–84. Evangelista's speculations on this point were corroborated by Robert Bowman. Recounting discussions during a trip to the Kremlin in 1987, Bowman explained: "The very first thing Arbatov (Gorbachev's right-hand man on relations with the U.S.) asked was 'What are we doing wrong? What could we do to end the arms race? . . . What about the Krasnoyarsk radar?' I said, well, there you can do some helpful things. Technically, it's a violation of the ABM Treaty because it's pointing in the wrong direction, but you and I, and just about everyone in the U.S.

military knows, it's only a technical violation of the ABM Treaty; it has no real ABM capability. It would be very useful if you let U.S. delegations come and look at the Krasnoyarsk radar, show them that it has no ABM capability. . . . Very shortly thereafter, they did allow an American delegation go to see the Krasnoyarsk radar" (Bowman, telephone interview with the author).

249. Powaski, *Cold War,* 123. Edgar Bottome points out that scientists such as Edward Teller "allegedly provided more scholarly and 'scientific' explanations for the existence of the missile gap" than did journalists and politicians. However, "[i]t is interesting to note that it would seem that not one journalist, politician, scientist, nor scholar had his reputation even slightly damaged for his part in the creation of the missile gap myth" (Edgar Bottome, *The Balance of Terror* [Boston: Beacon Press, 1986], 61). For the "smoking gun" document revealing the "missile gap" as an illusion, see Adam Yarmolinsky, memorandum to McGeorge Bundy, 15 March 1963, copy on file with the author. Document acquired with the help of Erik Doxtader.
250. Richard Lawrence Miller, *Heritage of Fear* (New York: Walker Press, 1988), 59.
251. Ronald E. Powaski, *The March to Armageddon* (New York: Oxford University Press, 1987), 227.
252. Noam Chomsky, *Deterring Democracy* (London: Verso, 1991), 23–24; see also Dana H. Allin, *Cold War Illusions* (New York: St. Martin's Press, 1994); Bottome, *Balance of Terror.*
253. Tom Gervasi, *The Myth of Soviet Military Supremacy* (New York: Harper and Row, 1986), 92–112.
254. Milton Leitenberg, "The Numbers Game or 'Who's on First?'" *Bulletin of the Atomic Scientists* 38 (1982): 28.
255. Gary Chapman and Joel Yudkin, *Briefing Book on the Military-Industrial Complex* (Washington, D.C.: Council for a Livable World, 1992), 3.
256. Miller, *Heritage of Fear,* 363.
257. See Herbert F. York, "Controlling the Qualitative Arms Race," *Bulletin of the Atomic Scientists* 29 (1973): 4–8; idem., *Race to Oblivion: A Participant's View of the Arms Race* (New York: Oxford University Press, 1970).
258. Guertner, "What is Proof?" 187.
259. Chomsky, *Deterring Democracy,* 29.
260. See Noam Chomsky, *World Orders Old and New* (New York: Columbia University Press, 1994).
261. Quoted in O'Neill, "Fear and Laughter," 35.
262. McGeorge Bundy, William J. Crowe, and Sidney D. Drell, *Reducing Nuclear Danger* (New York: Council on Foreign Relations, 1993), 10–11.
263. Valery Yarynich, "The Doomsday Machine's Safety Catch," *New York Times,* 1 February 1994, A15.
264. Quoted in Christoper Ruddy, "'Father of H-Bomb' Worried about Status of U.S. Defenses," *Pittsburgh Tribune-Review,* 12 January 1997, A1.
265. Ibid. See also Edward Teller, *Better a Shield than a Sword: Perspectives on Defense and Technology* (New York: Free Press, 1987), 119–20.
266. Edward Teller, *Better a Shield than a Sword,* 120.

President George Bush delivers a speech to Raytheon, Inc. employees at the Raytheon plant in Andover, Mass., on 15 February 1991. Rick Friedman/Black Star.

CHAPTER THREE

Placebo Defense: The Rhetoric of Patriot Missile Accuracy in the 1991 Persian Gulf War

Bullseye.[1]

—ABC News Anchor Sam Donaldson, 26 January 1991

Defenses against tactical ballistic missile defenses work and save lives. The effectiveness of the Patriot system was proved under combat conditions.[2]

—Secretary of Defense Richard Cheney, 21 February 1991

It is difficult to imagine a better validation of President Bush's redirection of SDI and continued approach to our negotiations with the Soviet Union [*sic*] than the clear lessons of the Gulf War.[3]

—SDIO Director Henry Cooper, 17 April 1991

On his hospital deathbed one day in late 1990, Lee Atwater reached underneath the covers and pulled out a crumpled envelope, which he passed to Secretary of State James Baker. As a loyal Republican and the strategic architect of George Bush's devastating "Willie Horton" victory over Michael Dukakis in the 1988 presidential election, the dying Atwater wanted to make sure that before passing away, he would leave the GOP with a winning strategy for electoral victory in 1992. "I have a plan," Atwater told his bedside visitor. "If Georgie blows it, open this. This is the surefire, ultimate election winner." Months later, with Bush reeling in the polls, Baker opened the envelope to discover Atwater's plan: a Hollywood-style video war, starring the U.S. armed forces, modeled after the Super Bowl. Acting on Atwater's suggestion, the Bush administration secretly hired a Hollywood producer and director, cast Saddam Hussein in the role of Hitler, secured world sponsors, and staged the 1991 Persian Gulf War.

This is the plot of Larry Beinhart's 1993 novel *American Hero,* which the author labels ostensibly as a "work of fiction." However, Beinhart clouds the status of his book by prefacing it with the following statement: "There are those who feel that fact and fiction are significantly less distinguishable than they used to seem to be."[4] Beinhart teases readers further by including in the final chapter of the book a list of thirty-nine factual "anomalies," many supported by citations from legitimate news sources, that cast doubt on the official story of the Gulf War.[5] "[I]t's legitimate to regard the official story about why and how [the Gulf War] happened as a hypothesis, an unproved theory," writes Beinhart in the final chapter, "just as many, many people regard the official story of the assassination of John F. Kennedy as flawed."[6]

Why would *Publisher's Weekly* state that "the best tribute one can pay the book is that, wacky as the thesis seems, it makes more sense than the actual war itself"? And why would *Kirkus Reviews* laud Beinhart's book as "gorgeously plausible," seconding critic Carl Hiassen's suggestion that "*American Hero* could, in fact, be a 'brilliantly disguised documentary'?"[7] With such hints of plausibility, these reviewers reflect a deep public suspicion regarding the status of the Gulf War. Did the Gulf War really take place? French writer Jean Baudrillard expressed his skepticism in a series of articles that appeared before, during, and after the war. Baudrillard suggested that the Gulf War was a mass exercise in postmodern simulation, more virtual reality video game than traditional military conflict, in which "[o]ur strategic site [was] the television screen."[8]

While Baudrillard's bold suggestion may strain credulity, there is no denying that the 1991 Gulf War was a tightly scripted, smoothly orchestrated, and well-edited affair, with the U.S. military setting down exact deadlines and timetables to demarcate discrete phases of the conflict and dictate the flow of events. This theme of an "on schedule" war was prominent in President George Bush's 15 February 1991 address to the nation (hereafter "Raytheon address"), as well as in the steady barrage of CNN news coverage backing up Bush's arguments with corroborating images and reportage.[9] These narratives invited viewers to participate as victors in the story of Operation Desert Storm, by reveling in the West's high-tech triumph over Saddam Hussein.

The factor that most seriously threatened to disrupt the American-mandated war script was Iraq's nightly Scud missile attacks on Saudi Arabia and Israel. The Scud attacks, which began on 18 January 1991 and continued throughout the war, exacted a heavy psychological toll on the Israeli civilian population, generating anxiety-ridden calls for Israeli Defense Force (IDF) retaliation against Iraq. Knowing that such retaliation would complicate the war plan by widening the conflict and alienating key members of the U.N. coalition,[10] American military planners sought feverishly to dissuade the IDF from escalating the war.

The United States attempted to accomplish this rhetorical task in part by providing Israel with Patriot ballistic missile defense batteries. It was hoped that Patriot defense, by neutralizing the Scud threat to Israeli cities, would calm frayed Israeli nerves and help stave off calls for IDF intervention against Iraq. The strategy worked; nightly video clips of the Patriot in action had what MIT defense and arms control scholar Theodore A. Postol calls "a magical effect on the public's perception of events."[11] Network news reporters enhanced the effect by providing dramatic, real-time commentary on Patriot missiles apparently foiling Scud attacks.[12]

Television coverage of the Patriot in action was not only instrumental in preserving the coherence of the American war script; such coverage also provided unprecedented documentation of the first-ever performance of a ballistic missile defense system under wartime conditions. Perceived Patriot success "was a heaven-sent gift for the Bush administration, building support for its Gulf War policy and for Raytheon [Patriot's manufacturer] and the defense industry in general," according to *Boston Globe* reporter Randolph Ryan.[13] In the aftermath of the war, missile defense advocates seized upon this evidence to argue that a sea change had occurred in the long-running debate over BMD feasibility. They contended that Patriot's remarkable effectiveness proved BMD was no futuristic illusion, but had instead become a proven battlefield option.

The claim of Patriot's Gulf War accuracy has become a staple argument of missile defense advocates starved for favorable BMD feasibility data. During the Gulf War, the claim of Patriot accuracy was used by military officials to assert narrative control over the flow of events and thereby preserve the integrity of a triumphant victory script. After the war, the claim of Patriot accuracy was used by opportunistic military hawks to justify massive postwar jackpots for weapons contractors who had engineered the arsenal of "smart" weapons used in the Gulf War.

In late 1991, the army's assertion of nearly unqualified Patriot success was called into question. A group of physicists, led by Postol, closely scrutinized the video evidence of Patriot-Scud engagements and found the army's claims of Patriot accuracy unfounded. When Postol attempted to publish these findings, missile defense advocates resorted to severe tactics, including attempts at post-publication classification, intimidation, and reliance on secrecy as a strategic tool in public debates over Patriot accuracy.[14] Eventual publication of Postol's article triggered a debate that still lingers to this day, even after U.S. military officials have conceded that Patriot's Gulf War accuracy was inflated during the war, and after several congressional investigations have issued findings in support of Postol's argument that Patriot successes shown on television were illusory.[15]

This chapter explores how Patriot worked rhetorically to bolster the American diplomatic position during the war, as well as the way in which Patriot's perceived Gulf War effectiveness shaped postwar missile defense debates. This analysis is accompanied by an examination of the scientific and political controversy over Patriot's Gulf War accuracy. President Bush's Raytheon address serves as the focus of textual analysis in the opening stages of part one, where the rhetorical significance of Patriot's wartime image as a "Scudbuster" is examined. Because Bush's message of Patriot invincibility resonated with the stream of hyperreal video images released by military officials during the Gulf War, it will be appropriate to couple textual analysis of Bush's speech with consideration of how the Pentagon's "Nintendo war" framework transformed war audiences into bedazzled consumers of the official war script, solidifying Bush's position as narrative commander. With the rhetorical significance of Patriot missile accuracy established, part two turns to the post–Gulf War dispute regarding Patriot's performance. In this analysis, journal articles, congressional testimony, correspondence, official reports, and interviews are consulted to elucidate the arguments advanced by each side in the dispute over the validity of data offered to prove Patriot's battlefield effectiveness. The chapter closes with critical commentary focusing on the normative issues and political stakes implicated by the vigorous post–Gulf War controversy over Patriot's performance.

Patriot's Rhetorical Role as "Scudbusting" Hero

The meaning of the Gulf War was shaped to conform with a carefully fashioned script, a narrative frame that structured perceptions and dictated the flow of events. In an immediate sense, this script was chiseled by George Bush during his official appearances and speeches conducted during the war. However, former defense industry insider Carol Rosin says that the script for the Gulf War was hatched many years before Bush even became president. Rosin first worked with Wernher Von Braun (inventor of the V-2 rocket) and held the position of corporate manager at Fairchild Industries in

Germantown, Maryland, from 1974 to 1977. The story of her ultimate resignation from Fairchild in 1977 provides a glimpse into what may have been the early roots of the Gulf War.

> I was sitting in a room called the War Room, the conference room at Fairchild. . . . On the wall was a list of a number of names of potential enemies against whom we were eventually going to build the space-based weapons. There were names that most of us had never heard before, names like Saddam Hussein, Khadafy, etc. . . . *The Gulf War was planned in this meeting.* At this meeting, they were saying that there would be a war in the Gulf region, because it was the hottest spot, the easiest place to have it, we were prepared. During that war was when we would dump the old weapons, test the new ones, and start the rationale for the newer ones. . . . I stood up in the meeting, I was the only woman, and the youngest one in the room, and I said "excuse me, I may have missed something here, but am I hearing that you're planning a war to dump weapons, test the new ones, and prepare for space-based ones, and give the rationale to the Congress, the public, and the allies, when there's $25 billion in the SDI, and it doesn't have anything to do with security, it's just to perpetuate and develop a space-based weapons program, or have I missed something?" Nobody answered. And I repeated the question. And nobody answered. And then they went on with this conversation they were having as though I had not said anything. My hair stood up on end. I stood up in the room, and I said that if this is true, unless someone can correct me, you need to consider this to be my 60-day resignation . . . and as I walked out of the room, that began my exit process from Fairchild Industries.[16]

Rosin's startling account fits with important historical aspects of the Gulf War experience. The war did turn out to be a proving ground for a new generation of "smart" weapons, many which were guided to their targets with the aid of information sent from space-based satellites. World television audiences watching such pyrotechnics were led to believe that the U.S.'s high-tech arsenal was performing flawlessly in its first ever dry run in the desert. However, viewers reached this conclusion based on carefully filtered and screened misinformation. "[M]ost people had no direct experience or knowledge of these phenomena," writes cultural critic Douglas Kellner, "so their pictures of the events in the Gulf were a product of the Bush administration discourse and media frames through which the crisis was constructed."[17]

In the Bush administration's official discourse, as well as the media reportage of the war, the Patriot missile system was highlighted prominently. The following discussion of Patriot's role in official Gulf War narratives promises to yield understanding of the Bush administration's key rhetorical themes as they played out in the context of a hyperreal, "Nintendo war" framework of public discourse.

BUSH'S RAYTHEON ADDRESS

On 15 February 1991 President Bush traveled to Andover, Massachusetts, paying a visit to the Raytheon plant where Patriot missiles were manufactured. Bush's visit to Raytheon was highly symbolic, since his sojourn came at a time in the Gulf War when Patriot missiles were taking center stage. The nerves of anxious Israeli citizens in Haifa and Tel Aviv had been rubbed raw by a steady barrage of Iraqi Scud missile attacks. Those critical of the war effort had began to doubt that the United States and its allies would be able to maintain command over military and political events, which seemed to be spinning out of control.

Bush's reception at Raytheon provided a comforting break from the harsh realities of war. During his visit to the plant, he entered a scene described by journalist Rick Atkinson as "gaily decorated with yellow ribbons, patriotic buntings, and hundreds of American flags."[18] President Bush inspected the assembly line that produced the Patriot air and missile defense system, then sat in a model of the Engagement Control System, the command post for Patriot missile batteries. Following the tour, he proceeded to the fabrications building, where at 1:45 P.M. he delivered a nationally televised address. "And look," Bush began, "I view it as an honor to be here, to come to Raytheon, the home of the men and women who build the Scudbusters."[19] Indeed, Bush's presence at the Raytheon plant was a testament to the importance of the Patriot missile as a persuasive crutch in the administration's unfolding Gulf War narrative. As Representative Frank Horton (R-NY) reflected, "[t]he story of the Patriot missile in the Persian Gulf is a good one. It is a story of heroes, ingenuity, of cooperation."[20] In the Raytheon address, Bush told the triumphant story of Patriot, weaving the tale into the larger structure of the basic organizing themes of official Gulf War discourse, punctuating his rhetoric with a flurry of fist shaking.

Perhaps the most crucial of these themes was script control. For Bush, it was absolutely essential that the United States dictate the sequence and timing of events in the war. The original 15 January 1991 deadline for Iraq's withdrawal from Kuwait provided a temporal guideline for the war script, and in the Raytheon address Bush stated, "we're going to continue to fight this war on our terms, on our timetable, until our objectives are met."[21] Bush was determined to maintain full temporal control over the flow of events, elevating the struggle over the war's timetable to the status of an intense personal struggle with Saddam Hussein. "[W]e will control the timing of this engagement, not Saddam Hussein," he declared.[22]

With script control firmly established, Bush could cast coalition soldiers in the role of proactive initiators of carefully planned and sequenced maneuvers, leaving Iraq to react passively to the coalition-mandated agenda. Saddam Hussein's bids to interfere with this process by altering the course of events were nullified before they could even materialize. His offer to withdraw from Kuwait, issued on the very morning of Bush's Raytheon address, was declared by the president to be "dead on arrival."[23] Responding to reporters, Bush had labeled Hussein's peace offer a "cruel ploy . . . it's totally unacceptable to everybody."[24] Like a Hollywood director hewing tightly to an iron-clad production timeline, Bush brushed aside potential derailments, announcing sternly that "[t]he war is going on schedule."[25]

Maintaining script control in this rhetorical situation was a difficult task, because of the inherent uncertainties and multiple levels of contingency built into the unpredictable war environment. Iraqi forces set fire to Kuwaiti oil fields. Major antiwar demonstrations flared up in Egypt and Jordan. Russian diplomats intervened as peace brokers. In the face of these uncertainties, President Bush relied heavily upon the claim of Patriot missile accuracy to maintain control over the unfolding storyline. Bush cited the Patriot as a key tool protecting coalition script control, since Iraq's Scud missiles, as instruments of psychological warfare, clearly possessed the capability to disrupt the flow of events. As International Institute for Strategic Studies scholar Roland Dannreuther explains, the Scud attacks "reinforced the illusion that Iraq, not the allied forces, was dictating the scope of the war."[26] In response to this threat, President Bush reasserted control of the conflict timetable by dwelling on the marvel of Patriot success. "[W]hen you go home at night, you can say with pride that the success of Patriot is one important reason why Operation Desert Storm is on course and on schedule," said Bush.[27] Asking the audience to "imagine what course this war would have taken without the Patriot,"[28] Bush invited listeners to complete his enthymeme by imagining any number of a variety of doomsday scenarios: gas attacks and mass civilian casualties, Arab-Israeli war, or even nuclear escalation.

With Bush in the director's chair, the war was framed as a "clear case of good versus evil."[29] Invoking what international relations scholar David Campbell calls a "political discourse of moral certitude,"[30] Bush likened Saddam Hussein to Adolf Hitler, tagging the Iraqi leader as a "brutal dictator . . . without regard for human decency."[31] The model script here was World War II, what Campbell calls the last "unambiguously 'good war' in U.S. memory."[32] "We have such a clear moral case, it's that big. It's that important. Nothing like this since World War II. Nothing of this moral importance since World War II," Bush insisted.[33] Locating Bush's rhetorical moves here in the tradition of presidential "crisis promotion," critic Denise Bostdorff points out how Bush's early (and largely unsuccessful) oil-based justifications for the war eventually gave way to more strident, crisis-laden rhetoric.[34]

Making moral arguments in this charged framework, Bush counterposed demonization of the enemy with celebration of the ingenuity and resourcefulness of the rescuing state. This strategy permitted Bush to use the war as an occasion to trumpet the awesome technological powers of the United States, pointing to such prowess as a symbol of progress and democracy offsetting the depraved image of Saddam Hussein.[35] In the Raytheon address, Bush pointed to the example of the Patriot missile as a condensation symbol for American ingenuity, technical achievement, and personal accomplishment through hard work. In Bush's formulation, the manufacture of the Patriot missile was an unvarnished American success story, replete with a dramatic beginning (the heroic effort of Raytheon employees to rush the system into production in time for the war) and a stirring climax. "Patriot is 41 for 42: 42 Scuds engaged, 41 intercepted!" Bush exclaimed.[36] Atkinson reported that after this line, "the adoring crowd roared it's approval. 'U-S-A! U-S-A!' they chanted, and the incantation echoed from the high ceiling with the deep, ferocious timbre of a war cry."[37]

Emboldened, Bush pressed on: "[Too often we hear about how] our kids, our children, our schools fall short. I think it's about time that we took note of the success stories, of the way the brave young men and women who man the Patriot stations perform such complex tasks with *unerring accuracy*."[38] Because the army had had a paltry supply of Patriot missiles in its inventory in the first week in August 1990, a crash program had been undertaken to increase production when tensions in the Gulf mounted. In the Raytheon address, Bush lauded this effort as a key instance where American dedication to the war effort succeeded in wresting the initiative from Iraq. "Just days after

Saddam Hussein *took the offense* against an undefended Kuwait, the people of this plant went into overdrive and *took the offense*," Bush said, sounding like a football coach. "And since mid-August, it's been an around-the-clock effort, three shifts a day, seven days a week. And I know many of you gave up your own Thanksgiving and Christmas even to be right here, to keep these lines moving," he said (149, emphasis added).

Bush described how earlier in the day, when he was sitting in an Engagement Control System on display in the Raytheon plant, he had become even more impressed with Patriot's "split-second accuracy" (149). For Bush, this impression extended beyond mere appreciation for technical achievement; in his eyes, the remarkable phenomenon of Patriot's pinpoint accuracy reflected a profound personal achievement of missile defense scientists and engineers: "Patriot works, and not just because of the high-tech wizardry. Patriot works because of patriots like you" (149). Perhaps this was the sort of flag-waving flourish that had been envisioned by the Carter administration officials who changed the SAM-D missile defense system's name to "Patriot" in a 1976 maneuver designed to stoke excitement surrounding bicentennial celebration of the American Revolution.[39]

Turning his attention to the military aspects of Patriot performance on the battlefield, Bush stated that his intention in the Raytheon address was to pay tribute not only to the "high tech workers like yourself" who built Patriot but also the "highly skilled servicemen and women who operate Patriot in the field."[40] As the highest of the "high-tech" weapon systems on display during the Gulf War, Patriot was touted as paradigmatic of a new form of advanced warfare, in which technical sophistication promised to be the sine qua non of military superiority. Speaking while standing next to the Raytheon assembly line, Bush offered a vision: "[W]e are witnessing a revolution in modern warfare, a revolution that will shape the way that we defend ourselves for decades to come" (149).

One of the most important themes growing out of this new military paradigm was the citation of Patriot's Gulf War performance as validating evidence for the general concept of ballistic missile defense. For years, critics had asserted that antimissile defenses did not work, Bush told the audience of Patriot employees. "They were wrong, and you were right. Thank God you were right," he said. (149). Patriot's effectiveness in intercepting Scuds, Bush asserted, was "proof positive that missile defense works" (150). In the first-ever test of a missile defense system under actual wartime conditions, Bush intimated, Patriot had passed with flying colors.

In Bush's view, it was appropriate to count Patriot's performance as a significant breakthrough, an event fundamentally altering the nature of the public debate over missile defense. Some critics maintained that "results from the test range wouldn't stand up under battlefield conditions," but as Bush said, "You knew they were wrong, those critics, all along. And now the world knows it, too" (150). With their incredible engineering feat, Bush suggested that Raytheon employees had demonstrated to the world that it was now realistic to assume that ballistic missiles could be neutralized by defensive systems. "Because of you, the world now knows that we can count on missile defenses," Bush announced (150).

Echoing a line of argumentation developed by Ronald Reagan nearly eight years before, Bush argued that the Patriot system had allowed the people of Saudi Arabia and Israel to receive material protection against ballistic missiles, transcending the thin and ephemeral guarantee of deterrence. "Thank God," Bush said, the people of Israel and Saudi Arabia had "more to protect their lives than some abstract theory of deterrence. Thank God for the Patriot missile. Thank God for that missile" (150).

As physicist Richard Garwin reflected, both appeals had a substantial and immediate effect on popular perceptions:

> During the Desert Storm operation against Iraq in 1991, it was announced by President George Bush that the effectiveness of the Patriot missile in intercepting Scuds was almost 100 percent. . . . [O]nce the words are out of the mouth of the President, there is a substantial establishment devoted to establishing their truth or reality, as was the case following the announcement of the Strategic Defense Initiative by President Ronald Reagan on March 23, 1983.[41]

Patriot(ic) News Reporting

Bush's laudatory tributes to the Patriot missile, and his overall narrative framework for the war, resonated against the backdrop of daily news reports. The stream of video footage featuring high-tech U.S. weapons in action lent support to Bush's suggestion that the Gulf War was a harbinger of a new military era, in which the American technological edge would bring unquestioned battlefield superiority. As Douglas Kellner observes, such ubiquitous footage endowed officials "with power as well as credibility, providing an aura of veracity to whatever claims they would make, which were seemingly grounded in technological omnipotence and evidence too compelling to doubt."[42]

Specifically, Bush's claims of Patriot missile effectiveness were validated strongly by media coverage of the war. Robert Lichter, editor of *Media Monitor*, reports that during the war, "the Patriot missile was the media darling, with 91 percent positive ratings."[43] According to Congressional Research Service analyst Steven Hildreth, "[g]lobal media reporting, including live camera coverage throughout Desert Storm, portrayed Patriot's performance against Iraqi missiles as a technological marvel."[44] The dramatic video clips of Patriots making apparent intercepts of incoming Scuds were buttressed by official military statements certifying Patriot accuracy.[45] The long string of official statements maintaining Patriot accuracy supplied in military press briefings,[46] coupled with other propaganda measures,[47] conditioned members of the press to expect, perceive, and report Patriot proficiency. As Hildreth explains, such reporting "shaped the public perception of Patriot's high level of effectiveness in the Gulf War."[48]

The media's pro-Patriot orientation came through in shaded play-by-play war commentary. Richard Blystone of CNN described the Scud as "a quarter-ton of concentrated hatred," while another CNN reporter hailed Patriot as "one of the first heroes of the Gulf War," and reported that it "uses quick thinking to outsmart the enemy."[49] *USA Today* described Patriot as "three inches longer than a Cadillac Sedan de Ville." NBC's Tom Brokaw called Patriot "the missile that put the Iraqi Scud in its place."[50] There was even evidence of the media's infatuation with Patriot in the lobby of the main media hotel in Dhahran, Saudi Arabia, where "We Love You" was scrawled on the side of a model Patriot missile.[51]

Favorable media coverage of Patriot and other high-tech weapons systems (such as smart bombs, sea-launched cruise missiles, and stealth fighters) in the Gulf War was, to a large extent, guaranteed because of the lack of contrary information available to journalists. Subject to strict military guidelines that kept noncombatants away from the vast majority of military action, nearly all reporters were forced to cover the war from remote hotels in Dhahran, Saudi Arabia. Dependent on official military "press pool" briefings for information, these reporters had no choice but to pass along the military's official accounts as authoritative news.[52] "I guess you could call it censorship by lack of access," said Walter Porges, an ABC News vice president.[53] With even limited press pool access tied to good behavior, most reporters willingly submitted to the military's guidelines and restrictions. As Australian journalist

McKenzie Wark noted at the time, "in going in to battle with each other for ratings, the media happily surrender to the demands placed on them by the military machine."[54]

Transformed from independent reporters into military public relations agents, many journalists lost sight of their status as watchdogs during the Gulf War.[55] This phenomenon filtered down even to pronoun usage. "The use by journalists of 'we' to mean U.S. military forces was constant," explains media analyst Jim Naureckas, "so that one seemed to hear of CBS taking out half the Iraqi Air Force . . . and even Walter Cronkite manning Patriots: 'We knocked one of their Scuds out of the sky' [Cronkite exclaimed at one point]."[56] Commentary and reaction from U.S. media sources operating at home reflected a similar partisanship; skeptics of the war effort were vilified frequently as unpatriotic. For example, editor Joe Reedy of the Kutztown, Pennsylvania, *Patriot* was fired for writing an antiwar editorial for the newspaper.[57] This instance was typical of a general pattern of exclusion which kept oppositional voices and critical perspectives out of print and off the air during the war.[58]

The lack of critical media reportage during the Gulf War stemmed in part from the fact that the Pentagon's virtual reality was the only game in town—there were precious few counterrealities available to journalists seeking to question the veracity of official military accounts. Pentagon control over news feeds enabled military officials to frame their public statements against a backdrop of carefully screened, hyperreal images portraying military maneuvers as clean, surgical operations. In a critical analysis of the Gulf War, Baudrillard draws from his central theoretical concepts of "hyperreality" and "simulation"[59] to advance the bold thesis that "[w]ar itself has entered a definitive crisis."[60] The Gulf War, writes Baudrillard, "is unreal, war without the symptoms of war, a form of war which means never needing to face up to war, which enables war to be 'perceived' from deep within a darkroom."[61] Although Baudrillard's breezy ontological conclusions regarding the nature of war continue to prompt quizzical double takes from scores of skeptical readers, his analysis highlights the importance of appreciating exactly how Gulf War discourses were shaped by Operation Desert Storm's hyperreal context.[62]

The next section explores elements of this hyperreality, in order to develop further the rhetorical context framing Bush's Raytheon address. This exploration sheds light on the manner in which the illusion of Patriot effectiveness took hold after Bush's address, as well as the ways that the Pentagon's information-control machinery locked in "Nintendo"-style news reporting that reinforced the basic themes of Bush's argument.

"Nintendo" Warfare

For U.N. coalition soldiers, the battlefield experience of the Gulf War recalled childhood visits to the video arcade. In the first thirty-eight days of the conflict, coalition fighter pilots made bombing runs in the dead of night, using infrared vision and computerized navigation aids to make their way through the desert and to their targets, "not real locations but map coordinates displayed on a VDU," as Benjamin Woolley, author of *Virtual Worlds,* described.[63] Alienated from the direct reality of the battlefield, there was little to distinguish the coalition pilots' experience from training runs made in simulation machines. "Shells burst in silence; explosions have no source," wrote journalist Robert Fisk during the war. "A fighter-bomber will attack a distant target, bathe the terrain in fire and twist away in the sky without the slightest sound," he reported.[64]

During the brief ground phase in the final five days of the conflict, soldiers on the desert floor also engaged the enemy in virtual reality warfare. "[T]roop movements were formations of pixels in computer-enhanced, false-colour satellite images," wrote Woolley.[65] American M1-A1 tanks tracked and locked onto remote targets from afar, then delivered precision-guided munitions from distances of up to 3,300 yards.[66] Operation Desert Storm represented a new kind of warfare, what James McBride called a "Nintendo War," with strategy and tactics plotted on computer screens and executed on remote video displays.[67]

Front-line reports of action contained eerie echoes of a major theme of Beinhart's novel *American Hero,* that the war was scripted on the model of the Super Bowl. A chief warrant officer for the 82nd Airborne announced the onset of hostilities with the remark "[i]t's time to quit the pre-game show."[68] A pilot returning from one of the first bombing runs over Baghdad exclaimed, "[w]e've scored a touchdown and no one was home!"[69] Another pilot warned against early optimism, cautioning "[w]e had one good morning. You sting 'em quick, you're winning 7–0, but it's not over."[70] When the expected Iraqi counterattacks never came, Major General Barry R. McCaffrey of the 24th Infantry Division said the war resembled "an eighth-grade team playing a pro football team."[71] Explaining the American decision to call off the war after Iraq's retreat from Kuwait, Presidential Special Assistant Richard

Haas invoked the football analogy once again: "[W]e didn't want to be accused of piling on once the whistle had blown."[72] According to Lewis Lapham, editor of *Harper's* magazine, media coverage of the Gulf War borrowed many of the rhythms and conventions of play-by-play football commentary featured prominently in blockbuster American television shows such as *Monday Night Football.*

> The Pentagon produced and directed the war . . . with a script that borrowed elements of "Monday Night Football." . . . The synchronization with prime-time entertainment was particularly striking on Super Bowl Sunday. ABC News intercut its coverage of the game in progress in Tampa with news of the bombing in progress in the Middle East, and the transitions seemed entirely in keeping with the spirit of both events. The newscasters were indistinguishable from the sportscasters, all of them drawing diagrams in chalk and talking in similar voices about the flight of a forward pass or the flare of a Patriot missile. The football players knelt to pray for a field goal, and the Disneyland halftime singers performed the rites of purification meant to sanctify the killing in the desert. . . . Just as it never would occur to Frank Gifford to question the procedures of the National Football League, so also it never occurred to Tom Brokaw to question the ground rules of the war.[73]

While coalition soldiers were remote spectators to the damage wrought by their high-tech weapons, the surreal nature of Gulf War placed the world audience in a perceptual position even further removed from the direct action. Because the media relied heavily on military-furnished remote video for their reports, the hyperreal experience of the battlefield was transferred to civilian television viewers. "[J]ust as the military found itself located in video hyperspace for the conduct of the war, the media found itself a coproducer of this virtual reality," writes Campbell.[74] Images recorded from the nose cones of smart bombs as they homed in on ground targets were replayed again and again on television. "These images literally took the TV viewers into a new high-tech cyberspace," what Kellner describes as "a realm of experience with which many viewers were already familiar through video and computer games."[75] The University of Oklahoma student who announced "I'm gonna pop some popcorn and watch the war"[76] captured the sense of political numbness created by mass exposure to the hyperreal press pool images of gee-whiz weapon gadgets performing trick after trick. As bedazzled television viewers witnessed this virtual reality video warfare presented as live news, the technological phantasmagoria whipped the world citizenry into a sense of suspended postmodern vertigo.

In the context of the Gulf War, the bizarre postmodern carnival described by Baudrillard was not a pre-given or inevitable feature of the reality. Instead, it was a constructed feature, manufactured in large part by a military research apparatus invested heavily in advanced communication technology, and also in part by political and military leaders who positioned themselves as exclusive information sources and monopoly providers of finished news products. "The Gulf war was 'total television,' an entertainment form that merged media and military planning. The pentagon and its corporate suppliers became the producers and sponsors of the sounds and images, while the 'news' became a form of military advertising," argue Kevin Robins and Les Levidow.[77] "Seen on network TV, the video-game images were crucial in recruiting support for the U.S.-led attack," they explain.[78]

Knee-jerk commitment to the idea of an unavoidable postmodern "condition" risks reifying hyperreality as an always pre-given phenomenon. This perspective obscures the fact that hyperreality can be constructed strategically, as well as exploited opportunistically, to further rhetorical objectives. The Gulf War was "calibrated for controlled thrills, anxiety, and relief from its opening laser-guided, son-et lumière spectacular and its initial fears of massive U.S. casualties to its triumphant helicopter descent on the U.S. Embassy in Kuwait," writes commentator Tom Engelhardt. "That scene was indicative of the production's carefully designed nature," he explains.[79]

From the vantage point of CNN viewers, the Gulf War appeared to be a sterile and precise military campaign. However, behind the Pentagon's decisive view charts and Hollywood-style video clips, there was massive carnage. Widespread destruction wrought by allied bombing was far from a hyperreal experience for the people of Iraq. As a result of the attacks, "the devastation to the industrial infrastructure of most of Iraq and portions of Kuwait, along with an unprecedented environmental disaster brought about by oil spills and Iraqi-set fires," notes international relations scholar George Lopez, "will take a toll on people and resources for years to come."[80] In addition to the thousands of direct civilian casualties from allied bombs, many more were affected dramatically by the razing of Iraq's civilian life-support infrastructure.

> The United Nations Children's Fund (UNICEF) stated immediately after the ceasefire that Baghdad "is a city essentially unmarked, a body with its

> skin basically intact, with every main bone broken and tendons cut." . . . There was little rubble, and civilians were spared, but their life support systems—electricity, water, transportation, communications—were disabled. To some, this is the very definition of strategic. In the words of Lieutenant Colonel Daniel Kuehl, USAF, Retired, it was "the progressive entropic dislocation of the innards and connective tissue of the Iraqi society."[81]

These comments are stark reminders that academic debates over the ontological status of military conflict in the "Nintendo" warfare era unfold on a plane far removed from the lifeworld of Iraqi civilians on the receiving end of U.S.-led "surgical strikes." Perhaps the appropriate answer to Baudrillard's famous question "Did the Gulf War take place?" is a follow-up question: For whom did the Gulf War take place? Insight into the ways in which discourse and perceptions during the Gulf War were structured and organized through hyperreal media frames helps explain how it was possible for remote television viewers to see the conflict as benevolent surgery, while ground-zero civilians experienced bombing in a very different, and far more violent, register.

Media critic Philip Taylor argues that the Gulf War's video format alienated viewers from the brutality of warfare: "This was a war fought by professionals and seen by the public from a distance."[82] During the war, military officials shot video footage that featured gruesome images of war violence. For example, so-called turkey shoots of retreating Iraqi soldiers were documented on video clips shot by cameras mounted on the front of Apache helicopters. Unsure about whether to release such video evidence to the media press pool, Pentagon officials showed segments to selected reporters in a screen test. As one reporter recounted, the screen test "showed frightened, disoriented Iraqi infantrymen being shot to pieces in the dark by U.S. attack helicopters. One by one they were cut down in the middle of the night by an enemy they could not see."[83] Invoking a popular football metaphor to describe the scene, John Balzar of the *Los Angeles Times* said the tape showed Iraqi soldiers "as big as football players on the TV screen. . . . A guy was hit and you could see him drop and he struggled up. They fired again and the body next to him exploded."[84] Censors eventually ruled the video "too brutal for general audiences," and the public never witnessed the events.[85] Retrospective accounts of the 2 March 1991 Battle of Ramaylah contained similar accounts of one-sided slaughter. In this battle, retreating Iraqi troops ran into the U.S. Army's 24th Infantry Division, commanded by Maj. Gen.

Barry R. McCaffrey. Maj. Gen. McCaffrey was the same high-ranking officer who characterized some of the Gulf War's ground battle as a pro-football team running up the score on eighth graders. According to reporter Seymour Hersh, audio tape-recordings of conversations conducted in battle show that McCaffrey's own troops were horrified at what they saw, including one episode where U.S. soldiers fired high-powered machine guns into a crowd of more than 350 disarmed Iraqi prisoners.[86] Such allegations are explosive, given that the Battle of Ramaylah occurred after a cease-fire had been negotiated, and that findings of war crimes on the part of U.S. troops would potentially implicate General McCaffrey, who currently sits as Director of the Office of National Drug Control Policy in the Clinton administration. In a pre-emptive media campaign conducted prior to release of Hersh's article in *The New Yorker*, McCaffrey accused Hersh of "journalistic stalking" and released a number of documents defending the appropriateness of his actions.[87]

In a *New York Times* letter to the editor, former CBS anchorperson Walter Cronkite commented on the Hersh-McCaffrey row: "The controversy over the actions of Gen. Barry R. McCaffrey's 24th Infantry Division after the cease-fire in the Persian Gulf war points up again the serious fallacy of the Pentagon's censorship policy in that conflict. That policy severely restricted the right of reporters and photographers to accompany our troops in action, as had been permitted in all our previous wars. This denial prevented the American people from getting an impartial report on the war."[88]

Apparently, Cronkite's letter had an effect on the *New York Times* editorial staff, who wrote two days later that "As Walter Cronkite . . . noted in a letter to *The Times* earlier this week, the Pentagon's effort to restrict coverage of the war denied the American people an immediate and full account of the battles American forces fought in Kuwait and Iraq. More comprehensive coverage might long ago have clarified whether General McCaffrey's order to attack was appropriate."[89] Clearly, the question of whether the 24th Infantry Division committed war crimes in the "last battle of the gulf war"[90] deserves close consideration. However, Cronkite's commentary points to a question that is potentially even more important: how come it is so hard for us to tell? As Taylor concludes, during Operation Desert Storm, allied forces proved that it was possible to fight wars "in the television age without allowing too much of war's 'visible brutality' to appear in the front rooms of their publics."[91] Similarly, Dana Cloud argues that the perceptual distance from military operations created by video news coverage shifted public war discussion to an interpersonal frame that focused attention on "yellow ribbon therapeutic" news at the expense of "harder" news about battle developments, casualties, and prisoners.[92]

With these reflections, Taylor and Cloud provide critical commentary that couples textual analysis of public arguments with appreciation of the powerful ways that the Gulf War's hyperreal space enabled and constrained deliberative exchanges. The next section of this chapter embarks on a similar vector of commentary focusing on the postwar controversy over Patriot's Gulf War performance. In the aftermath of the conflict, many of Bush's Raytheon address themes shifted from widely accepted truisms to hotly contested propositions, as a host of critics disputed official accounts of war events and deconstructed the Pentagon's hyperreal video evidence. Analysis of the process through which discourse surrounding Bush's initial claims evolved into a full-fledged public controversy promises to elucidate important contours of the debate.

Postwar Controversy over Patriot Accuracy

Following the war, many commentators picked up on and extended the key themes developed in Bush's Raytheon address: Good had prevailed against evil, the United States had demonstrated its unsurpassed technological capability, the Patriot missile had performed flawlessly, and the concept of missile defense had been validated triumphantly. Echoing Bush's suggestion that the Patriot was the cornerstone of the United States' new high-tech military revolution, Lt. Gen. Charles Horner suggested that "[n]o one should underestimate the value of the Patriot system in this war. . . . In the historical analysis and studies of this war, Patriot will be one of the key systems which influenced the outcome."[93] This position received support from a parade of rosy expert assessments trumpeting Patriot's wartime performance, each of them corroborating Bush's exuberant wartime assertions of Patriot effectiveness. On 13 March 1991 Maj. Gen. Richard Beltson said Patriot units were "receiving world-wide recognition for their unprecedented success in Operation Desert Storm."[94] Less than a week later, Gen. Carl Vuono, chief of staff of the army, declared "[t]he now world-famous Patriot PAC-2 missile . . . has achieved spectacular results."[95]

Conservative commentators such as the Heritage Foundation's Baker Spring reveled in the apparent display of American high-tech weaponry: "The early days of the war against Iraq confirm that high-technology weapons are

America's trump card against Saddam Hussein. No weapon used so far in Operation Desert Storm has performed so surprisingly well as has the Patriot missile system."[96]

In general, commentators noted that Bush's Raytheon address theme of a "revolution in modern warfare" seemed to be validated by the Gulf War's pyrotechnics. "This led to the launch of the RMA [Revolution in Military Affairs]," notes war studies scholar Lawrence Freedman.[97] Immediately after the war, Secretary of Defense Richard Cheney wrote to Congress that the Gulf War "demonstrated dramatically" the possibilities of a high-tech military revolution.[98] Amidst this bullish chorus of praise, Raytheon, Inc., manufacturer of the Patriot missile defense system, saw its stock jump in value from $68.50 to $81.88 per share.[99]

Following through on the themes laid out in Bush's Raytheon address, missile defense advocates claimed further that the Patriot's spectacular Gulf War performance constituted general evidence of missile defense feasibility, and warranted major investments in new missile defense projects. "After the war, policymakers throughout the Government continued to assess Patriot as a highly effective missile defense system. This support helped justify budget requests for additional improvements in the Patriot system, funding increases in the Strategic Defense Initiative (SDI), and plans to proceed with a limited strategic missile defense of the United States," observes Hildreth.[100] As veteran arms control official Albert Carnesdale points out, "the widespread perception is that Patriot was very effective. As a result, the Gulf War has given a political boost to SDI."[101] Defense analyst Marvin Feuerwerger corroborates this line of reasoning, stating that "[a] major impetus to ATBM programs has been the perception that the Patriot missile worked in the Gulf War."[102] Asked to list the six most significant recent changes in the world of ballistic missile defense, Carnesdale cites as number one "the political push given to all forms of ballistic missile defense by the perceived success of the Patriot missile system in the Persian Gulf War."[103]

Missile defense advocates were able to package evidence of the Patriot's Gulf War performance as scientific fact, while simultaneously exempting such evidence from the rigorous burdens of proof usually viewed as necessary entailments of scientific claims. These advocates contended that the Patriot had succeeded in a crucial, unprecedented technological experiment. The Gulf War "was a laboratory for the American military's new weapons and fighting doctrines," wrote Michael Gordon and General Bernard Trainor.[104]

With the Patriot's success in this laboratory environment, Republican politician Patrick Buchanan exulted that "the United States has shown it can attack and kill ballistic missiles. . . . The [SDI] debate is over."[105] "The Patriot is now known to work in a real-life situation. For the modern generation of aerospace weapons, where fantasies often outpace technological reality, this is a major accomplishment," declared defense analyst Wayne Biddle.[106] According to the Gulf War Air Power Survey, the war was not merely "a conducive environment for the successful application of air power," but also an opportunity "so ideal as to approach being the best that could be reasonably hoped for in any future conflict."[107] Campbell notes that by pointing to these ideal "laboratory" conditions, "[t]he Pentagon sought to objectify representations of engagements that obfuscated their constructed and contested character."[108] Drawing on the persuasive power of scientific realism, missile defense advocates presented Patriot performance data as closed to further interpretation, thus drawing attention away from the intensely political dimensions of discussions about Patriot effectiveness.

The Illusion Evaporates

Several months after the cessation of hostilities in the Persian Gulf, MIT defense and arms control scholars Theodore Postol and George Lewis became interested in the issue of Patriot's Gulf War performance when Lewis noticed that when replaying television video clips in slow motion, allegedly successful Patriot intercepts of Scuds appeared to be misses. Postol, Lewis, and their colleagues found similar phenomena in nearly all of the Patriot-Scud encounters they could find captured on video. Most of this footage was publicly available in promotional videos packaged in red, white, and blue boxes that contained triumphant commentary by announcers touting the awesome technological prowess of the U.S. military.[109] As Postol described one typical video, in hyping Patriot's contribution to the war effort the audio track would declare, "never before has a system worked so well," and then, on video, "they'd show you two Patriot-Scud engagements where the Patriot missed by half a kilometer. We're not talking about small misses. Enormous distance misses" (see Fig. 4).[110]

The MIT researchers supplemented this video research with analysis of wartime ground damage reports published in Israeli newspapers. If the Patriot was truly effective in neutralizing Scuds, one would expect that the amount of damage done by incoming Scuds, substantial in the first five days of the conflict, would have been reduced markedly in the period following deployment of the Patriot. However, ground damage reports did not reflect such mitigation. Reported damage was, in fact, greater when Israel was being shielded by Patriot. During the period of defense the number of apartments reported damaged tripled and the number of reported injuries increased by almost 50 percent per Scud attack.[111]

There are a number of factors that may have accounted for the increase in ground damage during Patriot deployment. First, Patriot interceptors that missed their Scud targets could have fallen to Earth and caused damage on their own. Given that each Scud launch was met by at least two or three Patriot interceptors, even with a 100 percent Scud neutralization rate, for each Scud launched by Iraq, several loose Patriots were likely to sail past their targets and dive to the ground (see Fig. 5, Fig. 6). "In cases where the Patriot interceptor still contains burning rocket propellant and hits this ground at a small elevation level," wrote Postol, "this may cause greater ground damage than the impact of an intact Scud."[112] Thus, "significant damage from a small number of Patriot impacts could have measurably added to the total net ground damage during the period of Patriot defense operations," he explained.[113] Second, partial Patriot intercepts may have cut Scuds into various large pieces which then fell in multiple locations, causing damage at several different sites (see Fig. 7, Fig. 8).[114] Third, bugs in the Patriot software may have caused incomplete or failed intercepts. In fact, the most lethal Scud attack of the entire war took place when a Patriot computer software error prevented it from tracking and engaging an incoming Scud. Twenty-eight American soldiers were killed and ninety-eight were wounded.[115]

Postol et al. supplemented their video and ground damage analysis with interviews of Israeli scientists and military officers associated with the Patriot deployment. These interviews confirmed suspicions that the Patriot had not performed as well as originally believed. For example, a "very senior Israeli officer," with "both a detailed understanding of air defense operations and first-hand knowledge of the Israeli defensive operations against Scuds" indicated that Patriot deployment in Israel was marred by several significant problems.[116] Among these problems were "difficulties associated with the Patriot's automatic computer operations, the coordination of independently operating fire units, misinterpreted intercept data by Patriot crews, the unexpected breakup of the attacking Al-Husayn [modified Scud] missiles, and timing errors in the interceptor's fuzing devices."[117]

Postol et al.'s analysis of video evidence, scrutiny of ground damage reports, and interviews with top Israeli scientists and military officials challenged fundamentally one of the most widely held lessons of the Gulf War, that Patriot performed as an impeccable "Scud-buster."[118] After the research team's revelations, "the certainty soon began to crumble, even by official accounts," explained *Science* columnist James Glanz.[119] As Postol reflected incredulously, "[a]pparently most people, no everyone, I guess, who saw this on television, everyone just didn't realize what they were seeing."[120] Patriot "is a story of how we projected what we wanted to believe onto the TV screen," reflected Representative John Conyers (D-MI).[121]

How could such a massive audience have been misled in this way? Common sense would dictate that missile defense interceptors take out their targets by smashing into them directly. However, Patriot interceptors were designed to fly to the neighborhood of a targeted incoming ballistic missile, self-detonate, and scatter a stream of lethal metal debris in the path of the enemy target. During the Gulf War, many viewers of Patriot-Scud engagements probably interpreted these interceptor self-detonations as direct hits on Scuds.[122] Postol et al.'s video analysis showed that in many engagements the Patriot interceptor was either behind or not even "in the neighborhood" of the incoming Scud at the time of self-detonation, and that the resultant spray of debris did little or nothing to alter the trajectory of incoming Scuds. Lacking a sophisticated technical understanding of BMD intercept dynamics, journalists and Gulf War television viewers were not in a position to make such detailed discriminations in real time, and were instead forced to rely on scripted interpretations provided by U.S. military officials.

Raytheon's Reply

In late 1991, the journal *International Security* agreed to publish an article by Postol that presented his research findings on Patriot accuracy in the Gulf War. When Raytheon, Patriot's private manufacturer, learned that such an article was slated to be published in the winter issue of *International Security*, its company representatives contacted the journal and demanded reply space in the same issue, so they could challenge Postol's claims and defend the reputation of their missile defense system. Although the editors of *International Security* did not grant an immediate right of reply, they did agree to allow Raytheon to attach a letter to each of the winter issues sent to regular *International Security* subscribers,[123] and agreed to publish the company's full response (along with a rebuttal by Postol) in the next (summer 1992) issue. With these unusual publishing events, the extraordinary public debate over Patriot's Gulf War effectiveness was joined.

The Raytheon letter was authored by Robert Stein, manager of Advance Air Defense Programs for the Raytheon Corporation. In taking direct aim at Postol's analysis, Stein defended the Patriot's record by citing positive evidence of the Patriot's Gulf War performance, challenging the integrity of Postol's data and analysis, offering explanatory rationales for the system's shortcomings, and pointing out psychological and political benefits of perceived Patriot success. Since Stein's presentation represented the first detailed and complete public defense of Patriot missile performance against Postol's challenges, it would be appropriate to consider each of his main arguments in detail.

In countering Postol's analysis, Stein cited a number of sources supporting Patriot's Gulf War proficiency. Virtually all of these citations were classified "SECRET" by the U.S. government. For example, for proof of Patriot effectiveness, Stein referred readers to the "order of magnitude" assessments issued by U.S. Army Brig. Gen. Robert Drolet. These assessments stated that "[i]n Saudi Arabia, Patriot successfully engaged over 80 percent of the TBMs within its coverage zone [and] in Israel . . . Patriot successfully engaged over 50 percent of the TBMs in the coverage zone."[124] However, Stein noted that the rationale behind these official assessments could not be disclosed, because Drolet based his findings on the Patriot "scorecards" that were classified SECRET by the U.S. government. Borrowing one of Edward Teller's most effective cold war techniques of persuasion, Stein lamented: "Thus, although this author would like to use these figures to refute erroneous statements such as Patriot's performance 'resulted in an almost total failure,' this cannot legally be done" (2).

Stein also cited an "independent assessment" of Patriot performance by the Army's Ballistics Research Laboratory (BRL) as proof of Patriot accuracy: "The team performed its assessment by examining the holes in the ground left when the pieces of the Scud missiles impacted" (9). By correlating the deduced energy transferred to the ground with analysis of the missile fragments, the "BRL team" attempted to reconstruct Patriot-Scud engagements. While Stein did not disclose the specific findings of the BRL study, he did indicate that the "Army's 80 percent TBM warhead kills or duds in Saudi Arabia is based on this investigation" (8).

The final major source cited by Stein to prove Patriot effectiveness was a "joint USA/IDF report" which contained analysis of ground damage in

Israel. Although this report was also "classified SECRET," the "released number" summarizing its findings regarding Patriot effectiveness in Israel was given as "over 50 percent" (15). The methodology of this study was similar to that used by the BRL team; Israeli researchers "examined craters, pieces of Scuds," and attempted to reconstruct the likely outcomes of Patriot-Scud engagements.

Stein's second approach was to bolster his case by leveraging the argumentative weight of secret evidence. For example, Stein privileged the epistemic authority of classified sources and implied that reviews of Patriot performance not based on such secret evidence lacked sufficient detail when he wrote, "[t]hose who have performed a detailed review of Patriot's performance in the Gulf and had access to all the facts" disagree with popular statements alleging Patriot failure (2). In addition, Stein quoted defense analyst Charles Zraket's statement that the assertions of Patriot critics "are incorrect, distorted, and designed to discredit [Patriot]. . . . These critics do not have access to the classified performance data from Saudi Arabia and Israel" (2–3). Stein quoted Zraket as saying that the analyses of Patriot critics were "pure conjecture at best" (3)

Stein's third strategy was to plead for more permissive standards to be used in evaluations of the Patriot's Gulf War performance. Such leeway was justified, Stein reasoned, because of the unusual nature of the Scud threat that the Patriot was asked to neutralize. Iraq's Scuds had a higher velocity, were stealthier, and were more difficult to track than any other ballistic missile that the Patriot had been designed for or tested against (3).[125] Even partial success, Stein implied, should warrant enthusiastic endorsement of Patriot, given that the system was put into a tough situation by being asked to engage a novel target. Patriot failures, on the other hand, should be excused. For example, in rationalizing Patriot's 25 February 1991 breakdown which failed to prevent nearly one hundred American casualties in Dhahran, Stein wrote, "[o]n occasion, a series of unanticipated and seemingly unrelated events will come together and create a tragic situation" (13). Further, Stein argued that the stress and unpredictability of the battlefield environment made precision measurement of the Patriot's performance difficult: "[T]he detailed performance data that are available from a flight test are not available under conditions of actual combat" (5). With this line of argument, Stein implied that it would be unreasonable for one to expect Patriot proponents to furnish clear and unambiguous scientific evidence demonstrating Patriot performance.

The final major strategy Stein utilized involved an attempt to shift the focus of debate away from the issue of actual Patriot performance to the benefits flowing from perceived Patriot effectiveness. "Patriot's success and very credible performance can be measured by the events as they occurred," wrote Stein. "The coalition did not falter. Israel did not have to mount offensive actions against Iraq, and was able to stay out of the war," he explained (1). By invoking this "real-time" standard of evaluation, Stein attempted to ratchet the debate over Patriot performance to a higher level of abstraction. According to this logic, although the Patriot may not have actually worked, the perception that it did had beneficial effects. Such effects could be evaluated, Stein suggested, quite independently of concrete evidence such as ground damage or video footage. According to this line of argument, Israel did not enter the war, and a conflagration of all-against-all did not materialize, thus the Patriot was effective.

Stein turned to psychological evidence for proof of the Patriot's worth in this capacity: "The world was fortunate that there was a Patriot. . . . Many of those who were on the scene [in the Gulf War] and were held hostage to Saddam Hussein's nightly attempts at spreading a rain of terror have attested to this fact" (21). Stein quoted a young Israeli student who sent a letter of thanks to the "Patriot soldiers" manning a Patriot missile battery outside the windows of his school. Because of these troops, the child wrote in his letter, "I don't feel anxiety anymore" (21). These testimonials from laypersons constituted significant evidence for Stein. "They were not scientists or engineers," Stein wrote, "but their eyes had provided them with all of the analysis they seemed to need" (21). Versions of this same argument could be found in a Raytheon press release issued in response to a WGBH-TV *Frontline* report critical of the Patriot's Gulf War accuracy. Arguing for the psychological benefits of perceived Patriot effectiveness, Raytheon officials asserted that final judgment in this case should rest with officials from Israel, Saudi Arabia, and Kuwait. "These countries understand first hand the effectiveness of Patriot, it's [*sic*] ability to be easily integrated with their existing air defenses and the piece [*sic*] of mind provided from knowing that the U.S. government and U.S. industry will continue to improve the system in light of changing threats and provide continued logistics support for the life of the system," stated Raytheon's press release.[126]

The Debate over Use of Classified Evidence

In the summer 1992 issue of *International Security,* Postol and Stein continued the dispute over the Patriot's Gulf War performance. Stein submitted a letter to the editor, and Postol replied. Punctuating the value of the Patriot's perceived effectiveness in wartime, and adding fuel to the dispute with a dose of condescension, Stein wrote, "the events that are so central to this discussion occurred during a war. To the men, women, and children who became prey to Saddam's TBMs of terror, this was far more than an academic debate."[127] In his reply letter published in the same issue, Postol issued a detailed defense of his original article and decried Stein's black box style of argumentation: "[D]espite pages of discussion, Stein presents no analysis or data in support of his claims for high Patriot intercept rates. Instead his arguments are based on citations of classified reports, and on public statements of individuals who themselves provide no data and who also cite classified reports."[128]

While Postol challenged Stein's use of classified evidence, Raytheon officials challenged Postol on the grounds that (contrary to his assurances) his article contained classified information. As a defense contractor with a facility security clearance, Raytheon was required to report all suspected clearance breaches to the Department of Defense. Thus, upon receipt of a draft of Postol's article in November 1991, Raytheon Security agents had conducted a classification review and had forwarded the results to the Intelligence Directorate of the U.S. Army Defense Investigative Services (DIS) on 11 November 1991.[129]

Acting on this tip three months later, after publication of Postol's article in the winter 1991/92 issue of *International Security,* DIS agents contacted Postol to inform him that his article contained classified information.[130] This contact was the first round of a remarkable public tug-of-war between Postol and the government intelligence community over the proper boundary separating classified and public information on missile defense. This struggle broke open another fascinating layer to the unfolding argument over the Patriot, adding the issue of the proper scope of secret intelligence to the agenda of discussion.

As a "Type A Consultant" with government security clearance, Postol was authorized to receive classified information, but in the case of the controversy over the Patriot's Gulf War performance, he told DIS investigators that he had deliberately eschewed reliance on classified sources in preparing his manuscript for *International Security.* "I had intentionally not used my security access to obtain any data on Patriot performance. . . . This had been my policy so that I would not be constrained in public discussions by the knowledge that certain data might be classified," Postol explained.[131] However, when DIS invited Postol to attend a classified meeting to further discuss his article, Postol's strategy was put in jeopardy. By attending such a classified meeting, Postol would have moved discussion of his work on the Patriot into the closed intelligence community, and his security agreement would have prevented him from making further public statements on the issue. Thus, Postol refused to attend the secret meeting proposed by DIS, maintaining that all of his data were drawn from openly available public sources: television and print news, interviews, and public descriptions of the Patriot system.[132] DIS responded by declaring Postol's entire article secret, a step that essentially amounted to the issuance of a postpublication gag order. DIS "informed me that because I would not agree to have a classified meeting to go over the particulars of this article, that this article was secret, and that I had to treat everything in the article as secret and, therefore, could not talk about any part of it," Postol said.[133]

Seeking relief from Congress, Postol described his predicament to the Committee on Government Operations at a 1992 House of Representatives hearing on government secrecy after the cold war. "Today, as I sit before you, I cannot talk about my article," Postol said. "[I]f I understand the claims of the DIS, they believe that I am obligated to give up my first amendment rights to save their organization the effort of analyzing and understanding other people's claims about my article," he testified.[134] Following his testimony, Postol talked with members of the news media. "The Army and Raytheon are now using the DIS, which appears to be more than an unwitting partner, to suppress my speech on the subject of Patriot performance in the Gulf War," Postol explained.[135]

Acting on Postol's allegations, Rep. John Conyers (D-MI) asked the Department of Defense to review the matter. Nina Stewart, deputy assistant secretary of defense, convened an expert panel to conduct a technical review of DIS's claims regarding Postol's article, to determine "whether [the article] contains classified information; whether it could have been written using open sources as claimed by Dr. Postol; and whether access to specific classified information would have been required to develop the conclusions reached."[136] Finding that "no further action is warranted," the panel concluded that "Dr. Postol's experience and background, combined with open source data may have enabled him to hypothesize or synthesize his conclusions."[137] Subsequently, the army dropped its attempt to classify Postol's article. "I am very pleased the gag order against Professor Postol has been lifted and that the Department of Defense is not going to attempt to classify a published academic journal article. Using the classification to suppress public debate on the performance of the Patriot Air Defense System is just not acceptable," said Conyers.[138]

However, following this reprieve, pressure on Postol continued. Soon after the favorable DOD finding, the army contacted the Office of Technology Assessment to review his security clearance,[139] and Raytheon began circulating a video clip implying that Postol had doctored video data in his Patriot research.[140] "Raytheon does their best to squash every hint of a question" about the Patriot, said Joseph Cirincione, deputy staff director of the U.S. House Subcommittee that conducted hearings on the missile in 1992.[141] "They move to squash it quickly and cleanly, discredit the people, push the product, and increase their sales," he explained.[142]

Congressional Investigation

Initial escalation of the Postol-Stein dispute soon brought national media attention to the debate, and with the controversy spilling over into public debates about post–Gulf War missile defense, a number of government agencies conducted investigations designed to ferret out the truth and lift the "fog of war." These investigations were difficult, in that they required analysts to assess the quality of the government's official classified information and compare it to the publicly available data cited by Postol, without disclosing any of the classified information in the process.[143] Members of the House Committee on Government Operations sought to evaluate the army's data showing Patriot accuracy in a classified briefing at the Army Missile Command. According to an anonymous source cited in *Inside the Pentagon*, the army's presentation was "appalling. The Army was defensive and very reluctant to answer questions. Their methodology for scoring Patriot kills is scandalous."[144] *Inside the Army* reported that "Congressional experts were uniform in their opinion that the analysis was totally inadequate."[145]

In classified interviews with the Patriot system manager, the General Accounting Office (GAO) found similar serious weaknesses in the quality of the army's data. GAO investigator Richard Davis discovered that the army "scorecard" data (which initially indicated an 80 percent/50 percent Patriot success rate in Saudi Arabia and Israel), were largely based on probable kill indications from Patriot computer guidance systems.[146] The Patriot system manager told GAO that he was "not surprised" that investigators were turning up evidence of Patriot failure that contradicted this guidance system data.[147] In the GAO's investigation of Patriot's system manager, investigators learned that the guidance system data "was not intended to be used as an analysis of Patriot's performance. [The system manager] said it was intended as a tool to keep himself and others at the Air Defense School abreast of events in the Gulf; therefore, he made no attempt to analyze the data."[148] The informality of the scorecard (also referred to as "spreadsheet") data was evident to GAO. Its investigators found that "the spreadsheet was actually an inaccurate summary of information obtained through telephone calls to various units in the Gulf, army staff offices in the Pentagon, or the Patriot project office."[149] Investigators also found "many gaps and inconsistencies" in the scoresheets and supporting records.[150]

Congressional Research Service (CRS) investigators found other weaknesses in the army's scorecard data. Much of the army's data turned out to be based on "Patriot unit reports of Scud engagements," that is, shot-by-shot accounts of whether the soldiers in the Patriot command centers believed they had successfully intercepted Scuds.[151] However, such evidence was found to be of dubious value, given the public statements made during the war downgrading the ability of human "after action" reports to accurately assess Scud-Patriot engagements. For example, army skepticism was quite evident during a 25 January 1991 press briefing, in which General Thomas Kelly had the following exchange with a reporter:

> Q: The videos of the latest Scud attack from Tel Aviv looks as if some Patriot missiles are taking off and impacting. Do you know anything about that? Can you shed any light on that?
> A: Whether Patriot missiles hit some of those incoming Scuds?
> Q: No, it looks as though Patriots are taking off and then crashing and exploding.
> A: I've not seen the videotape. Again, let me tell you why it takes awhile to sort through what happens in these Scud attacks . . . a lot of things happening very quickly, and it's almost impossible—especially with cloudy skies—for observers just to stand there and know what and what fell where. . . . So it's very, very hard to tell immediately, or within even a few hours, even after the event, precisely what happened and what fell where.[152]

With this explanation, General Kelly clearly questioned the reliability of human after-action accounts as data for Patriot performance, stating that it was "almost impossible" for ground observers to discern the outcome of

Patriot-Scud engagements. At the same time, General Kelly's comments also reflected evidence of army dependence on video data. Kelly said he was not prepared to answer reporters' questions, because "I've not seen the videotape." It is difficult to reconcile this comment with the fact that, following the war, the army elected not to utilize video data in their assessments of Patriot accuracy.

To elucidate further the evidentiary limitations of the army's data, Hildreth found other instances where human after-action reports of weapons system performance were unreliable. For example, in the 1988 *Vincennes* incident involving the accidental U.S. downing of an Iranian civilian airliner, "[t]he physical evidence available in this case—AEGIS radar recording tapes—proved conclusively that the recollections of officers directly involved were wrong on basic facts, such as whether the Iranian plane was ascending or descending as it flew toward the ship."[153] As former Secretary of Defense Les Aspin commented on the incident, "I think it raises very serious questions as to all the other reports that we have ever done [regarding other incidents], whether in fact that is what ever really happened."[154]

Scrutiny of the army's classified evidence produced other evidentiary anomalies. The classified BRL study, cited by Stein as produced by a "team" of Israeli experts,[155] turned out to be the work of a lone BRL engineer who worked in Saudi Arabia for five days in February 1991 and nineteen days in March 1991, visiting the sites of Scud-Patriot impactage "days or weeks after an impact when craters had often been filled and missile debris removed."[156] Further, the GAO found that the BRL study contained data for only about one-third of the Saudi engagements, although it had been cited by the army for all engagements.[157]

The overall assessment provided by congressional investigators was that the quality of the army's classified evidence showing Patriot Gulf War effectiveness was poor: "[T]he data that would be needed to conclusively demonstrate how well the Patriot performed during Operation Desert Storm does not exist and there is no way to conclusively tell how many targets the Patriot killed or failed to kill."[158] While the initial official assessment of the Patriot's effectiveness was that the system successfully intercepted 80 percent of Scuds fired at Saudi Arabia and 50 percent of Scuds fired at Israel,[159] a panel of senior army officials downgraded this assessment by 10 percent during congressional testimony, responding to charges by the GAO that the army's "original assessments . . . just could not be supported by the data that they said they used to come up with that assessment."[160] Under questioning, Brig.

Gen. Robert A. Drolet, author of the army's official assessment of Patriot effectiveness, stated that the service had a "high degree of confidence" in only 40 percent of the kills. In a further retreat, Drolet qualified that all the army meant by "successful intercept" was that "a Patriot and Scud had passed in the sky."[161] In post–Gulf War interviews, Israeli officials also backpedaled on the IDF's original assessments of Patriot accuracy.[162]

These concessions lent credibility to claims that the Patriot's Gulf War effectiveness was a manufactured fiction. "With significant support of the Bush Administration," argues Israeli scientist Reuven Pedatzur, "Raytheon, the manufacturer of the Patriot missile, produced a deceptive picture of the performance of the Patriot missile."[163] Despite admitting that its initial assessments of Patriot accuracy were invalid, the army stalled further attempts to clarify the Patriot's Gulf War record by failing to disclose key classified evidence and resisting calls for arbitration by an independent scientific panel, such as the American Physical Society (APS),[164] even though such arbitration would appear to be particularly appropriate in light of the findings by independent physicists such as IBM research fellow Richard Garwin that cast further doubt on official claims of Patriot accuracy.[165]

Recently, Jeremiah Sullivan, a physicist and former director of the Program in Arms Control, Disarmament, and International Security at the University of Illinois, Urbana-Champaign, headed up an APS review panel that evaluated the Postol et al.-Raytheon dispute over the Patriot's Gulf War accuracy. Because the Pentagon still refuses to release classified evidence on Patriot performance (almost a decade after the Gulf War), Sullivan's APS review panel could not compare the army's classified evidence against Postol et al.'s analysis of the video evidence and ground damage reports. However, Sullivan's team did assess the dispute between Postol et al. and critics such as Raytheon vice president Robert Stein and physicist Peter Zimmerman. Stein and Zimmerman questioned Postol et al's methodology of video analysis,[166] but as *Science* writer James Glanz explains, the APS panel definitively supported the technical soundness of Postol et al.'s video analysis, and discredited the objections raised by Stein and Zimmerman: "Accepted for publication at the journal *Science and Global Security*, the team's report analyzes all of the technical criticisms raised against the video evidence. It concludes that those criticisms are 'without merit' and goes on to identify 'an absolute contradiction' between the Army's scoring of Patriot performance and that video record."[167]

Despite the findings of several independent reviews supporting Postol et al.'s analysis, controversy over the Patriot's Gulf War accuracy continues to simmer after years of claims, counterclaims, and official attempts to gag dissent. Domestically, advocates of a Star Wars-style national missile defense still point to the Patriot as a shining example of empirical missile defense success.[168] Internationally, Patriot sales have been brisk, with Germany, the Netherlands, Israel, Japan, Saudi Arabia, Kuwait, and Taiwan all recently buying systems.[169] Since the Gulf War, Raytheon has generated $2.4 billion in foreign sales.[170] This remarkable success in marketing the Patriot to foreign customers has been in part due to the ability of missile defense advocates to maintain the narrative coherence of the triumphant story of the Patriot system, even in the wake of serious criticism. Missile defense advocates have extended the lines of argumentation set down by President George Bush in his 1991 Raytheon address, employing the rhetorical strategies examined in this chapter to keep the credibility of the triumphant story intact.

A Tough Placebo to Swallow

Although recent publication of some critical views on the Patriot's Gulf War performance shows that the Pentagon's postwar deception program has not been completely effective, it has had sufficient traction to alter fundamentally the trajectory of public dialogue on missile defense in this country and abroad. The following pages analyze this phenomenon. First, the vexing dilemmas embedded within the logic of "placebo defense" are explored. Second, the political impact of perceived Patriot effectiveness is investigated in order to throw light on the manner in which the Gulf War legacy has shaped recent trends in missile defense politics. Through an examination of the ways in which postwar strategic deception on the Patriot impacted the pattern of public argument on missile defense, it will be possible to arrive at a more contoured normative assessment of the value of the deceptive practices employed in pro-Patriot advocacy campaigns.

Dilemmas of Placebo Defense

In post–Gulf War dialogues about the Patriot's Gulf War accuracy, one trump card frequently employed by missile defense advocates is the argument that even if the Patriot did not actually work, it was still politically effective as a symbol. With such a strategy, missile defense advocates skirted the issue of actual performance. Even if not leakproof, the Patriot was good enough to stop Israeli retaliation, said Representative William Dickinson (R-PA). "[I]n providing even a limited defense to Israel," he said, "the Patriot allowed Israel to stay out of a war and thus save the coalition from fragmenting and ensured a militarily united front."[171] Representative Dickinson's justification for strategic deception on the grounds of political expediency echoed similar arguments offered during the cold war by Star Wars enthusiasts such as Daniel Graham and Edward Teller. Where Soviet audiences were manipulated to overestimate the effectiveness of Star Wars in the cold war, Israeli citizens were led to believe that the Patriot was protecting them flawlessly in the Gulf War.

Representative George Atkins (R-NY) invoked the instructive metaphor of a medical placebo to explain this perspective on the value of the Patriot: "[T]he primary benefit of Patriot is a political and psychological one, which I would say militarily is like a placebo."[172] Advocates invoking the placebo defense justification for Gulf War strategic deception emphasize that the mere perception of Patriot effectiveness bore substantial fruit during the war. Indeed, it appeared that the wartime perception of Patriot accuracy had a beneficial calming effect on Israeli civilians. "Thanks to widespread public perception of Patriot effectiveness, Israeli defense officials were apparently better equipped to resist public pressure to intervene against Iraq," argued Horton.[173] "Both critics and advocates believe the use of Patriot was an important psychological and political factor during the Gulf War," wrote Marvin Feuerwerger, senior strategic fellow at the Washington Institute for Near East Policy.[174] Even Postol acknowledged the benefits of this deception. Commenting on the fiction of Patriot effectiveness during the Gulf War, Postol said, "since war is both a physical and a psychological phenomenon, it had an important political leverage that was useful."[175] "During the war," Postol explained, "I think there was a good argument for not revealing the fact that the [Patriot] system was not functioning."[176]

When thought of in terms of a benign medical procedure, placebo defense has strong appeal as a sensible instrument of realpolitik statecraft. However, the effectiveness of placebo strategies is limited by the fact that placebos lose much of their effectiveness when patients become aware of the fact that they are being deceived—placebos depend on deception to achieve their effects. When the placebo fiction is exposed, the psychological efficacy of the remedy is eroded. Speaking of the psychological and political role of the Patriot in this vein, Feuerwerger wrote, "[g]iven the lessons that the Israeli and other publics may have drawn from the Gulf War experience, no one can guarantee that Patriot or similar systems will be able to perform a similar role in the future."[177] While strategic deception of enemies in wartime may be justified as a component of military doctrine, when such a strategy is extended to allied publics, the resultant skepticism may very well render such an approach unsustainable.

The placebo metaphor suggests that the benefits of wartime strategic deception on missile defense can be consistently realized only if world audiences maintain confidence in the U.S. military as a source of reliable information. Like the patient who loses trust in his or her doctor upon learning that all the doctor has in her medicine cabinet is salt tablets, such audiences are likely to question official American declarations when they are advanced in repetitive patterns of deception, disclosure, and detraction. Eventually, alternating cycles of deception and cover-up may stimulate such profound general cynicism that audience trust could evaporate completely. In this scenario, military officials could encounter difficulties in convincing even sympathetic allied publics that the most basic claims should be believed.

Should military officials seek to dodge this credibility spiral by covering up wartime deception with long-term information control strategies designed to keep the details of war permanently under wraps, such strategies are likely to be dogged by another horn of the placebo defense dilemma: the trap of self-deception. Misrepresentation of BMD capabilities could boomerang on U.S. officials, inducing ill-informed commanders to make battlefield decisions that mistakenly put allied soldiers in harm's way. Such permanent deception strategies could also erode the principle of civilian military control, undermining the ability of citizens and their elected officials to make sound decisions about military matters. These dilemmas of placebo defense will be explored more thoroughly in chapter five, where such issues will be revisited in light of recent developments in missile defense politics.

Public Spheres under Peacetime Assault

During the Gulf War, military officials manufactured factual support for President Bush's triumphant war script with tight media control and deft exploitation of video evidence. After the war, missile defense advocates and private contractors defended the claim of Patriot accuracy by utilizing secrecy strategically in public argumentation, attributing scientific objectivity to classified (yet unfounded) claims, and harassing scientific opponents with frivolous security clearance investigations. When appraising the costs and benefits of post–Gulf War strategic deception on the Patriot, it is important not to overlook the fact that the rhetorical campaign of missile defense advocates in this case represented a full-scale assault on public spheres as sites of open discussion. While this country has long excused such practices under emergency wartime conditions, if efforts to muzzle critical voices and neutralize the public's argumentative competence become routine peacetime practices, our system of democratic decision making is put in jeopardy. As a cost of post–Gulf War strategic deception on the Patriot, this unraveling of the democratic fabric can be more fully appreciated by understanding the dynamics and ramifications of missile defense advocates' techniques of blocking open discussion in public spheres of deliberation.

After the war, missile defense advocates employed multifaceted strategies to distort the content of public debate on missile defense. By relying heavily on the classification system to shield their data from scrutiny in the field of open argumentation, they precluded detailed public comparison of competing evidence and protected their data from the rigors of scientific peer review.[178] By publicly overstating the level of Patriot effectiveness, advocates infused false information into public debate.[179] By intimidating critics holding security clearances, advocates chilled contributions to the debate from the most qualified professional scientists.[180] By declining to submit their arguments for objective review by an independent panel,[181] and by "burying" an independent congressional evaluation of the technical controversy, advocates further restricted the flow of critical evaluative commentary on the Patriot's Gulf War performance.[182] "It's important to have an accurate understanding of Patriot's effectiveness against Iraqi Scud warheads in Desert Storm. It's important to public debate, it's important to defense planning purposes, and, therefore, I would argue that a credible account should be part of the public record," says Hildreth.[183] Unfortunately, systematic distortion of information in the public debate over the Patriot precluded the cultivation of this level of public understanding. According to Representative Conyers, "[t]he misinformation supplied by the administration and Raytheon has given the American people and Congress profoundly mistaken impressions about the Patriot."[184]

By protecting the Pentagon's internal data on the Patriot's performance behind the curtain of classification, missile defense advocates evaded burdens of proof in argumentative discourse and created leverage for specious attacks against Patriot critics. While such moves might have been legitimate if the government could have demonstrated that disclosure of the information on Patriot would have constituted, in the words of the Supreme Court, a "grave, direct and immediate" harm to national security, no such justification was ever provided in the case of the Patriot, even in the face of repeated official calls for declassification. In fact, persons with security clearances who were in a position to evaluate the secret data found little reason to keep Patriot performance data under wraps. "I received [a] classified briefing on this system," said Representative Stephen Neal (D-NC), "[a]nd I, frankly, can see nothing in that classified briefing that should not be made public."[185]

In the case of the controversy over the Patriot's Gulf War performance, the government's reliance on the classification system to shield its data from public scrutiny appeared to be a clear example of classification abuse, since the unclassified evidence actually turned out to be superior in quality to the data kept secret by the government. "[T]here is absolutely no need for classified information to be used in the public debate [on the Patriot]. I think the amount of data, and the kind of data that we have is actually better than any of the classified data that was collected by the U.S. Army," said Postol.[186] Further, because the precise details of the Patriot system were deducible from public sources, information about the Patriot's design was not "protectable," and hence not an appropriate candidate for classification.[187] As one Pentagon official put it, the "toothpaste was already out of the tube," yet the army continued to protect its data from scrutiny.[188]

The army's classification strategy skewed public debate on the Patriot by shifting the burden of proof to critics and placing those who dared to challenge official statements of Patriot effectiveness under suspicion for violation of secrecy rules. An example of how the army's use of classified information worked to shift the burden of proof can be found in 1992 congressional hearings convened to investigate the Patriot's Gulf War performance. Conyers, chair of the committee conducting one of these investigations, stated at the outset that the objective of his committee's inquiry was to determine whether the army's official statements on Patriot effectiveness were valid. However, because classification rules prevented witnesses and committee members from openly debating the connection between the army's data and its claims, attention shifted to the issue of whether the data provided by the Patriot's critics could support the claim that the Patriot was completely ineffective.[189] Missile defense advocates used refutations of such straw-person arguments as substitutes for positive demonstrations of Patriot effectiveness.[190] By going negative and attacking the claims of opponents, missile defense advocates were able to take advantage of the reversed burden of proof made possible by their reliance on classified data.

Another strategy utilized by missile defense advocates to use the classification system to control the tenor of debates about the Patriot involved the harassment and ad hominem attack of critics holding security clearances. On the one hand, missile defense advocates impugned the credibility of Patriot critics who did not base their contentions on the army's classified evidence. On the other hand, however, they suggested that critics who had access to the classified data were on thin ice by even participating in public debate. An exchange between Representative John Kyl (R-AZ) and Postol during congressional hearings demonstrated the widely divergent perspectives on the appropriate burden of proof in such classification questions. While Postol maintained that analysis derived from openly published information should not be subject to prepublication review by authorities, Kyl argued that holders of security clearances should always bear the burden of subjecting their public statements to advance review.

> Mr. KYL: Did you ever seek a security review before publication of your article?
> Mr. POSTOL: No, I didn't.
> Mr. KYL: Wouldn't it appear to you to be prudent, if you had obtained information in classified meetings, whether you thought that this was the source of the information in your article or not, to seek such a review?
> Mr. POSTOL: No. I don't think the system should be allowed to review things that have been obtained by unclassified work without access to the classified data in question, and where analysis independent of the system is done. No, quite emphatically.[191]

Representative Kyl's position on this matter was echoed by the DIS agent who eventually cleared Postol of any breach of secrecy rules. In the very next paragraph following her exoneration of Postol, Stewart wrote: "[I]t would seem reasonable to expect that Dr. Postol's broad experience with classified DOD weapon systems, his many years holding a security clearance, and his briefing in conjunction with signing a classified information nondisclosure agreement, should have led him to conclude that a security review by the government before publication of his article would be most prudent."[192]

This placement of security clearance holders under permanent discursive review for all contributions to public debate positions the government as gatekeeper of all public comments made by an important population of technically knowledgeable sources. With this "check with us first" policy, the government vastly exceeded its legitimate authority to control information. "Security procedures and the rules for enforcing protection of classified information have been stretched to include suppression of alternative thought," Postol stated, "not to mention rights of speech guaranteed under the first amendment."[193]

As former assistant for weapons technology in the Office of the Chief of Naval Operations in the Reagan administration, Postol was involved directly in a host of detailed defense analyses utilizing classified information.[194] "I have a full and deep appreciation and respect for the importance of secrecy in military planning, and I have no reservations about the government's need to be able to classify information. However, I have serious reservations about the implications of abuse of the classification system, as such abuses pose one of the most serious and overriding threats to democracy and its institutions," Postol said.[195] From the perspective of independent scientists such as Richard Garwin, postwar strategic deception on Patriot accuracy constituted a major threat to the democratic process.

> Perhaps the claim of a highly successful Patriot system kept Israel from responding militarily against Iraq, which would have complicated the military and political situation, to say the least. But in the U.S. democracy we would be deceiving not only our citizen-bosses but also our leaders themselves if we did not tell the truth in such matters. Although Winston Churchill remarked that "in wartime, truth is so precious that she should always be attended by a bodyguard of lies," that bodyguard is stifling not only to democracy but also to the national security unless used only where strictly necessary. In the case of Patriot as an antimissile system, insufficient provision was made for gathering information on its effectiveness.[196]

One direct consequence of this short-circuiting of democratic deliberation in post–Gulf War debates on Patriot accuracy was distortion of the budget appropriations process. With the Patriot as the centerpiece of the new U.S. high-tech war machine, its performance in the Gulf was linked to the wider fortunes and prospects of America's so-called Revolution in Military Affairs (RMA), with its first trial run in the Gulf War.[197] Accordingly, the widespread perception of the Patriot's wartime effectiveness was a major factor driving support for maintaining high military budgets in the threat-reduced post–cold war milieu. As sociologists Gregory McLauchlan and Gregory Hooks argue, this phenomenon was one key element in the translation of the post–cold war "peace dividend" into a post–Gulf War "war dividend."

> [M]ilitary pronouncements and media coverage of the role of the U.S. Patriot antimissile system against Iraqi Scud missiles claimed it an unqualified success, though a subsequent analysis revealed that the large majority of Patriots failed to intercept their targets and did not reduce ground damage in Israel. . . . Thus, the claims for and public perception of success of the paradigmatic high-tech weaponry of the Gulf War were the products of military public relations and uncritical media coverage. . . . Though detailed considerations of the outcome of the Gulf War is beyond the scope of this article, there are indications that it has been significant in recalibrating the military R & D agenda for the post–cold war era and by blocking the substantial redirection of resources to civilian pursuits anticipated by a "peace dividend." Following the war, defense industry publications and the business press were awash with articles speculating on a Gulf "war dividend" for the defense industry and military laboratories. As anticipated by *Fortune,* "Even if the cold war is really over, the strategic arms race, largely involving nuclear weapons, could give way to an equally costly conventional arms buildup aimed at keeping ahead of Third World threats."[198]

The lingering fiction of the Patriot's Gulf War effectiveness appears to have had substantive impacts on American military spending patterns. When Bush established in his Raytheon address that the Patriot was the pivotal weapon system in the United States' new high-tech arsenal, he paved the way for defense contractors and military planners to cash in on perceived Patriot effectiveness at budget appropriations time.[199] In recent budget cycles, their extraordinary successes have reversed the momentum of the cold war peace dividend and triggered a new round of conventional arms spending.[200] Apart from the issue of whether such defense spending patterns are appropriate in the current international threat environment, the fact that these spending decisions were made with flawed data is a matter of serious concern.[201] This concern deserves a full hearing in retrospective discussions regarding the merits of strategic deception on the Patriot's Gulf War accuracy, as well as prescriptive discussions on the proper future course of missile defense research and development.

Conclusion

In his 1993 novel *American Hero,* novelist Larry Beinhart suggests provocatively that the Gulf War was rigged by the GOP and Hollywood to improve

President Bush's chances in the 1992 election. While most readers are likely to find Beinhart's hypothesis as far-fetched as Jean Baudrillard's audacious proposition that the Gulf War "did not take place," this chapter has shown that one major aspect of the war, perceived as factual at the time, turned out to be fictional. There was no evidence to prove that the Patriot missile defense system knocked out "41 of 42" Scuds during the war, as President Bush claimed in his 15 February 1991 address to employees of the Raytheon Corporation.

Although vigorous postwar scrutiny of video evidence and ground damage reports have discredited Bush's rosy assessment of Patriot performance, vestiges of this fiction continue to live on in official missile defense discourses. The fiction has been nurtured and protected through strategic utilization of the classification system and systematic efforts to block balanced public review of the Patriot's Gulf War performance. With these measures of strategic deception, missile defense advocates have been able to maintain the narrative coherence of the key rhetorical themes advanced in Bush's Raytheon address, such as the idea that missile defense has crossed a crucial feasibility threshold and should now be considered a high-tech archetype of the bold "Revolution in Military Affairs."

Has the illusion of Patriot effectiveness been a useful fiction? This is a complicated question, given that there appeared to be clear benefits flowing from perceived Patriot effectiveness during the Gulf War. Many commentators have staked out a middle position on this point, holding that strategic deception on the Patriot was useful during the war because it dissuaded Israel from widening the conflict, but also stipulating that the ensuing peacetime deception was an unwarranted insult to this country's commitment to democratic decision making. One difficulty with the wartime exception for strategic deception is that such a strategy overlooks the co-dependence of wartime and postwar deception. To make good on the commitment to democratic decision making, the bifurcated normative framework requires that military officials follow up wartime deception with postwar disclosures of manipulation and the detraction of inaccurate statements. When repeated, this deception-disclosure-detraction cycle is likely to diminish U.S. credibility as a reliable source of information.

Wartime strategic deception thus introduces a vexing dilemma for missile defense advocates. Coming clean after war may repay the debt to democracy, but it may also cancel the effectiveness of strategic deception in future conflicts. Conversely, peacetime strategic deception may maintain the credibility of official sources of information, but it also may constitute a major erosion of democratic principles and signal dangerous movement toward a permanent national security state that operates in a constant state of emergency. In chapter five, I speculate on this latter possibility, exploring the possibility that "Fortress America" has evolved into a modern day Sparta, supported by a "citadel culture" of perpetual military buildup. The analysis in this chapter shows such national security imperatives can be used to manufacture hyperreal deliberative spaces that serve as exclusive sites for public learning about military affairs. In such spaces, military officials enjoy wide maneuvering room to package data favorably and manipulate popular perceptions with riveting video clips. Where Livermore researchers rigged X-ray laser and HOE experiments to generate the false perception of Star Wars' feasibility in the 1980s, in the 1990s army officials manufactured the illusion of Patriot effectiveness by rigging the reporting and interpretation of a large-scale, high-tech experiment in military performance, the Gulf War.

Although Beinhart's suggestion that the entire war was really another Lee Atwater political campaign spot for President Bush seems improbable, his ruminations on this point have proved to be quite prescient. In 1998, another series of events unfolded to muddy further the boundaries separating life from art. The plot of the 1997 film *Wag the Dog* involved an executive scheme by presidential aides to shift public attention away from an Oval Office sex scandal. This scheme used hoopla from a trumped up military invasion of Albania as a political distraction. When *Wag the Dog* was released on video, President Clinton found himself mired in the midst of his own lurid scandal involving sexual dalliances with a White House intern. Members of the media dogged Clinton relentlessly after the scandal broke, even following Clinton from Washington, D.C., to Martha's Vineyard, where reporters camped out on the periphery of the president's vacation compound.

Waiting for news, reporters killed time in mobile units parked in Clinton's driveway. On 20 August 1998 some reporters happened to be watching *Wag the Dog* on video when Clinton emerged from his vacation house to make the dramatic announcement that the United States had bombed Sudan and Afghanistan in a counterterrorism operation.[202] After Clinton shared news of the bombing campaign, entertainment writer Michael Fleeman asked "[i]s life imitating art again?"[203] Although the official justification for such attacks was that they were necessary retaliation against terrorists for the 7 August 1998 bombing of American embassies in Tanzania and Kenya, Clinton's skeptics had their doubts. "Look at the movie *Wag the*

Dog. I think this has all the elements of that movie. Our reaction to the embassy bombings should be based on sound credible evidence, not a knee-jerk reaction to try to direct public attention away from his personal problems," said Representative Jim Gibbons (R-NV).[204]

Speculation that Clinton had manufactured the 1998 military conflict to take the heat out of a hot scandal kitchen spurred *Wag the Dog* video rentals and sales. Immediately after Clinton's announcement of the airstrikes against Sudan and Afghanistan, there was a run on *Wag the Dog* videos at stores like Cox Video in Dallas.[205] Industry analyst Paul Dergarabedian says "the real-life scandal may have helped [*Wag the Dog* at] the box office."[206] With this strange confluence of events, "[t]he phrase 'wag the dog' has come to mean more: that there's no telling what politicians might do in their efforts to manipulate the public," says commentator Bob Fenster.[207]

If citizens choose to assume a more skeptical posture and try to counter such manipulative schemes in the future, they will need to find ways of resisting the massive institutional inertia that has already gathered behind the Patriot and other BMD systems. This is no easy task, especially because institutions have learned to utilize postmodern strategies that construct hyperreal communicative spaces facilitating top-down control of public discussion and anesthetizing the citizenry with fantastically absorbing television images. Understanding of the ways that military officials employ hyperreality to bolster schemes of strategic deception in public argument is a crucial resource that can inform efforts by citizens to counter such manipulation. My analysis of the controversy over Patriot missile accuracy in the Gulf War is offered here as one example of how it is possible to explicate and challenge strategic deception in official missile defense advocacy. It is hoped that there is value in this analysis as an exemplar of rhetorical criticism that is grounded in textual artifacts of public controversies, yet is aware of the complex semiotic destabilizations brought about by hyperreal experiences in the postmodern age.

Notes

1. Donaldson used this term to describe a Patriot-Scud engagement over Dhahran, Saudi Arabia on January 26, 1991 (quoted in Randoloph Ryan, "Patriot Propaganda," *Boston Globe*, 21 March 1992, 9). Analysis of the film footage after the Gulf War revealed that in this particular engagement, Patriot missed its mark. Like virtually all other television commentators, ABC's Sam Donaldson reported each Patriot-Scud engagement in terms that constructed the fiction of uncanny Patriot effectiveness.

2. Cheney made this statement in response to a question by Senator Dan Coats (R-IN) during a congressional hearing on 21 February 1991 (Richard Cheney, statement, *Department of Defense Authorization for Appropriations for Fiscal Years 1992 and 1993.*, 102d Cong., 1st sess., Senate Hearings, Committee on Armed Services, 21 February 1991 [Washington, D.C.: GPO, 1991], 15).
3. Cooper made this statement in testimony to Congress on 17 April 1991 (Henry Cooper, statement, *Hearing Before the Subcommittee on the Department of Defense*, 102d Cong., 1st sess., House Hearing, 17 April 1991 [Washington, D.C.: GPO, 1991], 613).
4. Larry Beinhart, *American Hero* (New York: Ballantine, 1993), ii.
5. Ibid., 388–94.
6. Ibid., 387.
7. Quoted in ibid., 401.
8. Jean Baudrillard, "The Reality Gulf: Why the Gulf War Will Not Take Place," *Guardian* (London), 11 January 1991, 25.
9. For discussion of television's power to produce a "reality effect" for viewers, where images "show things and make people believe in what they show," see Pierre Bordieu, *On Television*, trans. Priscilla Parkhurst Ferguson (New York: New Press, 1998), 21–28.
10. After the Scuds hit, "[i]n the inner councils of the Bush administration, no problem worried officials more than what might happen if Israel entered the war" (Michael R. Gordon and Bernard E. Trainor, *The General's War* [Boston: Little, Brown and Company, 1995], 230). By crossing over Jordanian airspace on their way to Iraq, it was feared that Israeli fighter planes would draw Jordan into the conflict and ignite "the nightmare Middle East scenario of a war of all against all" (Gordon and Trainor, *General's War*, 231).
11. Theodore A. Postol, "Lessons of the Gulf War Experience with Patriot," *International Security* 16 (1991/1992): 119.
12. Narrating Patriot's attempted but failed intercept of a Scud missile near Dhahran on 26 January, ABC anchor Sam Donaldson said, "[t]here's an explosion, there's an intercept, there's another intercept . . . there's something falling . . . Uh-oh." Donaldson rationalized the ensuing ground explosion: "It could be a spent Patriot." In fact, there was no successful Patriot intercept, and the ground flash was a Scud explosion (see John Conyers Jr., "The Patriot Myth: Caveat Emptor," *Arms Control Today* 22 [1992]: 4).
13. Ryan, "Patriot Propaganda," 9.
14. See Christopher Anderson, "Classification Catch-22," *Nature* 256 (1992): 274; Theodore A. Postol, "Improper Use of the Classification System to Suppress Public Debate on the Gulf War Performance of the Patriot Air-Defense System," *Government Secrecy after the Cold War*, 102d Cong., 2d sess, House Hearing, Committee on Government Operations, 18 March 1992 (Washington, D.C.: GPO, 1992), 38–62.
15. Ultimately, congressional investigations were performed by the General Accounting Office, the Congressional Research Service, and the House Committee on Government Operations. In the hearings conducted as part of the latter investigation, Brig. Gen.

Robert Drolet conceded under intense questioning that the military's initial estimates of Patriot's Gulf War accuracy were inflated.

16. Carol Rosin, telephone interview with the author, 24 September 1999. Rosin went public with this story immediately after the meeting at Fairchild, relating the events to multiple radio, television, and live audiences.
17. Douglas Kellner, *The Persian Gulf TV War* (Boulder, Colo.: Westview, 1992), 56.
18. Rick Atkinson, *Crusade: The Untold Story of the Persian Gulf War* (Boston: Houghton and Mifflin, 1993), 303.
19. George Bush, "Remarks to Raytheon Missile System Plant Employees in Andover, Massachusetts, February 15, 1991," *Public Papers of the Presidents of the United States: George Bush, 1991*, vol. 1 (Washington, D.C.: GPO, 1992), 148.
20. Rep. Frank Horton, statement, *Performance of the Patriot Missile in the Gulf War*, 102d Cong., 2d sess, House Hearing, Committee on Government Operations, 7 April 1992 (Washington, D.C.: GPO), 5.
21. Bush, "Remarks to Raytheon," 150.
22. Ibid., 150.
23. Ibid,, 149.
24. George Bush, "Exchange With Reporters in Andover, Massachusetts, on the Iraqi Offer To Withdraw From Kuwait, February 15, 1991," *Public Papers of the Presidents of the United States: George Bush, 1991*, vol. 1, (Washington, D.C.: GPO, 1992), 148. Bush's dogged attempts to discredit the legitimacy of the new Iraqi peace offer caused him to assume some curious rhetorical contortions. In an exchange with reporters, he said the members of the coalition rejected the offer "because there wasn't anything new or significant" ("Exchange," 148). But in an address to Raytheon employees just an hour later he stated that the peace offer was a "cruel hoax" because it contained significant new elements: "Saddam Hussein has added several new conditions" (Bush, "Remarks to Raytheon," 148).
25. Bush, "Remarks to Raytheon," 149.
26. Roland Dannreuther, "The Gulf Conflict: A Political and Strategic Analysis," Adelphi Paper 264 (London: Brassey's/International Institute for Strategic Studies, 1992), 49.
27. Bush, "Remarks to Raytheon," 150.
28. Ibid., 150.
29. Quoted in Bob Woodward, *The Commanders* (New York: Simon and Schuster, 1991), 343.
30. David Campbell, *Politics Without Principle: Sovereignty, Ethics and the Narratives of the Gulf War* (Boulder, Colo.: Lynne Rienner Publishers, 1993), 17.
31. Bush, "Remarks to Raytheon," 149–50.
32. Campbell, *Politics Without Principle*, 21; see also James McBride, *War, Battering and Other Sports* (Atlantic Highlands, N.J.: Humanities Press, 1995), 46.
33. Quoted in Woodward, *Commanders*, 343.
34. Denise M. Bostdorff, *The Presidency and the Rhetoric of Foreign Crisis* (Columbia: University of South Carolina Press, 1994), 237–40.

35. This strategy reflects a broader tendency of rhetors to graft bipolar cold war categories of discourse onto contemporary political discussion, substituting outlaw "rogue" states for the menace of the Soviet Union (see Rebecca Bjork, *The Strategic Defense Initiative: Symbolic Containment of the Nuclear Threat* [New York: SUNY Press, 1992]; John Pike, "Star Wars: Clever Politics in the Service of Bad Policy," *F.A.S. Public Interest Report* 49 [1996]: 1–14).
36. Bush, "Remarks to Raytheon," 150. Further references to this speech will be made by page number in the text.
37. Atkinson, *Crusade*, 303.
38. Bush, "Remarks to Raytheon," 149, emphasis added.
39. See Frank N. Schubert and Theresa L. Kraus, "The Whirlwind War," Center of Military History Report, 1995, Internet, Federation of American Scientists Website, online at http://www.fas.org/spp/starwars/docops/wwwapena.htm, Appendix A. This was not the last effort by the Carter administration to take political advantage of Patriot. "In what was reputed to be a political ploy to achieve the backing of House of Representatives speaker Thomas P. 'Tip' O'Neill of Massachusetts, the Carter administration approved the army's production contract with Raytheon on the eve of the 1980 election. Though Carter's bid for reelection failed, the Patriot was well placed to take advantage of the generous defense funding policies of the newly elected administration of President Ronald Reagan" (Schubert and Kraus, "The Whirlwind War," Appendix A).
40. Bush, "Remarks to Raytheon," 149.
41. Richard L. Garwin, "Theater Missile Defense, National ABM Systems, and the Future of Deterrence," in *Post-Cold War Conflict Deterrence*, ed. National Research Council (Washington, D.C.: National Academy Press, 1997), 187.
42. Kellner, *Persian Gulf TV War*, 157. As Hallin and Gitlin document, "images of technology dominated the visual representation of the [Gulf War] conflict, taking up to 17 percent of total television time, considerably more than any other major category" (Daniel C. Hallin and Todd Gitlin, "The Gulf War as Popular Culture and Television Drama," in *Taken By Storm: The Media, Public Opinion, and Foreign Policy in the Gulf War*, ed. W. Lance Bennett and David L. Paletz [Chicago: University of Chicago Press, 1994], 154–55).
43. Robert Lichter, "The Instant Replay War," in *The Media and the Gulf War: The Press and Democracy in Wartime*, ed. Hedrick Smith (Washington, D.C.: Seven Locks Press, 1992), 227.
44. Steven A. Hildreth, "Evaluation of U.S. Army Assessment of Patriot Antitactical Missile Effectiveness in the War against Iraq," Report, *Performance of the Patriot Missile in the Gulf War*, 102d Cong., 2d sess., House Hearing, Committee on Government Operations, 7 April 1992 (Washington, D.C.: GPO, 1992), 23.
45. For example, in a Riyadh press conference on 30 January 1991 General Norman Schwartzkopf, commander-in-chief of the U.S. forces in Desert Storm, announced that at that point in the war, "of thirty-three [Scuds] engaged, there have been thirty-three destroyed" (quoted in Conyers, "Patriot Myth," 4). Testifying before Congress,

Secretary of Defense Richard Cheney stated on 21 February 1991, "Patriot missiles have demonstrated the technical efficacy and strategic importance of missile defense" (Statement, 15).

46. See Hildreth, "Evaluation," 31–76.
47. According to Gordon and Trainor, "the Patriot batteries in Saudi Arabia were displayed conspicuously to the press, the military's version of the Potemkin village the Russian czars used to build to impress foreign visitors" (*General's War*, 64).
48. Hildreth, "Evaluation," 23.
49. CNN broadcast, January 22, 1991, quoted in Hallin and Gitlin, "Gulf War as Popular Culture," 154.
50. Quoted in Jim Naureckas, "Gulf War Coverage: The Worst Censorship Was at Home," in *The FAIR Reader: An EXTRA! Review of Press and Politics in the '90s*, ed. Jim Naureckas and Janine Jackson (Boulder, Colo.: Westview, 1996), 33–34.
51. David Shukman, *Tomorrow's War* (New York: Harcourt Brace Jovanovich, 1996), 103.
52. According to David Paletz, "[t]he military gave uninformative or at best guarded briefings, denied unfettered access to the battlefield, deployed an access-limiting system of reporter pools, used public affairs officers to supervise interviews, engaged in story-delaying of copy, rendering stories obsolescent, and imposed outright censorship" (Paletz, "Just Deserts?" in *Taken By Storm: The Media, Public Opinion, and Foreign Policy in the Gulf War*, ed. W. Lance Bennett and David L. Paletz [Chicago: University of Chicago Press, 1994], 282). Media critic John MacArthur's description of "Operation Desert Muzzle" provides a similar account (John R. MacArthur, *Second Front: Censorship and Propaganda in the Gulf War* [New York: Hill and Wang, 1992], 146–98).
53. Quoted in John J. Fialka, *Hotel Warriors: Covering the Gulf War* (Washington, D.C.: Woodrow Wilson Center Press, 1991), 6.
54. McKenzie Wark, "Gulfest," *Australian Left Review*, February 1991, 7.
55. As former Reagan press aide Michael Deaver remarked during the war, "if you were going to hire a public relations firm to do the media relations for an international event, it couldn't be done any better than this is being done" (quoted in Naureckas, "Gulf War Coverage," 30). This sentiment was echoed by one SDI official, who said, "[i]f someone had asked me to go out and execute an advertising campaign to get the point across to the American people and the rest of the world . . . I could not have done it any better than the television has done" (quoted in Edward Reiss, *The Strategic Defense Initiative* [Cambridge: Cambridge University Press, 1992], 187). As reporter Malcolm Browne reflected, the Pentagon turned the average journalist into "an unpaid employee of the Department of Defense, on whose behalf he or she prepare[d] the news of the war for the outer world" (Malcolm Browne, "The Military vs. the Press," *New York Times Magazine*, 3 March 1991, 29).
56. Naureckas, "Gulf War Coverage," 29.
57. See Grant Mahan, "Hawk Publisher Bombs Dove Editor," *Washington Journalism Review* 13 (1991): 14–15.
58. A survey conducted by Fairness and Accuracy in Reporting found that of 878 on-air sources used by ABC, CBS, and NBC in the first two weeks of the war, only one was a

representative of a national peace organization. In contrast, seven players from the Super Bowl were interviewed for their perspective on the war (see Naureckas, "Gulf War Coverage," 35).

59. See Jean Baudrillard, "Simulacrum and Simulations," in *Baudrillard: Selected Writings,* ed. Mark Poster (Stanford, Calif.: Stanford University Press, 1988); idem, *Simulations* (New York: Semiotext[e], 1983). The allegory that Baudrillard cites to exemplify the essence of the concepts of hyperreality and simulation is a Borges short story in which cartographers set out to make a map of their empire so accurate that it exactly covers the space of the mapped territory. In this instance, "simulation is no longer that of a territory, a referential being or a substance. It is the generation by models of a real without origin or reality: a hyperreality" (Baudrillard, "Simulacrum," 166).
60. Baudrillard, "The Reality Gulf," 25.
61. Ibid.
62. The ontological skepticism projected in Baudrillard's analysis drew biting retorts from social critics such as Christopher Norris. Although Norris granted that Baudrillard was right to point out some of the "hyperreal" aspects of the Gulf War, such as real-time battlefield reporting and long-distance combat engagement, Norris contended that Baudrillard pushed the postmodern envelope too far when he denied the very existence of the war. "[T]hese latest notions fall in very readily with a 'post-modern' mood of widespread cynical acquiescence, a feeling that the war was indeed so utterly unreal . . . that nothing we could possibly think or do would have any worthwhile impact as a means of contesting the official (media-sponsored) version of events" (Christopher Norris, *Uncritical Theory: Postmodernism, Intellectuals, and the Gulf War* [Amherst: University of Massachusetts Press, 1992], 29). Norris's argument was that Baudrillard's series of articles exhibited the "depth of ideological complicity that exists between such forms of extreme anti-realist or irrational doctrine and the crisis of moral and political nerve" among potential war resisters (*Uncritical Theory,* 27).
63. Benjamin Woolley, *Virtual Worlds* (Cambridge, Mass.: Oxford University Press, 1992), 193.
64. Robert Fisk, "The Bad Taste of Dying Men," *Independent* (London), 24 February 1991, 19.
65. Woolley, *Virtual Worlds,* 193.
66. Shukman, *Tomorrow's War,* 171. This long-distance engagement was typical in the ground war. "Although some of the Iraqis stayed and fought, there had not been any classic tank battles. The largest armored confrontation—the 1st Armored Division's battle at Medina Ridge—saw M1 tanks pummeling the immobile tanks of the Iraqi covering force from a distance" (Gordon and Trainor, *General's War,* 417–18).
67. See McBride, *War, Battering and Other Sports,* 36. The Nintendo aspects of the battlefield experience were heightened by the rapidity of the Iraqi retreat: "In some cases American GIs were alarmed that the collapse of the Iraqi front lines was proceeding at such a pace that they would not have the opportunity to 'get into the fun' of combat" (McBride, *War, Battering and Other Sports,* 58).

68. Quoted in Guy Gugiotta and Caryle Murphy, "Warplanes Roar Off in Darkenss in 'Desert Storm,'" *Washington Post,* 17 January 1991, A27.
69. Quoted in Kari Points, "Reporting Conventions Mask Sexual Politics," *Media and Values* 56 [1991]: 19.
70. Quoted in Philip Shenon, "Allied Fliers Jubilantly Tell of Early Control of the Skies as Iraq Planes Fled," *New York Times,* 18 January 1991, A10.
71. Quoted in John Mueller, *Policy and Opinion in the Gulf War* (Chicago: University of Chicago Press, 1994), 162.
72. Quoted in Lawrence Freedman and Efraim Karsh, *The Gulf Conflict, 1990–1991: Diplomacy and War in the New World Order* (Princeton, N.J.: Princeton University Press, 1993), 405.
73. Lewis H. Lapham, "Trained Seals and Sitting Ducks," in *The Media and the Gulf War: The Press and Democracy in Wartime,* ed. Hedrick Smith (Washington, D.C.: Seven Locks Press, 1992), 258–59.
74. Campbell, *Politics Without Principle,* 16.
75. Kellner, *Persian Gulf TV War,* 157.
76. Quoted in McBride, *War, Battering and Other Sports,* 59.
77. Kevin Robins and Les Levidow, "Video Games and Virtual War: High-Tech Imaging Transforms Social Perceptions," *New Statesman and Society* 8 (1995): 29.
78. Ibid.
79. Tom Engelhardt, "The Gulf War as Total Television," in *Seeing Through the Media: The Persian Gulf War,* ed. Lauren Rabinowitz and Susan Jeffords (New Brunswick, N.J.: Rutgers University Press, 1994), 84.
80. George Lopez, "The Gulf War: Not So Clean," *Bulletin of the Atomic Scientists* 47 (1991): 30–36.
81. William M. Arkin, "Baghdad: The Urban Sanctuary in Desert Storm?" *Air Power Journal* 11 (1997): 6.
82. Philip M. Taylor, *War and the Media: Propaganda and Persuasion in the Gulf War* (Manchester: Manchester University Press), 276.
83. Quoted in Ramsey Clark, *The Fire This Time: U.S. War Crimes in the Gulf* (New York: Thunder's Mouth Press, 1992), 140.
84. Quoted in Jason DeParle, "Keeping the News in Step: Are Pentagon Rules Here to Stay?" in *The Media and the Gulf War: The Press and Democracy in Wartime,* ed. Hedrick Smith (Washington, D.C.: Seven Locks Press, 1992), 387.
85. See Clark, *Fire This Time,* 140.
86. See Michael R. Gordon, "Report Revives Criticism of General's Attack on Iraqis in '91," *New York Times,* 15 May 2000, A11.
87. See Associated Press Online, "Army Releases Gulf War Records," 11 May 2000, Internet, Online, Lexis/Nexis.
88. Walter Cronkite, letter to the editor, *New York Times,* 16 May 2000, A30.
89. "The Last Battle of the Gulf War," *New York Times,* 18 May 2000, A30.
90. Ibid.

91. Taylor, *War and the Media*, 276, 278. This video censorship also facilitated the Pentagon's strategy of inflating perceptions about high-tech weapons accuracy. "Senior officers have said they viewed extensive footage of bombs that missed targets or hit targets selected in error, such as the civilian building across the street from the Iraqi Interior ministry," journalist Barton Gellman observes, "but the Pentagon has released none of the footage" (Barton Gellman, "U.S. Bombs Missed 70 Percent of the Time," in *The Media and the Gulf War: The Press and Democracy in Wartime*, ed. Hedrick Smith [Washington, D.C.: Seven Locks Press, 1992], 198).
92. Dana Cloud, "Operation Desert Comfort," in *Seeing Through the Media: The Persian Gulf War*, ed. Lauren Rabinowitz and Susan Jeffords (New Brunswick, N.J.: Rutgers University Press, 1994), 159–62.
93. Quoted in Robert Stein, letter to the editor, *International Security* 17 (1992): 215.
94. Quoted in Hildreth, "Evaluation," 74.
95. Quoted in Hildreth, "Evaluation," 75.
96. Baker Spring, "Congress Bickered over Weapons Now Proving Themselves in the Gulf," Heritage Foundation Backgrounder no. 808 (Washington, D.C.: Heritage Foundation, 1991), 1–2.
97. Lawrence Freedman, "The Revolution in Strategic Affairs," Adelphi Paper 318 (Oxford: Oxford University Press/International Institute for Strategic Studies, 1998), 32.
98. United States Department of Defense, *Conduct of the Persian Gulf War, Final Report to Congress* (Washington, D.C.: GPO, 1992), 164.
99. See Eric Schmitt, "Army Is Blaming Patriot's Computer for Failure to Stop the Dhahran Scud," in *The Media and the Gulf War: The Press and Democracy in Wartime*, ed. Hedrick Smith (Washington, D.C.: Seven Locks Press, 1992), 214; Dan Charles, "Critics Home in on Patriot Missile," *New Scientist*, 2 February 1991, 39.
100. Hildreth, "Evaluation," 23.
101. Albert Carnesdale, statement, *The Impact of the Persian Gulf War and the Decline of the Soviet Union on How the United States Does its Defense Business*, 102d Cong., 1st sess., House Hearing, Committee on Armed Services, 16 April 1991 (Washington, D.C.: GPO, 1991), 476.
102. Marvin Feuerwerger, "Defense Against Missiles: Patriot Lessons," *Orbis* 36 (1992): 581.
103. Carnesdale, statement, 476.
104. Gordon and Trainor, *General's War*, 468.
105. Patrick Buchanan, "Evidence of SDI's Potential," *Washington Times*, 30 January 1991, G2.
106. Wayne Biddle, "The Untold Story of the Patriot," *Discover*, June 1991, 74–79.
107. Barry D. Watts, "Effects and Effectiveness," in *Gulf War Air Power Survey, Vol. 2, Operations and Effects and Effectiveness*, ed. Eliot A. Cohen (Washington, D.C.: GPO, 1993), 363.
108. Campbell, *Politics Without Principle*, 1.
109. In one of these videos, the voice of the narrator is dubbed over a driving rock music beat. The video opens with the pronouncement that the Persian Gulf War was "the ultimate technology war. . . . Never before has the hardware of American battle enjoyed

such popularity. Never before has America so keenly romanced its own armaments" (*Technology of War*, prod. *Popular Mechanics*, New York: Goodtimes Home Video, 1994, videocassette.)

110. Theodore A. Postol, interview with the author, 10 May 1995, Cambridge, Mass. Postol and Lewis examined more than 140 videotapes, covering thirty-one Scud-Patriot engagements.
111. Postol, "Lessons," 145. The report cited by Postol, published in a Tel Aviv newspaper, stated that thirteen uncontested Scuds damaged 2,698 apartments and injured 115 people, while attacks involving eleven Scuds after Patriot was deployed damaged 7,778 apartments, wounded 168 and resulted in at least one death (see David Hughes, "Success of Patriot System Shapes Debate on Future Antimissile Weapons," *Aviation Week and Space Technology*, 22 April 1991, 90). Because of the lack of official ground damage reports, Postol notes that there is uncertainty built into the damage assessments published in Israeli newspapers. Thus, "while it is not possible to draw any detailed conclusions from this data, it is clear that at this time there is no publicly available evidence to support claims that ground-damage and injuries in Israel were lowered as a consequence of Patriot defense operations" (Postol, "Lessons," 145). See also Conyers, "Patriot Myth," 8–10.
112. Postol, "Lessons," 146.
113. Postol, "Lessons," 150. "All told, four Patriot missiles struck Israeli ground during the war. It turns out in those missiles the 'safe and arm' mechanism, which is responsible for self-destruct, was defective. The path of some of the Patriot missiles is clearly visible in ABC TV video recordings" (Reuven Pedatzur, statement, *Performance of the Patriot Missile in the Gulf War*, 102d Cong., 2d sess., House Hearing, Committee on Government Operations, 7 April 1992 [Washington, D.C.: GPO, 1992], 120). "The number of Patriots that fell on defended territory causing direct casualties and extra damage has not been accurately counted, but one witness saw four Patriots hit Tel Aviv in one night" (Alexander Cockburn, "Dumb Bombs," *New Statesman and Society*, 14 June 1991, 17).
114. This hypothesis is supported by the fact that "after only ten days of war, Israel reported that about 1,000 dwellings had been damaged in missile attacks and about four or five buildings would have to be demolished" (Dan Morgan and George Lardner Jr., "Scud Damage Suggests Patriot Needs Refinement: U.S. Missiles Sometimes Fail to Destroy Iraqi Warheads in Midair Interceptions," *Washington Post*, 21 February 1991, A27).
115. See David Hughes, "Tracking Software Error Likely Reason Patriot Battery Failed to Engage Scud," *Aviation Week and Space Technology*, 10 June 1991, 26; Eliot Marshall, "Fatal Error: How Patriot Overlooked a Scud," *Science*, 255 (13 March 1992) 1347; Theodore A. Postol, "Optical Evidence Indicating Patriot High Miss Rates During the Gulf War," *Performance of the Patriot Missile in the Gulf War*, 102d Cong., 2d sess., House Hearing, Committee on Government Operations, 7 April 1992 (Washington, D.C.: GPO, 1992), 142–43.
116. Postol, "Lessons," 154.
117. Ibid.

118. This was President Bush's phrase, used in his Raytheon address (see Bush, "Remarks to Raytheon," 148).
119. James Glanz, "Patriots Missed, but Critics Hit Home," *Science* 284, 16 April 1999, 413–16.
120. Postol, interview with the author.
121. Conyers, "Patriot Myth," 1.
122. As GAO investigator Richard Davis explained during 1992 congressional hearings on Patriot performance: "[T]his is the layman's definition or explanation. When the Patriot system engages an incoming missile, the Scud in this particular case, it doesn't run into it. It was a proximity-type explosion. When you saw the explosion on the television, CNN or whatever, you were probably seeing the Patriot exploding. And when the Patriot explodes, it displaces fragments, and these fragments are supposed to knock out the incoming missile" (Richard Davis, statement, *Performance of the Patriot Missile in the Gulf War,* 102d Cong., 2d sess., House Hearing, Committee on Government Operations, 7 April 1992 [Washington, D.C.: GPO, 1992], 98).
123. Robert M. Stein, "Patriot ATBM Experience in the Gulf War," letter to *International Security* subscribers, 8 January 1992. I thank Dan Reiter for making his copy of the letter available to me.
124. Ibid., 2. Further references to this letter will be made by page number in the text.
125. Ironically, Patriot had trouble engaging Scuds because Iraq's crude variant (Al-Husayn) of the Scud had a tendency to break apart into several pieces during re-entry. Still potentially lethal, these fragments were much more difficult for the Patriot radar system to track and target for interception.
126. Raytheon, Inc., "Raytheon's Response to WGBH Frontline 'Gulf War,'" News Release (1996), Internet, Frontline Website, online at http://www.pbs.org/wgbh/pages/frontline/gulf/weapons/raytheontext.
127. Stein, letter to the editor, 225.
128. Theodore A. Postol, "Reply to letter of Robert Stein," *International Security* 17 (1992): 225.
129. Raytheon, Inc., "Answers to Questions," appendix, *Performance of the Patriot Missile in the Gulf War,* 102d Cong., 2d sess., House Hearing, Committee on Government Operations, 7 April 1992 (Washington, D.C.: GPO, 1992), 321. Dave Harris, spokesperson for the Army Missile Command, confirmed Raytheon's account. "They came forward to us and said, hey, we think there's classified information in this. We said, yes, we agree. They went to DIS" (quoted in Tim Weiner, "Army Cracks Down on Patriot's Critic," *Chicago Tribune,* 20 March 1992, 16). By that time, however, Postol's article had already been published, and in the words of Harris, "the toothpaste was out of the tube" (quoted in Weiner, "Army Cracks Down," 16).
130. Postol, "Improper," 39. Apparently, DIS felt that Postol had run afoul of "compilation" classification rules, "when nominally unclassified data is assembled in such a way that the end result is more secret than the sum of its parts" (Anderson, "Classification," 274). Compilation regulation is "an all-too-common extension of an already aggressive classification policy," says Steven Aftergood, a security expert with the Federation of American Scientists (quoted in Anderson, "Classification," 274).

131. Postol, "Improper," 40.
132. When Postol pointed out to the DIS investigator that his article contained more than one hundred references which could be checked to determine whether they were drawn from open sources, "the investigator complained that this would be a lot of work for the DIS" (Postol, "Improper," 40).
133. Theodore A. Postol, statement, "Patriot Wasn't Precise, Scientist Shows," narr. Dan Charles, *All Things Considered,* National Public Radio, 19 March 1992, online, Lexis/Nexis (an electronic information database with over 28,000 full-text sources online at http://www.lexis-nexis.com/lncc).
134. Postol, "Improper," 39–40.
135. Quoted in John Aloysius Farrell, "U.S. Ordered a Stop to Patriot Criticism, MIT Professor Says," *Boston Globe,* 18 March 1992, 3.
136. Nina J. Stewart, letter to Hon. John Conyers, *Government Secrecy after the Cold War,* 102d Cong., 2d sess., House Hearing, Committee on Government Operations, 18 March 1992 (Washington, D.C.: GPO, 1992), 72–73.
137. Stewart, letter to Conyers, 72.
138. Quoted in John Aloysius Farrell, "MIT Physicist Cleared in Patriot Analysis," *Boston Globe,* 28 March 1992, 3.
139. The specifics of this investigation were not made public, and the investigation did not result in a termination of Postol's security clearance. See Postol, "Improper," 191.
140. Raytheon commissioned Shaoul Ezekiel of MIT to produce the video. Ezekiel argues in the video that "Postol's use of commercial video was not an accurate measure of high-speed encounters like those between Patriots and Scuds" (quoted in "Professor Revives MIT Missile Flap," *Boston Globe,* 26 February 1996, 15). Two years after the video was circulated to the news media, representatives of Raytheon wrote a letter to the *Boston Globe* that criticized Postol's work as coming from the "MTV school of weapons analysis," using a methodology "repudiated by Professor Shaoul Ezekiel of MIT" (quoted in "Professor Revives," 15). Citing an admission from Ezekiel that he had not read Postol's research, Postol asserted that Ezekiel had committed academic misconduct and demanded a correction. Because Postol was also a professor at MIT, the matter was taken up by a faculty review committee at the university. Although a three-person committee cleared Ezekiel of any ethical wrongdoing in a 1 February 1996 report, it did find that Raytheon had published an "exaggeration" in its letter to the *Boston Globe,* and it praised Postol and his work, finding "[t]he truth emerged as it should" ("Professor Revives," 15). According to Philip Hilts of the *New York Times,* this finding by the faculty review committee "did give a nod in Dr. Postol's direction by saying that he had won the debate about the Patriot's lack of effectiveness" (Philip J. Hilts, "Dispute on Gulf War Missile Continues to Fester at M.I.T," *New York Times,* 24 February 1996, 11).
141. Quoted in David Chandler, "No Letup in War of Words," *Boston Globe,* 4 April 1994, 27.
142. Ibid.
143. For general commentary on the difficulties involved in sorting out the competing claims in the post–Gulf War dispute over the Patriot's performance, see Harry M.

Collins and Trevor J. Pinch, *The Golem at Large: What You Should Know about Technology* (Cambridge: Cambridge University Press, 1998), 7–29.

144. "Representative Conyers Wants Cheney to Probe Army Claims Regarding Patriot's Success," *Inside the Pentagon*, 12 March 1992, 14.
145. Margo MacFarland, "Congressional Investigators Charge Army Data Doesn't Back Patriot Success," *Inside the Army*, 9 March 1992, 1.
146. Davis, statement, 78. See also United States General Accounting Office, *Operation Desert Storm: Evaluation of the Air War*, PEMD-96–10, July 1996 (Washington, D.C.: GPO, 1996); Conyers, "Patriot Myth."
147. See Davis, statement.
148. "The Chief Engineer said that Patriot's fuze can sense its target and detonate up to six times the required miss distance, resulting in an extremely low or no probability of kill. However, the system would still record a kill" (U.S. GAO, *Operation Desert Storm*, 7).
149. Davis, statement, 78.
150. As the GAO notes, "[f]or example, there were several instances in which Patriot operators reported destroying more Scud warheads than there were missiles launched" (Davis, statement, 78).
151. According to Hildreth, "the Army's assessment of Patriot effectiveness relies heavily on human after-action reports" ("Evaluation," 27).
152. Quoted in ibid., 31.
153. Ibid., 27.
154. Quoted in ibid., 27.
155. Stein, letter to the editor, 9.
156. Davis, statement, 78.
157. Ibid.
158. U.S. GAO, *Operation Desert Storm*, 4.
159. See Robert A. Drolet, "PEO Air Defense Responds to Patriot Criticisms," *Inside the Army*, 9 December 1991, 5.
160. Quoted in David C. Morrison, "Patriot Games," *National Journal*, 25 April 1992, 1028.
161. Quoted in Morrison, "Patriot Games," 1028; see also David F. Bond, "Army Scales Back Assessments of Patriot's Success in Gulf War," *Aviation Week and Space Technology*, 13 April 1992, 64.
162. As Lewis and Postol write, "Moshe Arens, Israel's minister of defense during the war, was videotaped in an interview with Pedatzur in September 1993 that was later reported in the *New York Times*. When Pedatzur asked Arens how many Scuds Patriot intercepted, the defense minister replied that the number was 'minuscule.' Pedatzur then read to Arens the conclusion from a still-classified Israeli report—'There is no evidence of even a single successful intercept of an Al-Hussein by Patriot. There is, however, circumstantial evidence for one possible intercept.' Arens responded, 'It sounds correct.' In another videotaped interview by Pedatzur, Gen. Dan Shomron, chief of staff of the Israeli Defense Force during the war, similarly stated, 'To the best of my recollection, only one Scud missile exploded in the air as a consequence of a

Patriot explosion'" (George N. Lewis and Theodore A. Postol, letter to the editor, *Christian Science Monitor,* 8 September 1997, 19).

163. Reuven Pedatzur, "Failed at Time of Test: The U.S. Administration and the Patriot's Manufacturer Presented a Deceptive Picture of the Performance of Patriot during the War," *Ha'aretz,* 24 October 1991, trans., Internet, Federation of American Scientists Website, online at http://www.fas.org/spp/starwars/docops/rp911024.htm.
164. See Chandler, "No Letup," 28.
165. "My own judgment," wrote Garwin in 1997, "largely based on a close reading of the analyses of Dr. T. A. Postol and his critics, is that few, if any, warheads were destroyed or disabled by Patriot interceptors" ("Theater Missile Defense," 184).
166. See Peter D. Zimmerman, "A Review of the Postol and Lewis Evaluation of the White Sands Missile Range Evaluation of the Suitability of TV Video Tapes to Evaluate Patriot Performance during the Gulf War," *Inside the Army,* 16 November 1992, 7–9; idem, statement, *Performance of the Patriot Missile in the Gulf War,* 102d Cong., 2d sess., House Hearing, Committee on Government Operations, 7 April 1992 (Washington, D.C.: GPO, 1992), 149–76.
167. Glanz, "Patriot Missed."
168. As John Pike observed recently, "as is well known, Patriot, rather than being virtually perfect, failed for precisely the reasons that critics of missile defense said that active defense was going to be leaky all along, mainly, it had a hard time dealing with the inadvertent countermeasures of tank fragments, and the software reliability proved to be a fatal flaw in the system. Unfortunately, there are an impressive array of institutional forces that are driving American policy in a direction of ignoring all of the true lessons that we should have learned from Operation Desert Storm" (John Pike, "Star Wars: Clever Politics in the Service of Bad Policy," *F.A.S. Public Interest Report* 49 [1996], Internet, Federation of American Scientists Website, online at http://www.fas.org/faspir/pir0996.html#starwars.)
169. See e.g. "Raytheon Awarded Deal to Sell Taiwan Patriot Antimissiles," *Wall Street Journal,* 1 July 1994, B4. On the day that Raytheon announced inking of the $600 million Patriot deal to Taiwan, the company's stock jumped 37.5 cents per share ("Raytheon Awarded," B2).
170. "Raytheon Awarded," B4. Italy, the United Arab Emirates, and the United Kingdom have also expressed interest in the Patriot since the Gulf War (David Hughes, "Saudi Order to Keep Patriot Line Open as Performance in Israel Is Questioned," *Aviation Week and Space Technology,* 25 November 1991, 38).
171. Rep. William Dickinson, statement, *The Impact of the Persian Gulf War and the Decline of the Soviet Union on How the United States Does its Defense Business,* 102d Cong., 1st sess., House Hearing, Committee on Armed Services, 16 April 1991 (Washington, D.C.: GPO, 1991), 418.
172. Rep. George Atkins, statement, *Performance of the Patriot Missile in the Gulf War,* 102d Cong., 2d sess., House Hearing, Committee on Government Operations, 7 April 1992 (Washington, D.C.: GPO, 1992), 214. The term *placebo* is "the first person singular of the future indicative of the Latin verb placere 'to please'" (Klaus Schonauer, *Semiotic*

Foundations of Drug Therapy: The Placebo Problem in a New Perspective [Berlin: Mouton de Gruyter, 1994]). In medical terms the placebo is defined as "a therapeutic procedure . . . which is given deliberately to have an effect, or . . . which unknowingly has an effect on a patient, symptom, disease, or syndrome, but which is objectively without specific activity for the condition being treated" (Arthur K. Shapiro, "A Historic and Heuristic Definition of the Placebo," *Psychiatry* 27 [1964]: 52–58). There can be a further distinction drawn between the placebo (i.e., "a biomedically inert substance given in such a manner to produce relief") and the placebo effect (i.e., "the resulting effect upon the patient") (Howard Brody, *Placebos and the Philosophy of Medicine: Clinical, Conceptual, and Ethical Issues* [Chicago: University of Chicago Press, 1980, 25–44]). In medical experimentation, placebos (in the form of inert pills) are given to control groups in order to maintain the integrity of double-blind drug trials. Clinically, physicians prescribe placebos to individual patients in cases where there appears to be no active drug available as an effective treatment (see Michael Jospe, *The Placebo Effect in Healing* [Lexington, Mass.: Heath Publishers, 1978]; Louis Lasagna, Frederick Mosteller, John von Felsinger, and Henry K. Beecher, "A Study of the Placebo Response," *American Journal of Medicine* 16 [1954]: 770–79; and Arthur K. Shapiro, "A Contribution to the History of the Placebo Effect," *Behavioral Science* 5 [1960]: 109).

173. Horton, statement, 5–8. See also Stein, "Patriot ATBM," 20–21.
174. Feuerwerger, "Defense Against Missiles," 584; See also David Chandler, "Sorting Out the Claims: A Q & A," *Boston Globe*, 4 April 1994, 25.
175. Postol, interview with the author.
176. Ibid.
177. Feuerwerger, "Defense Against Missiles," 584.
178. As Hildreth observes, "[t]he Army's claims of Patriot effectiveness in Desert Storm are classified, as is most of the material supporting its claims. This complicates debate over details" ("Evaluation," 25). "[T]he U.S. Army has thus far not declassified the best data in its possession, including its shot-by-shot assessments and the correlations of those assessments with damage reports. It would be extremely interesting to compare the reports of damage on a nightly basis to see whether more damage was produced on nights when the Army claimed Scuds were not intercepted than on those nights when it claimed success" (Feuerwerger, "Defense Against Missiles," 587).
179. As Conyers reflects, "[w]e now know that the public and the Congress were misled by definitive statements of success issued by administration officials and representatives of Raytheon, the Patriot's prime contractor, during and after the war. The high office of the individuals who made such statements, including President George Bush and General Norman Schwartzkopf, commander of the Desert Storm forces, added to these claims' credibility and acceptance" (John Conyers Jr., Response to letter of Frank Horton, *Arms Control Today* 23 [1993]: 29).
180. As Postol recounts, "[a] series of events following my April 1991 testimony and publication of my article in the Winter issue of *International Security* suggest that inappropriate attempts have been made to silence the debate over Patriot's Gulf War

performance. Among these events was an investigation by Defense Investigative Services (DIS) following an effort by the Army to classify my Winter 1991/92 article after its publication" ("Reply," 240). See also Tim Weiner, "Army Squelches Article—After It's Published," *Philadelphia Inquirer,* 19 March 1992, 1.

181. The Army and Raytheon spurned the proposal by the American Physical Society to conduct an objective review (see Chandler, "Sorting Out," 27).
182. As Representative John Conyers Jr., chairperson of the House Committee on Government Operations, recounted efforts of Raytheon and GOP missile defense advocates to impede publication of the committee's final report evaluating Patriot's Gulf War effectiveness, "Mr. Horton sought, in his own words, to 'bury' a Committee report on the performance of the Patriot missile. The Raytheon Company mounted a ferocious lobbying campaign against the report in the last few days of the 102nd Congress" (Conyers, "Response to letter," 29).
183. Hildreth, "Evaluation," 42.
184. Conyers, "Response to letter," 9.
185. Rep. Stephen Neal, statement, *Performance of the Patriot Missile in the Gulf War,* 102d Cong., 2d sess., House Hearing, Committee on Government Operations, 7 April 1992 (Washington, D.C.: GPO, 1992), 96.
186. Postol, interview with the author. Some contended that Postol's evidence was superior merely on the grounds that it was not classified. Daniel Fisher of Harvard University, member of an American Physical Society review panel which offered to provide an objective assessment of the controversy over the Patriot's Gulf War accuracy, argues that since Postol's evidence was subjected to peer review, it should be given more epistemic weight than Raytheon/army evidence, which has not been subject to peer review because of its classified status (see Chandler, "No Letup," 28).
187. Postol, "Improper," 47.
188. Quoted in Weiner, "Army Squelches," 16.
189. See, e.g., Horton, statement, 5–8.
190. Quoted in Postol, "Improper," 56.
191. As Postol corrected the GAO in a letter to Comptroller General Bowsher, "[t]he GAO review states that 'Professor Postol states that during the war the Patriot was not as effective as claimed by the U.S. Army, did not destroy any of the Scud missile's warheads, and did not reduce the damage in the areas of Israel it was sent to protect. . . .' All three of the above statements ascribed to me by GAO are incorrect. My position has been that *there is no evidence that indicates* that the Army's claims of Patriot effectiveness are likely to be correct and that the publicly available evidence strongly suggests that Patriot's performance was poor, that *we have been unable to find any evidence* in our study of press video taken during Desert Storm that even one Scud missile warhead was destroyed by the action of Patriot, and that *there is no evidence* that the action of Patriot either reduced the number of casualties or damage per Scud attack in Israel" ("Optical Evidence," 135–44, emphasis in original).
192. Stewart, letter to Conyers, 72.
193. Postol, "Improper," 46.

194. These analyses included issues relating to Soviet ABM capabilities, the Closely Spaced Based MX, U.S. SDI, the TRIDENT weapon system, and nuclear war planning (Postol, "Improper," 44).
195. Ibid.
196. Garwin, "Theater Missile Defense," 187.
197. In the euphoric aftermath of the war, military hawks and defense contractors recognized "the opportunity to revive the political capital of a cold war arsenal" and in the process reverse the tide of receding defense appropriations that had taken place since 1989 (Gregory McLauchlan and Gregory Hooks, "Last of the Dinosaurs? Big Weapons, Big Science, and the American State from Hiroshima to the End of the Cold War," *Sociological Quarterly* 36 [1995]: 749–76). The experience of the Gulf War provided an occasion for advocates of military spending to argue that weapons designed for fighting the cold war could be useful when deployed against new post–cold war adversaries: "By demonstrating the awesome power of conventional high-tech weapons that had been largely deployed only in an 'imaginary war' with the Soviet Union, it could be shown that these would also be strategically critical for use against new enemies" (McLauchlan and Hooks, "Last of the Dinosaurs?"). In a parade of post–Gulf War congressional hearings, Pentagon partisans parlayed the widely perceived success of Operation Desert Storm into massive increases in defense spending that ratcheted up the military budget to near cold war levels. Advocates relied explicitly on apparent Gulf War successes to justify huge spending increases in high-tech weaponry. Based on the widespread accounts of technological triumph in Operation Desert Storm, "subsequent Pentagon procurement budgets invested heavily in these extremely costly weapons systems, with nearly $60 billion allocated to 'smart bombs' alone. . . . Procurement costs for weapons systems loaded with dazzling experimental technologies of untested reliability are a major factor in keeping defense budgets close to cold-war levels in a world where no other global superpower threatens American security in the way the Soviet Union used to" ("Get Smarter on Smart Weapons," editorial, *New York Times,* 11 July 1996, A22).
198. McLauchlan and Hooks, "Last of the Dinosaurs?" 766–67.
199. The perceived effectiveness of the Patriot's Gulf War performance led to a groundswell of budget support for a variety of ground-based kinetic-kill missile defense systems such as Extended Range Intercept Technology (ERINT), THAAD, Arrow, Ground Based Interceptor (GBI), and Endoatmospheric/Exoatmospheric Interceptor (E2I) (Vincent Kiernan, "Beyond the Patriot," *Discover,* June 1991, 80–82). "Practically every new antitactical missile program from Arrow to Extended Range Intercept Technology (ERINT) envisions Patriot as part of the overall architecture in terms of ground-based sensors and fire control systems" (Hughes, "Success," 90).
200. As the GAO noted in its study of the effectiveness of so-called smart weapons in the Persian Gulf War, "The performance of aircraft and their munitions, cruise missiles, and other air campaign systems in Desert Storm continues to be relevant today as the basis for significant procurement and force sizing decisions. For example, the Department of Defense (DOD) Report on the Bottom-Up Review (BUR) explicitly

cited the effectiveness of advanced weapons used in Desert Storm—including laser-guided bombs (LGBs) and stealth aircraft—as shaping the BUR recommendations on weapons procurement" (U.S. GAO, *Operation Desert Storm,* 1). With high-tech weapons driving procurement decisions, the U.S. military budget has approached cold war levels in recent years. By signaling an intention to support significant increases in missile defense appropriations in 1991, Congress triggered a competition among the service branches that carried echoes of the Wizard-Nike BMD interservice rivalry of the 1950s. As journalist Bill Gertz noted in the wake of the Gulf War, "[t]he debate over Star Wars has been replaced by policy disputes . . . among the military services over which programs will be funded and who will get a bigger share of congressional appropriations" (Bill Gertz, "Son of Star Wars," *Air Force Magazine,* September 1994, 52; see also Lauren Spain, "The Dream of Missile Defense. So Where's the Peace Dividend?" *Bulletin of the Atomic Scientists* 51 [1995]: 49–53).

After the Gulf War, the army, navy, and air force each advanced bold proposals for their pet missile defense programs. The army backed upgrades in the Patriot system and development of THAAD. Proposing that missile defense could be best conducted from mobile ships, the navy urged funding for their sea-based Lower and Upper Tier programs. Convinced that airborne interceptor platforms offer the best prospects for effective missile defense that can cut off incoming enemy missiles early in flight, the air force recommended the Air-Borne Laser to Congress as the superior missile defense option. Gronlund, Lewis, Postol, and Wright offer an excellent comparative analysis of these competing systems (Lisbeth Gronlund, George N. Lewis, Theodore Postol, and David Wright, "Highly Capable Theater Missile Defenses and the ABM Treaty," *Arms Control Today* 24 [1994]: 3–8). Other informative descriptions can be found in Gertz, "Son of Star Wars"; Spain, "Dream of Missile Defense"; and United States General Accounting Office, *Ballistic Missile Defense: Evolution and Current Issues,* NSIAD-93–229, July 1993 (Washington, D.C.: GPO, 1993).

201. In addition to calling the Patriot's Gulf War accuracy into doubt, in 1996 the GAO filed a report that questioned the performance of virtually all of the highly touted "smart weapons" deployed during the Gulf War. GAO investigators found that "smart bombs," which had been credited widely as responsible for the wildly succesful air war, were not only used infrequently (less than one quarter of the tonnage delivered by air), but, when used, were often rendered ineffective by clouds, rain, fog, smoke, and high humidity. In the final assessment, only 8 percent of the delivered munitions tonnage was guided, but such weapons represented 84 percent of the total munitions cost. As the GAO researchers explained in the summary of their report, "Desert Storm demonstrated that many systems incorporating complex or advanced technologies require specific operating conditions to operate effectively. These conditions, however, were not consistently encountered in Desert Storm and cannot be assumed in future contingencies. Lastly, many of DOD's and manufacturers' postwar claims about weapon system performance—particularly the F-117, TLAM, and laser-guided bombs—were overstated, misleading, or inconsistent with the best data . . ." (U.S. GAO, *Operation Desert Storm,* 4–5). Closer postwar inspection of the Gulf War record also led a

Pentagon research team to doubt that Operation Desert Storm sorties destroyed any of Iraq's mobile Scud missile launchers or degraded Iraq's chemical or nuclear armaments (see Bruce B. Auster, "Sobering Lessons from the Gulf," *U.S. News & World Report,* 10 May 1993, 52).

202. See Susan Feeney, "Timing Poses Questions about President's Motivation: Parallels to Movie 'Wag the Dog' Spur Talk," *Dallas Morning News,* 21 August 1998, A1.
203. See Feeney, "Timing Poses Questions."
204. Quoted in Fleeman, "Is Life Imitating," 208.
205. Bob Fenster, "Life Imitates Art: Who's Playing Wag the Dog?" *Arizona Republic,* 21 August 1998, B7.
206. Michael Fleeman, "Is Life Imitating Movie, Again? 'Wag the Dog' Back in Spotlight," *AP Online,* 21 August 1998, Internet, Elibrary Website, online at http://www.elibrary.com.
207. Quoted in Fleeman, "Is Life Imitating," 205.

U. S. President Richard Nixon (seated, left) signs the ABM Treaty with a Soviet arms control negotiator in Moscow on 26 May 1972. Reprinted permission of Corbis photographs.

CHAPTER FOUR

Whose Shoe Fits Best? Dubious Physics and Power Politics in the TMD Footprint Controversy

Riding a cresting wave of pro-missile defense sentiment that was swelled by the Patriot's apparent Gulf War wizardry, Congress passed the 1991 Missile Defense Act. This legislation declared it an official goal of the U.S. to "provide highly effective theater missile defenses," and mandated that the "Secretary of Defense shall aggressively pursue development of advanced theater missile defense systems, with the objective of downselecting and deploying such systems by the mid-1990s."[1] While this package of new theater missile defense initiatives included plans for an upgrade of the Patriot (called PAC-3, short for Patriot Advanced Capability Level-3), there were also provisions made for research and development of a new missile defense system called THAAD (Theater High Altitude Area Defense).[2] On paper, THAAD was billed as a system that could defend limited areas from medium- or short-range missiles even more effectively than the Patriot. With faster interceptors and more powerful radar, THAAD was designed to minimize "leakage" by allowing the defense multiple chances to intercept incoming missiles in "shoot-look-shoot engagement opportunities."[3]

Soon after the announcement came that the Clinton administration planned to pursue THAAD aggressively, concerns were raised that the new system was illegal. Critics argued that THAAD's high-end design exceeded performance limitations set down in the 1972 Antiballistic Missile (ABM) Treaty. On the surface, such claims strained credulity, given that the ABM Treaty clearly does not cover theater missile defense (TMD) systems such as the Patriot (it prohibits only widespread testing and deployment of strategic missile defense systems designed to counter long-range rockets fired across whole continents).[4] Like the Patriot, THAAD was designed to intercept short- and medium-range ballistic missiles such as the Scuds used by Iraq in the 1991 Gulf War. However, apparent post–Gulf War engineering advances endowed THAAD with potential capability to intercept long-range strategic missiles as well as short-range Scuds, blurring the boundary between theater and strategic defenses codified in the ABM Treaty.

Attempting to find a way to pursue THAAD without undercutting the ABM Treaty, Clinton administration officials defended THAAD in the early 1990s on the grounds that although it would be highly effective against short-range missiles, it was still legal under the ABM Treaty because the system could not stop a long-range rocket attack from Russia or China, and therefore would not have "strategic" capability. This line of argument met substantial opposition in op-ed pieces, letters to the editor, congressional testimony, and arms control journals, where commentators questioned the administration's scientific calculations and warned of the grave dangers involved in eroding the limits on strategic defenses codified in the ABM Treaty.

Since 1993, this argument over the "demarcation" line separating legal theater missile defense systems from illegal strategic missile defense systems has spurred spirited and contentious debates in Congress and other public spheres. In each of these contexts, TMD demarcation discussions have been arduous as interlocutors have struggled to find common ground at the interface between complex questions of physics[5] and sensitive aspects of international politics.[6] The dispute over THAAD's ABM Treaty legality straddles this interface between physics and politics, and raises significant questions regarding the connection between scientific proof and public validation of scientific knowledge.

This chapter features an exploration of this controversy as it unfolded in the mid-1990s. Since the controversy played out in multiple discourses ranging from technical peer reviews to newspaper editorials, the multitiered critical approach I have used to investigate other missile defense controversies in this book is ideal for this case. Specifically, I focus on the interplay between public argument and scientific practice to discern how arguments drew from science, how science was shaped by arguments, and how these insights related to the substantial political, military, and social stakes in play. In a study that foregrounds the connections between eras and episodes of missile defense advocacy, there is much to be gained here, since commentators pointed to the mid-1990 TMD demarcation debate as a pivotal missile defense controversy that promised to shape future events powerfully. "All seem to agree," wrote Congressional Research Service analyst Stephen Hildreth in 1994, "that the outcome of the current debate will shape the future of U.S. TMD programs and the ABM Treaty, and hence, U.S. national security interests."[7]

Hildreth's hypothesis has interesting ramifications for the future of strategic deception in missile defense advocacy, in light of the multiple allegations levied against Ballistic Missile Defense Organization (BMDO) that it employed novel forms of strategic deception to gain the upper hand in mid-1990 TMD demarcation debates. For example, where Reagan-era Star Wars scientists inflated technical assessments to bolster the political case for missile defense, scientists in Clinton's THAAD program have been accused of deliberately *deflating* performance evaluations to portray the system as legal under the ABM Treaty. On another level, while SDI managers clamped down on scientific dissent in an attempt to stamp out political controversy in the 1980s, critics have claimed that THAAD managers *stimulated* scientific disagreement for political purposes in the 1990s. The THAAD case presents a ripe opportunity to explore strategic deception in missile defense advocacy as a protean phenomenon. Such exploration is likely to yield important background for understanding present-day missile defense discussions, since contemporary National Missile Defense (NMD) projects trace their lineage directly from the THAAD program.

This chapter explores the mid-1990s ABM Treaty demarcation controversy by combing through multiple sources and levels of public and scientific communication on the topic. Part one provides background on the technical arms control concepts and terminology that will help elucidate the political claims advanced by Clinton administration missile defense officials early in the dispute, as well as the process through which dissent arose in opposition to those claims as the issue evolved into a public controversy. Part two refocuses on strategic deception, as attention shifts to consideration of a particular dispute regarding TMD computer "footprint" methodology used to calculate THAAD's ABM Treaty compliance. Charges of strategic deception, subterfuge, and intimidation pervade the texts of this dispute, and the significant political stakes involved make this "footprint" debate a particularly appropriate case study for this book. With the basic outline of the arguments in the footprint controversy in place, part three moves on to study the texts that purport to judge the competing claims. IBM fellow Richard Garwin (a renowned physicist who was asked by BMDO to conduct a scientific review) has discovered evidence that corroborates critics' public charges of strategic deception on the part of a government-paid contractor in this case, so it is appropriate to consider the possible political ramifications of such findings. In this vein, part four features an analysis that places the TMD "footprint" controversy in the wider political context of ABM

Treaty demarcation negotiations and U.S.-Russian relations. Part five looks at recent developments in the late-1990s THAAD program in order to shed light on connections between the mid-1990s ABM Treaty demarcation disputes and current NMD debates. Such commentary is appropriate especially since NMD systems hoped for under the 1999 National Missile Defense Act are based directly on THAAD's "hit to kill" architecture.

Post–Cold War Missile Defense and the ABM Treaty

During a lull in the cold war, U.S. president Richard Nixon and Soviet secretary general Leonid Brezhnev signed the SALT I agreements. These accords, reached in Moscow on 26 May 1972, were designed to improve superpower nuclear stability in the midst of an escalating and increasingly costly mutual buildup of offensive nuclear weapons. Part two of the agreements, "Limitation of Anti-Ballistic Missile Systems," set down constraints on the development, testing, and deployment of defensive systems capable of intercepting strategic ballistic missiles. These limits, which came to be known as the "ABM Treaty," served an important arms control purpose, because superpower reductions in offensive nuclear forces were said to be possible only if each side could be assured that the other would not develop robust missile defense capability. "The ABM Treaty," explains Arms Control Association analyst Matthew Bunn, "fostered a predictable strategic balance by moderating fears of widespread missile defenses, allowing each side to plan its strategic forces with the knowledge that the other side could not put even a limited nationwide missile defense in place for at least several years."[8]

Rather than imposing a flat ban on all missile defense programs, the ABM Treaty allowed scientific research to continue and authorized deployment of limited defense systems. In time, debates over the exact meaning of these loopholes proved to be highly contentious, as disagreements about the legality of particular missile defense programs flared up perennially in arms control discussions. For example, in the 1980s, Reagan officials argued that even though their Star Wars research program was designed to yield eventually a nationwide "Astrodome" shield that could neutralize the Soviet Union's entire offensive missile force, such a system was still legal under a "broad interpretation" of the ABM Treaty. This strategy drew a flurry of protests from arms control advocates in both the United States and the Soviet Union, who argued that such a liberal translation of the ABM Treaty's legal loopholes constituted a unilateral "breakout" of treaty limits that threatened to undermine the basic foundation of superpower arms control as established in the SALT agreements.

The dispute over Reagan's "broad interpretation" of the ABM Treaty can be seen as a precursor to contemporary debates regarding the THAAD system's legality under the treaty. Where Reagan officials argued that Star Wars was treaty-compliant because it involved simply research, not deployment,[9] Clinton officials contended that as a "theater" system designed only to intercept short and medium-range missiles, THAAD was legal under the ABM Treaty loophole that allows deployment of such limited systems.[10] This position was advanced by John D. Holum, (President Clinton's arms control chief),[11] and by BMDO director Malcolm O'Neill during testimony given to Congress. In hearings held by the House Armed Services Committee on 10 June 1993, O'Neill outlined four technical factors that made THAAD's ability to intercept strategic missiles "doubtful."[12] In testimony before the Senate Foreign Relations Committee on 3 May 1994 O'Neill followed up on this claim by asserting definitively that THAAD fit within the legal parameters of the ABM Treaty.

> The Department is planning to develop and deploy theater/tactical missile defense systems to counter the projected threat to our forces abroad and to our allies. This mission is well within the purposes and objectives of the ABM Treaty. . . . Meanwhile, all of BMDO's testing and development activities remain compliant, *within the narrow interpretation* of the ABM Treaty.[13]

Just as the Reagan administration's official campaign to promote Star Wars as a legal ABM system triggered heated public controversy in the mid-1980s, O'Neill's claim of THAAD's ABM Treaty compliance met with substantial public opposition in the mid-1990s. In op-ed pieces, journal articles, congressional testimony, and letters to the editor, commentators expressed skepticism about the seriousness of the Clinton administration's commitment to the ABM Treaty, given THAAD's theoretical potential as a "strategic" missile defense system.

For example, David Wright, a senior scientist with the Union of Concerned Scientists, was quoted in the *Washington Post* as arguing that if the Clinton administration was allowed to interpret the ABM Treaty in such a way as to bring THAAD within its legal parameters, "this would effectively eliminate the ABM as a way to restrict strategic defenses."[14] In congressional hearings, John Rhinelander, one of the original negotiators of the ABM Treaty, testified that with the Clinton administration's strategy, "you really blow a very large hole in the ABM Treaty."[15] A *San Francisco Chronicle* editorial stated that "[k]nowledgeable critics" have cast doubt on THAAD's ABM Treaty legality by claiming that the system "would have substantial capability to shoot down strategic missiles."[16]

MIT physicists George Lewis and Theodore Postol were among the critics who challenged the Clinton administration's position on THAAD most roundly in public. In a letter published in the *Washington Post,* Postol wrote that the administration's strategy would "essentially eliminate [the ABM Treaty] as a means of limiting deployments of strategic-capable defensive systems."[17] In congressional testimony, Lewis argued that THAAD's theoretical capability (its "footprint") to defend against missile attack matched that of a strategic missile defense system, and therefore was illegal under ABM Treaty parameters. "This is clearly the footprint of a strategic defense system. This footprint is almost exactly the same size as the footprints for the Safeguard/Sentinel system that the ABM Treaty foreclosed," Lewis said.[18]

The Gronlund et al. Footprint Study

In April 1994 a group of physicists that included Postol and Lewis (as well as Lisbeth Gronlund and David Wright) published an article in *Arms Control Today* that utilized computer footprint methodology to support the contention that the planned THAAD system would possess dual capability to intercept both theater and strategic ballistic missiles.[19] A theoretical estimate of the capability of ballistic missile defense systems, computer-generated footprints consist of mapped areas demarcating the ground area that defensive systems can reasonably be expected to protect from incoming rocket attack (see Fig. 9) It will be useful to pursue a more nuanced technical understanding of Gronlund et al.'s "footprint" analysis in order to provide depth and texture to the broader political dispute over ABM Treaty compliance, since the Gronlund et al. analysis constituted the scientific basis of many public claims of skeptics who questioned the viability of the Clinton administration's position. More detailed knowledge of the technical subtleties in the Gronlund et al. footprint study will also pave the way for consideration of the controversy that subsequently arose when BMDO director Malcolm O'Neill ordered an official counterstudy to challenge the arguments advanced by Gronlund et al. in their *Arms Control Today* article.

There are a number of system parameters and situational factors that need to be taken into account in order to generate a reliable footprint. The radar detection range of the defensive system is an important factor, since longer detection ranges give the tracking and guidance components of a BMD more time to plan and execute intercepts. The speed, re-entry angle, and aerodynamic characteristics of incoming missiles are variables that help determine the vulnerability of target missiles and provide important information about the probability that an interceptor can neutralize its target. Further, the performance characteristics of the interceptor missiles themselves need to be factored into footprint calculations, as interceptor lethality, speed, and range play a prominent role in determining the expected proficiency of a defensive system in fending off enemy missile attacks.

Computers can be programmed to incorporate these factors into overall engagement models that generate footprints for particular defensive systems deployed in certain modes and locations. This computer war-gaming bears a resemblance to cold war analyses that modeled the casualty levels of nuclear exchanges, except that where much of the cold war computer modeling was concerned with predicting the likely levels of carnage arising from nuclear detonations over cities, computer footprint modeling assesses the degree to which such devastation could be prevented through active ballistic missile defense.

Utilizing computer footprint methodology, Gronlund et al. assessed the capability of the THAAD system to defend U.S. assets in a variety of threat scenarios. The footprints generated by their study suggested that once fully operational and effective, THAAD could be expected to defend considerable ground area against enemy missiles fired in a "theater" context (in the three thousand kilometer range). Based on this assessment, it would be reasonable to expect a deployed THAAD system to provide missile defense protection for forwardly deployed U.S. troops in foreign theaters of operation such as Europe or the Middle East.

However, Gronlund et al.'s analysis went further, to demonstrate that using footprint methodology it could be shown that the THAAD system would also have the capability to defend against much longer (ten thousand kilometer range) intercontinental ballistic missiles. This finding cast doubt on the classification of THAAD as a pure theater system. With capability to defend against much more distant threats, Gronlund et al. contended, THAAD could be expected to have "strategic" capability as well—that is, it could reasonably be counted on to provide some measure of protection to the

U.S. homeland against long-range ICBMs launched from Russia or China. Such a conclusion directly challenged the Clinton administration's pursuit of a TMD demarcation threshold to bring THAAD legally within the parameters of the ABM Treaty.[20] By suggesting that the THAAD system exceeded the limitations on acceptable missile defense capability set down in the treaty, Gronlund et al. complicated the administration's demarcation negotiation strategy and presented a perplexing rhetorical quandary to BMDO planners: how does one paint a system that will "walk like a duck and quack like a duck, and will eventually be a duck," as a duckling?[21]

Finding the Right Fit: The Debate over THAAD's Footprint Size

Gronlund et al.'s footprints came to the attention of BMDO director Lieutenant General Malcolm O'Neill during a 1994 briefing by government-paid analyst Keith Payne.[22] Interested in the differences he observed when comparing Gronlund et al.'s footprints with the footprints generated by Payne and other official BMDO analysts, Director O'Neill ordered further research into the footprint discrepancy. O'Neill's interest in further research on this matter might have been piqued by the fact that Gronlund et al.'s findings directly contradicted O'Neill's own claims regarding the politically sensitive issue of THAAD's ABM Treaty legality which he made in 1993 and 1994 congressional testimony.

Since the official task order for such research has not been released by BMDO, it is difficult to ascertain the precise research objectives laid down in this order. However, internal memos and interviews indicate that BMDO directly commissioned Sparta, Inc., a Rosslyn, Virginia-based think tank, to conduct the research.[23] As recipient of more than $10 million yearly in DOD contracts for largely secret research on ballistic missile defense,[24] Sparta has grown into one of the Ballistic Missile Defense Organization's de facto private intelligence agencies, a so-called SDI boutique.[25] To find the Rosslyn, Virginia, offices of Sparta, Inc., one needs to know more than the proper street address, 1911 Fort Myer Drive. Since there is no listing for Sparta on the marquee directory in the lobby of the building at this address, first-time visitors must call up from the public pay phone to discover on which floor to exit the elevator. Once inside Sparta's eleventh floor offices, visitors are registered in the receptionist's computer database and issued a black-and-white identification tag for the course of their stay. Visitors lacking security clearance are fenced away from "hot areas" where classified research and discussions take place.

Conspicuous signs reminding employees to "refrain from leaving classified information on desks" signal that Sparta is not only concerned with producing knowledge, but deeply entrenched in the business of hiding it as well.

Part of a cloistered network of private think tanks, government laboratories, and test ranges which make up the official "BMD community,"[26] Sparta works closely with top officials from BMDO ("bimdo," in Sparta parlance) to analyze key issues of systems architecture for BMD programs. For example, Sparta participated in a workshop conducted by BMDO's Data Validation Group (DVG) to develop official computer modeling methodology for measuring the theoretical capability of missile defense systems.[27] Sparta has also been commissioned to perform design studies for THAAD.[28] "We've grown as SDI has grown," said Jerry Kinney, the firm's vice president.[29]

The language used in Sparta's study on the Gronlund et al. matter intimates that the objective of the research, as ordered by BMDO, was to "review" Gronlund et al.'s *Arms Control Today* article and "understand why the conclusions they reached were at odds with the BMD community."[30] As Laura Lee, Sparta's lead researcher on the project, later recounted, "they [BMDO] gave us the [Gronlund et al.] article and said, 'Could you please explain to the General, so he understands, why you get two different footprints?'"[31] As Lee further clarified the approach of the study, "[w]e first tried to reproduce it [Gronlund et al.'s footprints] to understand their assumptions, and then explain to BMDO why we don't make those same assumptions."[32]

The Secret Lee et al. Counterstudy

To deepen earlier discussion of the technical nuances in the Gronlund et al. footprint study, it will be helpful to go into a more detailed investigation of the methodology and results of the Sparta counterstudy, because later charges of strategic deception on the part of missile defense critics relate directly to several fine points in Lee et al.'s work. The title of the Sparta study, "The Abuse of 'Footprints' for Theater Missile Defense and the ABM Treaty," signals immediately the tone of its conclusion: Gronlund et al. had "abused" footprint methodology in their *Arms Control Today* article. Completed in September 1994, the Sparta study alleged that Gronlund et al. had committed a number of serious technical errors in their research.

Before laying out these criticisms, the Sparta report successfully reproduced the Gronlund et al. footprint for a THAAD-like TMD deployed to defend in a tactical mode (i.e., against an incoming missile fired from less than three thousand kilometers). Mapping a footprint onto the Washington, D.C., area, the TMD's hypothetical defended area against an incoming tactical missile extended north up to the Pennsylvania border and south down to central Virginia. Sparta's confirmation of this theater footprint constituted theoretical corroboration of the idea that when engineered according to specifications, the THAAD system could be expected to perform its assigned mission of providing effective defense against short- and intermediate-range missile threats.

Next, the Sparta study attempted to reproduce Gronlund et al.'s second footprint, the one designed to demonstrate THAAD's theoretical capability against long-range (over ten thousand kilometers) intercontinental ballistic missiles (capability which would very likely run afoul of the legal limits prescribed under the ABM Treaty). Initially, the Sparta analysts noted that in order for a defensive system to have a chance against such long-range threats, it would need to track and fly to the incoming missile within a minimum "battlespace" time. In the case of TMD versus strategic ICBM, a battlespace time of twenty-two seconds was derived by taking into account ICBM velocity, detection time, and minimum intercept altitude.[33]

Using a radar detection range commonly associated with the proposed THAAD system (202 km for a 0.005 m^2 target), the Sparta study then contended that THAAD's long-range detection capabilities would require 30.6 seconds of battlespace time.[34] Because of this battlespace time discrepancy (30.6 versus 22 seconds), Lee et al. concluded that it was not possible for THAAD to generate a footprint when engaged in a strategic context. Unlike the first (theater) footprint, Lee et al. found that Gronlund et al.'s second (strategic) footprint could not be reconstructed (see Fig. 10). In concrete terms, the implication was that before the THAAD system could detect, track, and fly to a long-range ICBM, the incoming missile would be on its way down and past the last possible intercept point.

Lee et al. concluded that the only way such a footprint could possibly be generated would be if researchers radically simplified the modeling assumptions utilized in footprint methodology to increase the battlespace time available. For example, by ignoring atmosphere and gravity forces during the interceptor burn phase, researchers could have shaved required battlespace time to 28.8 seconds.[35] Further, by assuming infinite acceleration of the interceptor (when only average acceleration would be realistically possible), battlespace time could be cut to 20.4 seconds.[36] With these modeling advantages afforded to the defense, theoretically the THAAD system could lock on and meet an incoming ICBM in less than the battlespace time required (20.4 versus 22 seconds), yielding a defensive footprint.

The Sparta researchers reasoned deductively that the only way Gronlund et al. could have generated their strategic footprint would have been to adopt these "very, very simple" assumptions in their model.[37] This was quite a dramatic charge; the Sparta analysts alleged that Gronlund et al. had manufactured a footprint by conveniently ignoring gravity, atmospheric drag, and rocket deceleration. The weight of Sparta's charge was embedded in the argument's enthymeme: no legitimate physicist could bracket these obvious factors and still claim to offer serious analysis. Lee et al. claimed to have found a specific point where Gronlund et al. had "abused" the footprint process and illegitimately invoked the authority of science to bolster a political argument.

Sparta's second argument was that Gronlund et al. had ignored significant aspects of ballistic missile defense in the strategic context. Unlike the theater context, where the limited range of incoming missiles simplifies radar tracking, Lee et al. contended that in the strategic context such factors are less controllable by the defense, with the resultant uncertainty degrading the defense's capability significantly.[38] Specifically, Lee et al. suggested that missile defense in a strategic context is complicated by the fact that missiles may be incoming from a wide range of angles across the Earth's surface (threat azimuth) and the fact that missiles could also come in on any one of a large number of re-entry angles. Because Gronlund et al. did not sufficiently take these factors into account, Lee et al. contended that their strategic footprints were invalid. This finding of the Sparta study was amplified to influential audiences in the arms control community shortly after completion of the study in September 1994.

Information obtained under the Freedom of Information Act indicates that in March and April of 1995, BMDO funded and sponsored major official briefings in which Sparta researchers were permitted to present the findings of their "review" of Gronlund et al. without rebuttal or opposition. The first of these briefings took place at the National Defense University on 6 March 1995 and involved senior members of the Russian State Duma.[39] Since a major focus of this meeting was the issue of ABM Treaty demarcation,[40] the findings of the Sparta study very likely addressed some of the most central concerns of the Russian officials regarding U.S. TMD policy. In this "Missile Defense Discussion," organized by the National Institute for Public Policy for BMDO and conducted at Fort Lesley J. McNair, Sparta researchers were presented with the opportunity to outline the deficiencies they had found in the Gronlund et al. *Arms Control Today* article.[41]

In this briefing, official BMDO resources were utilized to create a forum where Sparta researchers could amplify their views to high-ranking foreign diplomats. The decision to hold such a briefing and to exclude Gronlund et al. signaled BMDO's judgment that the methodology of the Sparta study was sufficiently sound to warrant official amplification. Official support for these briefings also indicated that BMDO believed that the findings of the Sparta study were relevant to the ongoing Russia-U.S. arms control process.

BMDO sponsored a second major briefing designed to showcase the Sparta study on 3 April 1995 at the State Department in Washington, D.C. This briefing was conducted for a "big, large group" of high-ranking Arms Control and Disarmament Agency (ACDA) personnel.[42] Charged with carrying out U.S. arms control negotiations (including the ongoing ABM Treaty talks carried out through the Standing Consultative Committee), ACDA officials often engaged the issue of THAAD's legality under the ABM Treaty, thus the TMD footprint controversy was deemed a "very important" issue to the group.[43] One senior official estimated that representatives from as many as eight different ACDA groups attended the briefing.[44]

The Footprint Controversy Ignites

During much of the internal briefing process conducted in late 1994, the Sparta study was not cleared for public release, and therefore it was impossible for authors of the Gronlund et al. *Arms Control Today* article to learn that their work was being discredited in private meetings conducted by BMDO. However, when Keith Payne, director of the National Institute of Public Policy, made a public reference to the Sparta study in an op-ed piece published in the 15 January 1995 issue of *Defense News,* this reference opened up a new layer of the public controversy over ABM Treaty compliance policy.[45] In February 1995, Theodore A. Postol, a member of the Gronlund et al. research team at MIT, responded with a request to Keith Payne for a copy of the study. After Payne cleared the report for public release and sent a copy to MIT, Postol prepared a twenty-four-page rebuttal addressed to BMDO director O'Neill that was circulated throughout the Beltway BMD community.[46]

Postol claimed that the Sparta study contained "false and misleading information about the accuracy of technical work" published by Gronlund et al., and that "[h]ad the most basic quality control procedures been followed the results of the BMDO study would have been found to be in error."[47] Further, Postol stated incredulously that "I find it hard to understand how such a fundamentally flawed document could have been released by the BMDO. . . . [I]t is clear that public funds have not been spent well or appropriately."[48] Postol objected that the Sparta study was flawed on three levels: it incorrectly calculated Gronlund et al.'s radar detection ranges, it improperly overlooked early warning radar cueing as a significant aid to strategic defense, and the authors of the Sparta study violated norms of scientific peer review by failing to communicate with the researchers under review in order to clarify key methodological questions. It will be instructive to consider each of these objections in turn.

A technical paper by Gronlund, Lewis, Postol and Wright, attached to Postol's letter, argued that the Sparta study committed a basic calculation error by incorrectly determining the radar detection range of TMD systems engaged against long-range (strategic) targets.[49] The Sparta study, Gronlund, et al. argued, overlooked the iterative dimension of footprint calculation. Footprint size and radar search area are correlated; the larger the footprint of a TMD system, the more sky its radar must search. When the radar becomes overtaxed and cannot search all of the area needed to support a given footprint, the system cannot be expected to provide reliable defense. However, this does not mean that no defense is possible in such scenarios; there still remains the option of consolidating the size of the footprint in order to lessen the demands on radar. In a separate paper by Lewis and Postol, attached to Postol's letter, the MIT scientists argued that by shrinking the footprint, it is possible to reduce the slice of sky that needs to be searched, thus shifting the search area/anticipated footprint equation back into the defense's favor.[50]

Lewis and Postol pointed out that by using a "rote scaling" method, the Sparta team ignored the significant search flexibility afforded by this iterative optimization mechanism.[51] In the scenario where the needed search area exceeded the available radar resources in the strategic context, Sparta's model indicated that there was "no footprint possible." Gronlund, et al. contended that in jumping to this conclusion the Sparta researchers glossed over another alternative: that a smaller footprint might have been generated if the anticipated footprint size had been adjusted downward to ease the radar's search burden (see Fig. 11). The Sparta study "ignores the reduction in search area resulting from a smaller anticipated footprint for the low RCS targets. By not considering what the appropriate search area is, the study uses a search area that is much too large, calculates detection ranges that are much too small, and erroneously concludes that no footprints are possible," Gronlund, et al. argued.[52]

Postol's second major objection to the Sparta study was that it failed to take into account the advantages afforded to strategic defense by early warning radar cueing. During the Gulf War, U.S. space satellites provided Patriot missile batteries with data to enable early radar fixes on newly launched Scuds. When the Scuds went up, satellites detected the launch, and then pinpointed the trajectory of the Iraqi missiles. Although there is scant evidence that the Patriot system was able to convert this advantage into Scud intercepts (see Chapter 3), there is little doubt that the head start afforded by early space cueing reduced the radar demands on Patriot substantially.[53] Gronlund, et al. argued that this same sort of early warning radar cueing (e.g., provided by Defense Support Program satellites [DSP]) would increase significantly the search capabilities of a THAAD system deployed in strategic defense mode: "Thus by using early warning radars, the composite footprints against strategic targets with varying azimuths and reentry angles could in fact be quite large, even significantly larger than composite footprints against 3,000 km-range theater targets."[54]

In making this appeal to O'Neill, Postol played upon the general's own public position on this issue. In 1994, General O'Neill had told Congress "[t]oday's systems (TALON SHIELD and JTaGS) provide course cueing that will enhance footprints defended by the GBR/THAAD and SPY-I/Standard Missile systems."[55] General O'Neill also had anticipated that further remote radar advances would enhance the defense capability of THAAD even more. In this same testimony, he had asserted that "[f]uture systems such as Brilliant Eyes will provide higher-quality cueing information, enabling single-beam acquisition by the fire control radars and significantly larger defended areas."[56] This line of explanation was reinforced in testimony given to the Senate Armed Services Committee, in which O'Neill asserted that: "... the availability of space-based sensors, such as Brilliant Eyes, for missile tracking can significantly enhance the range of THAAD and Navy wide-area interceptors, by providing accurate and early midcourse cueing."[57]

In pointing out the iterative optimization mechanism and the radar search savings from early warning cueing, Postol isolated for BMDO director O'Neill two assumptions of Gronlund et al.'s model that permitted a footprint to be generated in the strategic context. Postol expressed disbelief that Lee et al. could have overlooked these crucial aspects of the Gronlund et al. methodology, and he contended that their oversight in this regard was symptomatic of a third major shortcoming of the study: failure to conduct legitimate scientific peer review. By attempting to "reconstruct" Gronlund et al.'s footprints using Gronlund et al.'s stated assumptions, Postol contended that the Sparta researchers had taken on important review burdens. Foremost among these burdens, argued Postol, was the obligation to discover fully and incorporate Gronlund et al.'s modeling assumptions prior to the attempt at footprint reconstruction. Only by duplicating Gronlund et al.'s methodology could a fair attempt to "reproduce" their footprints be undertaken, Postol contended.[58]

However, shortly after public release of the Sparta study, it became evident that not all of Gronlund et al.'s modeling assumptions had been incorporated into the analysis. Although the assumptions of space-based radar cueing[59] and iterative radar search optimization[60] were stated clearly in Gronlund et al.'s original *Arms Control Today* article, these assumptions were omitted in the Sparta study.[61] In place of these assumptions, Lee et al. included the "very, very simple" assumptions of no gravity, no atmospheric drag, and infinite acceleration. The upshot of this substitution, argued Postol, was that the Sparta study could not be considered a legitimate review of Gronlund et al.'s work.

> While the sole purpose of the BMDO study was to review our work, it is clear that such a review could not be accomplished without examining whether our "methods and assumptions" were appropriate and determining whether or not we correctly applied our "methods and assumptions" to the problem we analyzed. Hence, if the BMDO study applied different methods and assumptions . . . then it is being misrepresented as a review of our work.[62]

In addition to the fact that the Sparta study failed to incorporate several of Gronlund et al.'s key modeling assumptions into its footprint analysis, Postol contended that Lee et al. showed bad faith in failing to make a significant effort to communicate with the MIT group in order to gain a more detailed understanding of the modeling assumptions employed in its *Arms Control Today* article. "No attempt was made by the authors of the [Sparta] study or by BMDO contract monitors to contact us to check the accuracy of statements made about our work in the BMDO study. There was also no attempt by the BMDO-sponsored authors or BMDO staff to query us about discrepancies between our results and those found by the BMDO-funded analysts," wrote Postol.[63] Postol's reply thus constituted a sweeping and layered rebuttal of Sparta's secret counterstudy. On one level, Postol showed that the Sparta researchers had committed basic technical errors as a result of misreading the *Arms Control Today* article. On another level, Postol challenged the professionalism of the Sparta group by arguing that the misconstruals that had led to these basic errors could have been prevented had Sparta researchers contacted the MIT group for clarification prior to completion of their study.

The Twisting Trail of Footprint Arguments

The preceding analysis of back-and-forth argumentation exchanged by principal scientists in the Sparta-MIT footprint controversy highlighted several key points of stasis on which the dispute pivoted. The Sparta team claimed that strategic footprints generated by Gronlund et al. for a THAAD-like BMD system were based on unrealistic and overly optimistic assumptions regarding atmospheric conditions and physical dynamics. On the other hand, Postol claimed in rebuttal that the Sparta team had committed basic technical errors in reconstructing the Gronlund et al. footprints, and that Sparta researchers also had invalidated their study by engaging in practices inimical to professional norms of scientific peer review. The following section examines key texts that purport to judge these competing claims. This focus on judgment will shift attention back to the wider stakes involved in the Sparta-MIT footprint controversy, with the broader political problematic of ABM Treaty diplomacy coming back into play as a primary theme in the analysis. There are two texts particularly appropriate to frame this discussion: BMDO director Malcolm O'Neill's 3 May 1995 letter to Theodore Postol, and Richard Garwin's 10 October 1995 FILE Memorandum. O'Neill's letter has obvious textual significance as the major official statement from BMDO on this matter, while the Garwin FILE Memorandum deserves attention because its contents reflect the findings of an expert assessment of the controversy by a renowned physicist. The following section considers the O'Neill letter, examines Garwin's FILE Memorandum, and then concludes with critical commentary reflecting on the implications of these two texts.

On 3 May 1995 Lieutenant General O'Neill issued a two-page letter to Postol that outlined BMDO's official stance on the TMD footprint dispute and attempted to address Postol's criticisms of the Sparta study. In the letter, O'Neill offered three major arguments: 1) The findings of the Sparta study did not represent official BMDO policy, so Postol's dispute should be with Sparta, not BMDO; 2) The divergence between Sparta's and MIT's findings was not due to technical errors by Sparta, but instead was the natural outgrowth of a legitimate scientific disagreement; and 3) Postol's attacks on Sparta were unjustified because Lee et al. had worked in good faith to produce their study.

Occupying a seat of judgment invested with considerable institutional power, and addressed by Sparta and Postol as the primary audience for their scientific dispute, BMDO director O'Neill was situated at the locus of the controversy over THAAD's legal status under the ABM Treaty. It is interesting to sift through the text of his letter for insight regarding the official reaction to the scientists' argumentation, and to shed light on BMDO's assumptions regarding the role of scientific inquiry in the overall missile defense enterprise. A critical assessment of O'Neill's letter will provide an opportunity to work through these issues and to begin framing larger lessons from the experience of the Sparta-MIT footprint dispute.

In pursuing a strategy of disassociation, O'Neill sought to deflect Postol's criticisms of the Sparta study by pointing out that Lee et al.'s findings had not received official BMDO sanction. Although O'Neill admitted that his organization had commissioned the Sparta study, he added that "BMDO has not adopted the findings of the Sparta study," and that "[the Sparta study] neither states nor reflects U.S. or BMDO policy."[64] By opening up this distance between BMDO and Sparta, O'Neill cleared a space to pursue plausible deniability. Eschewing accountability for the specific findings of Sparta's work, O'Neill positioned himself to transcend the dispute and react to it as an interested but detached observer.

So positioned, O'Neill then proceeded to defuse the political pressure created by Postol's strident criticisms by removing himself from the seat of judgment. From O'Neill's point of view, the findings of the Sparta study lacked official status, so there was no pressing need for him to adjudicate the controversy. Freed from this burden to issue a definitive resolution of the dispute, O'Neill then proceeded to frame the TMD footprint controversy as a non-zero sum game where all sides could benefit from the open airing of scientific disagreement: "After a thorough review of your paper, your letter and the Sparta footprint analysis slides, I conclude that the difference of opinion between you and Dr. Lee is reasonable and reflective of healthy scholarly debate."[65]

In this same vein, O'Neill invoked the marketplace of ideas metaphor to illustrate his point that the arguments of both sides might usefully coexist. "True enough," O'Neill wrote to Postol, "your article and the Sparta study do not agree. That doesn't surprise me. In the marketplace for ideas there is room for vigorous and honest debate."[66] After disassociating himself from the seat of judgment and situating the argumentation within the framework of "open scholarly debate," O'Neill then proceeded to chide Postol for his professional attacks on Lee et al. O'Neill objected to Postol that "I do not agree with the personal and professional attacks you made in your letter against people who, in good faith, express a difference of opinion with you and reach different conclusions."[67]

One of the most basic insights of science and technology studies (STS) is that in scientific controversies, actors' arguments are colored by their political interests.[68] BMDO Director O'Neill's interests were clearly tied to continued political support for THAAD, one of BMDO's most promising weapons systems. Because THAAD's political future would have been severely clouded if the scientific claims advanced by Gronlund et al. had stabilized quickly, O'Neill's political interests were well served by the prospect of a long, drawn-out, and inconclusive scientific controversy over TMD footprint methodology. Uncertainty about THAAD's theoretical capability as a strategic BMD system resulting from such a protracted row promised to reduce the authority and credibility of computer footprint modeling as a scientific method of settling ABM Treaty compliance issues, thereby insulating THAAD from political criticism based on Gronlund et al.'s findings. Such controversy also promised to insulate O'Neill personally from political peril, by absolving him of the need to stand by the Sparta study or issue a definitive judgment on scientific dimensions of the dispute. "To the extent that science can be represented as indeterminate," explains sociologist Sheila Jasanoff, "political decision-makers absolve themselves of the need to toe the line on any particular scientific orthodoxy."[69]

An Independent Voice

Shortly after BMDO Director O'Neill provided an official response to the THAAD footprint controversy in his 3 May 1995 letter to Postol, Richard L. Garwin, renowned missile defense scientist and IBM research fellow, visited the Sparta offices for the purpose of conducting a review of the arguments in the case.[70] Given his sterling credentials as a brilliant and independent physicist having great familiarity with missile defense issues, Garwin was uniquely qualified for this task.[71] After interviewing Sparta analysts Laura Lee, Robert Bulk, and J. E. Lowder during his 28 September 1995 visit to Sparta's Rosslyn, Virginia, offices, Garwin drafted a FILE Memorandum that summarized his conclusions. The document was released privately to Sparta analysts on 8 October 1995 for advance review. After receiving the Sparta team's reply, Garwin attached it to his own memorandum, and circulated both widely.[72]

Garwin's investigation yielded the finding that basic technical errors in the Sparta study invalidated Lee et al.'s crucial claim that Gronlund et al. had used "very, very simple assumptions" in generating a footprint for a THAAD-like system against long-range ballistic missiles. Specifically, Garwin noted that the Sparta team ignored the "iterative optimization mechanism" included in the Gronlund et al. footprint model as an assumption that cuts radar search demands and gives a THAAD-like interceptor enough battlespace time to defend (theoretically) against long-range intercontinental ballistic missiles. This conclusion directly undermined the key claims made by Foil 104a/04 in the Lee et al. study, which stated: "The footprint shown . . . is impossible unless one assumes infinite acceleration, no gravity and no atmosphere."[73] In his review Garwin concluded that "[t]his statement is incorrect, according to what SPARTA told me. Foil/010 permits these results to be obtained without invoking not all, but one or more of these 'very, very simple assumptions' and now adds another assumption (tailoring of the search area) that would do it also."[74]

Since the Sparta team's stated methodology involved a commitment to "reconstruct" the Gronlund et al. model in order to check its validity, Garwin's finding that Lee et al. had omitted a key element of the model cast serious doubt on the credibility of the Sparta study. In fact, Garwin reported that during conversation, Sparta analysts conceded both that they had excluded the iterative optimization mechanism for "radar search area tailoring" (the Gronlund et al. iterative optimization mechanism) as an assumption in their attempts at footprint reconstruction, and that such an exclusion had had a definite effect on their conclusions. As Garwin recounted, the Sparta analysts agreed that incorporation of the iterative optimization mechanism as an assumption in the Gronlund et al. footprint model would have enabled the model to generate a valid footprint for THAAD-like systems defending against strategic missiles.

> I asked in particular about the assumption that was actually made by TAP [Gronlund et al.] and others—that it made no sense to scan an area larger than the footprint could be defended, so that for the small RCS case the scan solid angle would be reduced, compatible with the defended footprint to be obtained; launch-point cueing from DSP would make this possible.

> *Sparta stated that they had not considered this in their response to O'Neill* ("Abuses . . .") and they agreed that that *could have been listed as among the 'assumptions' that would make the difference.* I pointed out that I did not believe this was appropriately termed a "very, very, simple assumption."[75]

When Garwin asked why the Sparta researchers excluded Gronlund et al.'s iterative optimization mechanism in their attempt to reconstruct the footprints published in Gronlund et al.'s *Arms Control Today* article, he got the following reply: "SPARTA suggests that they did not have enough information on the methods used by the MIT group."[76] This is an odd explanation, given that a crucial dimension of sound scientific peer review in attempts to "reconstruct" or "replicate" experiments necessarily entails a commitment to seek out and acquire sufficient information about the original work under scrutiny in order to enable valid experimental replication, and that no evidence in the historical record suggests that the Gronlund et al. research team withheld evidence about their model. It would not have been difficult for the Sparta team to acquire the necessary information in this case, given that the iterative optimization mechanism for radar search area tailoring was an explicit modeling assumption stated in Gronlund et al.'s published work.[77] Garwin's investigation appeared to support the conclusion that in this case the Sparta analysts did not make a sufficient effort to gather information about the details of the Gronlund et al. footprint model to support a credible scientific review.

A more general conclusion regarding the reluctance of the Sparta team to engage in healthy dialogue on this matter can be drawn from the Sparta researchers' treatment of Garwin himself. Even though Garwin held top secret clearances (that were transferred for his visit to Sparta), Lee "declined to give . . . a copy" of foils describing the origin of the Sparta study, "to avoid the charts being criticized 'out of context.'"[78] This excuse is particularly curious in light of the fact that the meeting came "at the suggestion of BMDO,"[79] and was called presumably to provide Garwin with the necessary information required to reach an informed and independent judgment on the Sparta-MIT footprint controversy. Lee exhibited a similar lack of willingness to share research results when she denied permission to reprint a key viewgraph from the Sparta study for this book, justifying her decision to block publication of such material with the explanation that the study "is so old and probably no longer matters to anyone."[80]

Two years after completion of this review, Garwin published a chapter in a book on defense issues put out by the National Research Council.[81] In this chapter, Garwin reflected on the Sparta-MIT footprint controversy and reiterated emphatically his FILE Memorandum conclusion that the Sparta findings were wrong. Further, Garwin argued that official BMDO support for such "technical misinformation," as well as the willingness of top-level missile defense officials to convey such misinformation to Russia through diplomatic channels, had done substantial diplomatic damage to U.S.-Russian relations.

> Unfortunately, there is much misinformation, and even technical misinformation, provided to the Russian legislature, that could lead to substantial missteps by the United States and by Russia. For instance, a study paid for by the BMDO and released publicly in February 1995 has been claimed to counter the analysis of Professor T. A. Postol of the Massachusetts Institute of Technology and his colleagues that argues that THAAD has significant effectiveness against strategic ballistic missiles, if it is effective against missiles of 3,000-km range. Unfortunately, this BMDO-sponsored study has no 'study' behind it—just the briefing charts, as explained to me by BMDO staff and the contractor. Furthermore, the results are *wrong*, although it is more difficult to determine that they are wrong if there is no written analysis that can be evaluated.[82]

Garwin's ultimate conclusion was that the Sparta-MIT footprint controversy episode so thoroughly undermined BMDO's credibility as a source of technical information that in future debates in this area, "one should not trust the material published by BMDO, on which BMDO policy, that of the Department of Defense (DOD), and presumably U.S. national security policy are based."[83]

Sizing Up the Footprint Controversy

By concluding ultimately that the Sparta study was technically unsound, Garwin lent weight to political positions advanced in Gronlund et al.'s *Arms Control Today* article—that THAAD, when engineered to specifications, would violate the ABM Treaty. The results of Garwin's independent review thus cast doubt on the veracity of ACDA Director Holum's and BMDO Director O'Neill's official declarations of THAAD's ABM Treaty legality, made in congressional hearings during 1993 and 1994. Further, Garwin's findings raised troubling new questions about BMDO's penchant for funding and propagating technical misinformation for strategic gain. The THAAD case is a reminder that old habits die hard. A commitment to strategic deception endures in the missile defense bureaucracy today, more than a decade after the fall of the Berlin Wall. However, where previous cold war trickery campaigns inflated missile defense systems' capabilities, the post–cold war environment has given rise to a novel form of deception: strategic understatement. The flip-flops in advocacy performed by U.S. missile defense officials in this case left Russian arms control officials perplexed about the policy status of the Sparta study and more confused and suspicious than ever of U.S. intentions regarding the ABM Treaty. Coming at a time when Russia was already edgy about NATO expansion and when Russian arms control officials had made it abundantly clear that they regarded U.S. fidelity to the ABM Treaty as a litmus test for the prospects of post–cold war superpower arms control, the fallout from the Sparta debacle was particularly corrosive. The study not only squandered nearly $10,000 of U.S. taxpayers' hard-earned money, but by overtaxing Russian trust, it also unnecessarily depleted the arms control capital of U.S. ABM Treaty negotiators. Perhaps the most disturbing aspect of the study, however, is that as a larger symbol, it illustrates that systematically distorted science is still an inherent feature of the U.S. missile defense program. During the cold war, SDIO enlisted science to straightforwardly embellish missile defense capability in X-ray laser testing and the Homing Overlay Experiments. Today, BMDO-sponsored science performs more contoured rhetorical tasks, such as the carefully calibrated understatement of THAAD's strategic defense capabilities. The methods have changed, yet the result remains the same: misinformation and distortion of scientific data are the order of the day.

The remainder of this section is devoted to a more detailed consideration of these deceptive dynamics, as well as a broader analysis of the political stakes in play over the Sparta-MIT footprint controversy. First, a return to the contents of BMDO Director O'Neill's 3 May 1995 letter to Postol will provide an opportunity to reflect on BMDO's official role in the scientific debate. Second, the broader political consequences of strategic deception in this case will be explored, with correspondence by Alexei Arbatov, member of the Russian State Duma, serving as a text to ground analysis of the wider political significance of official attempts to propagate technical misinformation on THAAD.

Is This "Healthy, Scholarly Debate?"

BMDO Director O'Neill's 3 May 1995 letter to Postol announced BMDO's official stance on the Sparta-MIT controversy for the first time. Previous analysis in this chapter has highlighted the three main arguments in O'Neill's correspondence: 1) The findings of the Sparta study did not represent official BMDO policy, so Postol's dispute should be with Sparta, not BMDO; 2) The divergence between Sparta's and MIT's findings was not due to technical errors by Sparta, but instead was the natural outgrowth of a legitimate scientific disagreement; and 3) Postol's attacks on Sparta were unjustified because Lee et al. worked in good faith to produce their study.

While there is a surface coherence to O'Neill's three lines of argument, evidence of obfuscation and contradiction stews beneath the text. Although BMDO may not have adopted explicitly the findings of the Sparta study as official policy, the circumstances and character of official briefings indicate clearly that BMDO used its official status to create forums where Sparta researchers could present their views in an unopposed fashion to high-ranking audiences. Thus, while O'Neill forswore any official responsibility for the contents of the Sparta study in his letter to Postol, in practice it was official BMDO policy that provided institutional support for the study, and it was official BMDO policy that arranged for the unimpeded amplification of the study's findings to key official audiences. The tension between O'Neill's strategy of disassociation and BMDO's on-the-ground practice thus becomes manifest: BMDO appeared to have no reservations about promoting the Sparta study through official briefings, yet its director showed no interest in defending the study as a legitimate review in the face of vigorous external challenge.

Further, Sparta researchers were able to use their close association with BMDO to gain direct access to defense officials throughout the official BMD community. As Sparta's Laura Lee explained, this access translated into an exclusive right of first reply in cases where officials got wind of the Sparta-MIT controversy: "[After public release], we had to go through a lot. Every time Ted [Postol] flared it up, somebody somewhere would say, 'I want to see this.' And people would want to know the details. And so we'd show the charts, go through the diagrams and equations, approximations, and they'd go 'Oh, okay."[84] Intimating that this was a frequent occurrence, Lee went on to say, "I can't tell you the number of times I've dealt with this."[85]

The tension between O'Neill's strategy of deniability and BMDO's practical role in promoting Sparta's findings can be localized in two significant omissions in the text of O'Neill's letter to Postol. While O'Neill reiterated that BMDO's charge to Sparta was to "review" Gronlund et al.'s article, O'Neill ignored Postol's claim that the Sparta study did not even qualify as a legitimate scientific peer review, because it did not replicate Gronlund et al.'s modeling assumptions when attempting to reconstruct Gronlund et al.'s footprints.

Additionally, O'Neill failed to respond to Postol's charge that Lee et al. deviated from fundamental norms of scientific peer review by failing to contact Gronlund et al. prior to the completion of the Sparta study. This silence is deafening in light of Garwin's findings that the Sparta team made fundamental and careless errors in replicating the Gronlund et al. footprint model, such as ignoring key theoretical assumptions (e.g., the iterative optimization mechanism) built into the Gronlund et al. model that were disclosed clearly in published work.

O'Neill's second move, deferring judgment on the Sparta-MIT controversy by bracketing the dispute as a "healthy scholarly debate" takes on new light when assessed against the backdrop of Garwin's independent findings. By so naming the MIT-Sparta footprint controversy, O'Neill suggested that it was the interplay between the opposing positions that afforded him valuable insight into aspects of U.S. TMD policy vis-à-vis the ABM Treaty. "In the marketplace of ideas," wrote O'Neill, "there is room for vigorous and honest debate."[86] It was precisely this dynamic of debate that O'Neill claimed to find most illuminating about the Sparta-MIT exchange, and this fact was used to excuse him from any obligation to judge one side or the other as correct.

However, O'Neill chose not to reply to Postol's charge that Lee et al. sought to foreclose scientific debate by refusing to share information and by failing to make even minimal communicative attempts to clarify crucial aspects of the Gronlund et al. computer model. This silence takes on a more disturbing quality in light of Garwin's findings that this case was hardly a model of "open, honest," scientific debate carried on in the "marketplace of ideas." In fact, there was no debate in the first six months after completion of the Sparta study, because its contents were shrouded in secrecy. During this time, Gronlund et al. were not even aware of the fact that a major challenger had enjoined them in debate. Presumably, O'Neill never would have permitted this time period to elapse if he had been genuinely seeking to spur robust debate in the open marketplace of ideas. This tension is yet another manifestation of the contradictory dynamic latent in O'Neill's reply to Postol.

O'Neill's characterization of the controversy as an honest scholarly debate thus appears questionable when assessed in the context of the charges, countercharges, and third-party findings of systematic distortion in the peer review process of the TMD footprint controversy. Yet O'Neill relied heavily on this classification of the controversy to support his aloof rhetorical posture.

This pattern of argumentation mirrors earlier episodes of missile defense advocacy where military officials assumed malleable positions, putting a different spin on the same scientific evidence for various audiences in order to maximize persuasive leverage in BMD promotion campaigns. In the case of the TMD footprint controversy, BMDO's approach hewed to a similar double standard. In the internal "BMD community," a classic "constitutive forum,"[87] BMDO commissioned a secret study by a private SDI boutique and then amplified the findings of the study in official briefings. This internal strategy provided grounds for official missile defense advocates to discredit the contents of a threatening public article and help shore up support for THAAD. However, when the study drew intense criticism after its public release, BMDO denied sanctioning the study and declined to defend its findings in the public sphere.

Diplomatic Missteps and Superpower Suspicions

According to senior Russian arms control official Alexei Arbatov, in the mid-1990s the Gronlund et al. article became "widely known in Russia," and its results were "extensively used in the discussion about the problems associated with ABM Treaty modification."[88] It is easy to appreciate why Gronlund et al.'s argument would find such a wide Russian audience. Russians worried that headlong American pursuit of missile defense would nullify the deterrent value of their offensive nuclear forces and make them vulnerable to U.S. nuclear blackmail. The Gronlund et al. *Arms Control Today* article related directly to this concern, because its contents constituted the scientific basis for many public claims that BMDO was attempting to sneak strategic defense in through the back door, by cloaking strategic defense systems in the garb of a pure theater defense.

In the case of TMD footprints, Russian perception that American negotiators systematically employed a methodology that understated the capability of U.S. TMD systems against strategic targets worked to chill the negotiating climate and bolster the hand of arms control opponents in the Russian Duma. As chair of the Subcommittee on International Security and Arms Control in the Russian Duma, Arbatov exercised significant control over the fate of superpower arms control at the height of the TMD footprint controversy. "There is a considerable number of Duma members concerned about possible strategic capabilities of advanced theater defenses and this issue is very likely to be raised in connection with ratification of the START II Treaty. It is therefore very important for us to have clear and unambiguous understanding of the issues associated with discrimination of theater and strategic defenses," Arbatov wrote in 1995.[89]

A popular argument advanced by U.S. TMD advocates in response to these concerns was that on questions of ABM Treaty compliance, capability, not intention, was key to determining a given system's legality.[90] While such an approach may have been technically correct, it may not have been politically prudent, given the concerns voiced by Arbatov. Even if a TMD was technically legal under the ABM Treaty, Russian perception that such a system was being packaged to circumvent the treaty could have been as damaging to strategic stability as outright abrogation of the treaty.[91]

It was precisely this dynamic that led Russia to link its ratification of the START II arms control agreement not only to continued U.S. observance of the ABM Treaty, but also to satisfactory resolution of outstanding TMD demarcation issues. According to START I negotiator Yuri Nazarkin and former Arms Control and Disarmament Agency (ACDA) START specialist Rodney W. Jones, fear of U.S. TMD "weighs heavily in START II ratification deliberations. Members of Parliament and Russian defense experts have clearly linked the outcome of START II ratification to their concerns over the ABM Treaty."[92] Vladimir Lukin, head of the International Relations Committee, was the floor manager of the START II bill in mid-1990 deliberations held in the Russian Duma. Seconding Nazarkin and Jones's account, Lukin said "that the development of these [TMD] defenses could torpedo ratification [of START II]."[93]

These statements suggest that Russian linkage of START II to U.S. ABM Treaty compliance was more than a technical issue; it was also a rhetorical issue. Russian officials made it abundantly clear that their assessment of American intentions regarding missile defense deployment was as important as their determination of extant American capabilities as a factor determining whether to ratify START II. For instance, Pavel Povdig, arms control expert at the Moscow Centre for Disarmament, Energy and Ecology, argued that THAAD "will lay down a basis for the country's [U.S.'s] anti-missile defence system."[94] In a similar vein, Anton Surikov, general director of Russia's Institute of Defense Studies, contended that "[t]hese [TMD] plans are essentially another attempt at dragging the SDI idea in through the back door and they present a significant threat to strategic stability in the world"[95]

The gulf between American assurances of ABM Treaty compliance and Russian perceptions of covert U.S. ABM Treaty breakout was a rhetorical chasm that threatened to swallow up the START II Treaty. Unfortunately, this chasm was widened by Russian perceptions of U.S. strategic deception on TMD demarcation issues. Surikov's "back door" SDI hypothesis gained currency each time the U.S. deliberately underrepresented TMD capabilities to Russian demarcation negotiators.

Robert McFarlane touted the advantages of cold war deception on SDI by arguing that it led the Soviet Union to take on bloated military budgets, speeding up the collapse of the Soviet economy by five years. Ironically, U.S. strategic deception may turn out to have similar effects in the present context. As in the cold war, Russian suspicion of U.S. military intentions translates into significant argumentative ammunition for Kremlin hawks calling for boosts in defense spending. Where such massive enemy expenditures allegedly served U.S. interests in the late stages of the cold war by smothering the Soviet Union's economy and supposedly hastening the fall of communism, today similar spending patterns waste precious economic fortunes and subvert efforts by both Russians and Americans to move beyond debilitating superpower cold war antagonisms. This is no trivial concern, given that in the present milieu a revival of unbridled cold war passions on the part of major nuclear powers could trigger another Cuban missile crisis gone wrong. The world came uncomfortably close to outright nuclear war during the cold war, and the next time a harsh diplomatic stalemate like the Cuban missile crisis confronts hostile nuclear superpowers, there is no guarantee that the deterrence dice will come up lucky again. Should superpower antagonisms escalate to the level of a nuclear exchange, life on planet earth would be in grave jeopardy. Therefore, extraordinary efforts to stave off unnecessary superpower friction in the present milieu are prudent, if not imperative.

The zero-sum framework of cold war competition provided incentives for superpower adversaries to wreak havoc on each others' economies. Now the stakes are different as the superpowers attempt to navigate a transition toward a more cooperative and sustainable post–cold war posture. The success of such a transition depends in large part on mutual trust, since suspicion strengthens the hands of those in each country calling for a return to cold war-style military confrontation. Official U.S. policy statements recognize that strategic deception may entail huge risks in this delicate context. For example, in a recent policy address, U.S. Deputy Secretary of State Strobe Talbott prescribed that:

> [W]e should weave relationships and devise incentives that will encourage Russia to evolve as a democratic, secure, stable, prosperous state, at peace with its neighbors and integrated into a community of like-minded nations. But to do that, one challenge America faces, quite frankly, is to overcome Russian suspicions, Russian conspiracy theories, and Russian old-think.[96]

In this context, corrosive fallout from American strategic deception can be located on a rhetorical level. Should Russians begin to doubt fundamentally the credibility of U.S. promises and assurances, Talbott further suggested that we can expect them to begin translating our discourse in a more hostile manner, filtering ostensibly amicable American proposals for post–cold war partnership and cooperation through an adversarial, zero-sum lens.

> If the Russians overindulge their misplaced suspicions that we want to keep them down, then words like partnership and cooperation, translated into Russian, will become synonyms for appeasement, subservience, and humiliation at the hands of the West. The result then could be that we will indeed cooperate less, and compete more, on precisely those issues where it is in our common interest to cooperate more and compete less: arms control, environmental degradation, terrorism, regional conflict, and proliferation of weapons of mass destruction. . . . Russian policymakers—especially those still inclined to see their country's relationship with the United States as intrinsically a rivalry—may fall into the trap of defining what is in their national interest as pretty much anything that annoys, or causes problems, for us. If that reflex for scoring points against us in a zero-sum game become a kind of default feature in the software of Russian policy, it will only generate mistrust on our side. Suspicions of each other's motives could prove self-justifying, and pessimistic prophecies about the future of the relationship may be self-fulfilling.[97]

Mutual trust and cooperation are especially crucial in the context of nuclear stability, since, as Garwin observes, "American security, like that of all other nations, depends on political pressures and constructive diplomacy, imperfect though these tools are."[98] It is reasonable to expect such diplomacy to falter when Russian suspicions about the structure of the current U.S. missile defense program are allowed to fester. Therefore, there are grounds for grave concern raised by the finding of systematic and chronic strategic deception in missile defense advocacy.

THAAD's "Rush to Failure": On the Road to Nowhere?

It is possible to look back on the mid-1990s TMD demarcation controversy with a sense of irony. At the height of the debate, rosy projections of THAAD's theoretical capabilities raised significant questions about the system's legality under the ABM Treaty. However, at the same time that analysts such as Gronlund et al. were challenging THAAD on the grounds that its high-end engineering specifications ran afoul of ABM Treaty performance limits, THAAD was racking up a dismal track record for actually intercepting missiles on the test range. By the summer of 1999, THAAD had failed its sixth consecutive flight test (hardly a streak of futility one would expect from a "Treaty busting" system!) Conducted at the White Sands National Missile Range in New Mexico on 29 March 1999, this test was designed to demonstrate "integrated performance of the entire system," and it "incorporated an upgraded seeker on the missile and corrected numerous problems encountered on earlier intercept attempts."[99] When the THAAD interceptor was launched to destroy a "Hera" target missile, "instead of slamming into the target, the interceptor passed within 10 to 30 yards of it. . . . Efforts to determine what had gone wrong were complicated by the loss of telemetry data one minute into the interceptor's flight."[100] This flight test failure came on the heels of five previous failed experiments, where contaminated battery connectors, false alarms, software breakdowns, errors in flare deployment, and a host of other technical mishaps had resulted in missed intercepts.[101]

Although the beleaguered THAAD system experienced a bevy of technical woes, missile defense planners stood by the project and argued for approval of more funding and new missions. As the 1990s wore on and political support for National Missile Defense (NMD) grew, one of the new missions for the THAAD program was to serve as the "hit to kill" architecture of a Star Wars-style NMD system. This mission shift intensified the ABM Treaty controversy, since a THAAD-based NMD system would clearly break out of the treaty's original limits on strategic defenses. After the 1996 Missile Defense Act mandated review and possible deployment of NMD, discussions of "footprints" and the fine points of ABM Treaty demarcation gave way to more basic arguments about whether the U.S. should continue to honor the terms of the treaty at all.[102]

This new turn in the controversy took on a decidedly partisan color when forty-one House Republicans sued President Clinton personally for violating the terms of the 1996 Missile Defense Act. In their lawsuit, this group of Republicans (led by Sen. John Kyl, R-AZ) charged that the THAAD testing timetable approved by Clinton was too slow to meet the ambitious program milestones laid down in the 1996 Missile Defense Act.[103] Although the House members' suit turned out to be more of a publicity stunt than a serious legal challenge, relentless pressure for quick deployment of NMD led to an acceleration of the THAAD research and development timetable through the mid-1990s.

It now appears that this manipulation of the pace of scientific research in the THAAD program was a major cause of technical breakdown, responsible for the remarkable string of flight test failures racked up by Lockheed Martin (the company hired by BMDO to build the system). This was the finding of a 1998 report issued by an independent team of missile defense experts appointed by the Pentagon.[104] This team, headed by former Air Force Chief of Staff Gen. Larry Welch, produced findings that stirred memories of Reagan-era Star Wars science, where the 1985 Goldstone experiment (a Livermore X-ray laser test) failed after Reagan Star Wars officials rushed its scientific timetable for political reasons (see chapter 2). The Welch Commission noted that a similar "rush to failure" dynamic accounted for THAAD's poor field performance: "Specifically, the *perceived* urgency of the need for these systems has led to high levels of risk that have resulted in delayed deployments because of failures in their development test (DT) programs."[105]

While the Welch Commission found that political pressure for quick THAAD deployment was a major factor contributing to the program's dismal track record in field tests, it also made disturbing assessments of the overall level of technical competence among BMDO and Lockheed Martin program managers: "We felt that the program managers—both government and contractor—underestimated the degree of difficulty in achieving HTK [hit-to-kill]. The fact that the contractor simulations often predicted HTK 100 percent of the time gave us the impression that the contractors routinely underestimate the many things that can go wrong. We felt that this lack of appreciation for the complexity of the task continued after experience should have provided compelling evidence to the contrary."[106] These comments sounded much like earlier cold war review studies, conducted by such groups as the General Accounting Office (GAO), Congressional Research Service (CRS), and the Government Operations Committee, that warned about unjustifiably high levels of confidence in missile defense engineering on the part of SDIO and private missile defense vendors.

Today, the backbone of this bureaucracy is BMDO, the official government agency charged with overseeing missile defense research and development, with private contractors such as Sparta attached as offshoots from this spine. The elaborate and interconnected web of private think tanks and research consortia surrounding BMDO provides the organization with essential technical support and political influence. During the cold war, this military-industrial complex operated in scientific isolation and pursued secret missile defense research that papered over complex engineering problems with fabricated test results. Perhaps it should not be surprising to find similar instances of distorted science in BMDO's contemporary missile defense projects, given that the essential institutional characteristics of the U.S. missile defense bureaucracy have not changed significantly since the end of the cold war.

Although the 1993 name change from SDIO to BMDO was advertised by defense officials as a symbol of post–cold war rebirth for the U.S. missile defense bureaucracy, this symbolic adjustment turned out to be more of a cosmetic ploy than a genuine transformation. Because such a bureaucratic reorganization did not address the status and role of the network of private scientific advisors supporting the government's missile defense projects, it glossed over one of the root causes of systematic distortion in missile defense policy making and research.

Many of the private contractors, think tanks, and laboratories that worked on Ronald Reagan's Star Wars project are still involved in the official missile defense effort. These entities remain linked in a secretive support network that comprises what sociologist Chandra Mukerji has termed an "elite reserve labor force."[107] Largely dependent on Pentagon money for their existence, members of this private scientific network enter into contractual arrangements that oblige them to keep their research under wraps and refrain from sharing ideas with the community of public scientists.

Think tanks like Sparta operate as quasi-government agencies. They receive omnibus multiyear contracts for millions of dollars of open-ended research. When BMDO realizes a need for more information on a particular topic, it issues a "task order" that specifies a certain research project to be performed by a particular advisor in the official BMDO intelligence network. There is no bidding process involved to test whether other think tanks may be able to provide the information more effectively and economically. The task orders are presumptively secret and can be accessed by the public only through formal FOIA requests. This mechanism for private contracting permits BMDO to assemble a secret ring of official advisors that is virtually integrated into the administrative bureaucracy, yet is exempt from public accountability.

There is a good chance that physicists and arms control scholars in the open community would never have learned of the Sparta study, had Postol failed to press for its public release in 1995. Like the vast majority of other scientific work commissioned by BMDO, the Sparta study was "born secret," and could be released for public inspection only after passing a strict clearance process. Official DOD information-control policy provides numerous mechanisms through which the military's internal test results, studies, and memoranda can be shielded from public scrutiny.[108] The sprawling nature of this information-control protocol has created a distinct seam demarcating the secret and public discourse communities. Unfortunately, the more hermetically sealed this seam becomes, the less able are members of the public to engage vigorously in debates about defense policy.

The cozy military-contractor arrangement is particularly troubling in cases such as Sparta's, where multiple and cross-cutting task orders for program development and evaluation lead to conflicts of interest that invite biased research and reward fraudulent science. "[T]he infusion of SDI dollars has raised some questions. . . . Critics say that many smaller firms are in danger of becoming SDI addicts, creating a dangerous dependency in the event of political or international developments that could curtail or eliminate the program," wrote journalist Michael Isikoff.[109] "It's an enormous conflict of interest. There is no way these contractors can raise a skeptical eye. They look at SDI as an insurance policy that will maintain their prosperity for the next two decades," observed Senator William Proxmire (D-WI).[110] In Sparta's case, there are grounds for substantial concern that the company's ongoing contracts for THAAD engineering projects compromise the firm's status as an independent source of technical knowledge in this area. Along with Hughes Aerospace, TRW, and General Electric, Sparta was in on the ground floor of THAAD's engineering program. In 1992, Sparta researchers were awarded contracts to develop battle management software for THAAD's "dem/val" (Demonstration/Validation) phase,[111] and BMDO retained Sparta for numerous other THAAD engineering projects throughout the mid-1990s.[112] These contracts clearly endowed Sparta with a financial stake in political survival of the THAAD program, so it is curious indeed that BMDO selected Sparta researchers to conduct an independent assessment of the politically sensitive issue of THAAD's ABM Treaty status.[113]

The propensity of BMDO to cultivate close, reciprocal, and interlocking links with "SDI boutiques" such as Sparta raises serious questions about the ability of the BMD bureaucracy to carry out its work without breeding conflicts of interest. When allowed to persist, these conflicts of interest result in mutually reinforcing incentives for fraud, waste, and technical mismanagement on the part of contractors and government officials alike. Such close links also strain the legal boundaries of federal law. Congress enacted the Federal Advisory Committee Act (FACA) in 1972 to establish "a system governing the creation and operation of advisory committees in the executive branch of the federal government." As a legislative response to the growing dependence of the executive branch on external advice, the central purpose of FACA was to "control the advisory committee process and to open to public scrutiny the manner in which government agencies obtain advice from private individuals and groups."[114] FACA imposed a number of requirements on the advisory process, including advance notice of meetings, public access to records, and viewpoint balance in the composition of advisory bodies.[115] Given the tight-knit ties between government officials and private contractors in the BMD community, it is not surprising that missile defense research is one area where significant congressional oversight has been needed to monitor fidelity to FACA rules. For example, during the cold war, Congress convened hearings in 1988 to investigate the extent to which private missile defense advice to the government met the openness and notice requirements codified in FACA.[116] In the mid-1990s TMD demarcation controversy, Sparta's role as an official source of technical footprint analysis appeared once again to have stretched beyond the legal limits on official advice codified in FACA.[117]

FACA is a so-called sunshine law, designed to enhance governmental accountability by preventing conflicts of interest from corrupting the policymaking process. The need for such accountability in the missile defense context is evident, given the close integration of government-paid private advisors and engineering contractors into the official BMDO bureaucracy. Unless secrecy is explicitly counteracted in secretive science projects carrying major political significance, "peer review is more likely to fall prey to politics than to ensure that impartial standards of quality are maintained in policy-relevant science," Sheila Jasanoff explains.[118] In response to these conflict of interest concerns, one argument offered in defense of the strict compartmentalization of the scientific BMD community is that even though such a cloistered community might exclude certain viewpoints and foreclose wide-open discussion, enough space remains to allow for productive debate and disagreement. When I asked Sparta's Laura Lee to explain how she felt about excluding public figures such as MIT's Postol from secret discussions on THAAD footprint methodology, her answer reflected a sanguine optimism about prospects for achieving reliable and meaningful dialogue, even in the context of a closed scientific community.

> GORDON MITCHELL: One last question, and this is from the theoretical public argument perspective, taking the broader view. Some people say that for scientific truth to emerge, and to have high quality scientific research, you have to have an open scientific community. So while there's definitely a possible down-side to involving someone like [Ted] Postol, who from your perspective is not really making serious arguments and is just trying to stir up controversy, on the other hand, is there potentially a cost to excluding public physicists from participating in disputes like this?
>
> LAURA LEE: I see a difference between black programs and general secret programs.[119] In the general secret programs, there's a large community, and there's a lot of difference of opinion. There's a lot of debate; twenty-five different ways to calculate footprints. There's actually a footprint methodology paper produced that tells you how to calculate it; there's so much technical debate at the secret level. So I think we have a lot of different perspectives, and you can come up with the objective answers that way. . . . I think we have a lot of debate at the secret level. That keeps it open and honest. We get evaluated in our community; Sparta gets contracts, I only get funding when I do good technical work. Not when I raise issues. And Ted [Postol] will only get money when he makes a problem out of something, whether he's right or wrong.[120]

The danger in Lee's formulation is that insofar as the BMD community remains a cloistered conglomeration of relatively homogenous research units, scientific disagreement and debate in these circles will take place only in the register of restrictive and self-referential logics. Such regressive patterns of scientific inquiry are most likely to flourish in bureaucratic environments where internal dissent and skepticism are squelched to further political objectives. Unfortunately, there is a surfeit of evidence that such an intellectual environment exists currently in the official BMD community, where the flow of money tends to steer the course of technical work, experimental timetables are rushed to meet political deadlines, and rigorous scientific criticism has come to be seen as an inconvenient annoyance. On this topic, consider the comments of BMDO Director O'Neill, given in a 1995 interview. Asked by a *Defense News* reporter to comment on the state of relations between BMDO and those in the "development and acquisition branches in the military services," O'Neill provided the following response:

> Relationships between me and the services right now are fabulous. I couldn't have more support. Every time somebody makes trouble for me, the service leaders jump on them. That's because I control the dollars, and the dollars I control are benefiting the service. So when people start harassing me, the leadership tells them to get out of the way.[121]

Conclusion

Laura Lee continues to perform research for Sparta, but she recently set aside the TMD footprint issue and moved on to other projects. "There's nobody here [at Sparta] working in [the footprint] area anymore. We answered all the questions to our satisfaction, and any problems with the work have been resolved," she said in a recent telephone conversation.[122] How have the problems been resolved and the questions been answered? BMDO insiders may be the only ones who ultimately receive a full and fair accounting on these points. Until the entire missile defense enterprise is made more transparent, those in the public realm will continue to remain largely uninformed and essentially dependent on official sources of information. Under these conditions, there is every reason to expect that those in the missile defense bureaucracy will continue to resort to strategic deception as a routine strategy for the promotion of missile defense systems.[123]

This chapter's exploration of the mid-1990s public controversy over ABM Treaty demarcation illustrates that new strategies of deception have been developed to bolster missile defense advocacy in the post–cold war milieu. Specifically, in the case of the THAAD footprint issue, prominent independent scientists such as Richard Garwin have found that BMDO worked with Sparta to propagate "technical misinformation" as a way to influence official judgments on ABM Treaty compliance. The novel element of such a strategy is that it was employed to downplay THAAD's intercept effectiveness, in order to make the system more politically palatable. What other new forms of strategic deception pervade the missile defense enterprise? Is the Sparta study just the tip of the iceberg? For Garwin, the lesson of the Sparta-MIT controversy is that in future missile defense debates, "one should not trust the material published by BMDO, on which BMDO policy, that of the Department of Defense (DOD), and presumably U.S. national security policy are based."[124] If his conclusion is correct, then it is unlikely that members of the public will receive satisfactory answers to the tough questions that continue to dog BMDO while it remains a scientific island, ringed by a moat of secrecy and cut off from the wider physics community.

Technical breakdown and cover-up in the BMD community invite erosion of democratic decision-making processes, undercut arms control diplomacy, and frustrate superpower efforts to move beyond cold war antagonisms. Unnecessary secrecy always spawns mistrust, and in the present political context, the United States can ill-afford needless diplomatic friction. Public statements of Russian officials suggest that the BMDO's position on TMD footprint analysis may have sapped U.S. arms control credibility in ABM Treaty demarcation negotiations while also undermining support for ratification of START II in the Russian Duma. The success of the U.S. arms control agenda depends in large part on rhetorical persuasion, and persuasion depends on credibility. One challenge for U.S. diplomats in the new millennium will be to recover the lost credibility squandered during the mid-1990s Sparta-MIT footprint controversy.

Notes

1. Steven A. Hildreth, "Theater Ballistic Missile Defense Policy, Missions, and Current Status," CRS Issue Brief no. 95–585, 10 June 1993 (Washington, D.C.: GPO, 1993), 4.
2. THAAD "is a ground-based weapon system being developed by the Ballistic Missile Defense Organization and the Army to defeat theater ballistic missiles by colliding with them while in flight. The system supports the national objective of protecting U.S. and allied deployed forces, population centers, and industrial facilities from theater missile attacks" (United States General Accounting Office, *Issues Concerning Acquisition of THAAD Prototype System,* Letter Report GAO/NSIAD-96–136, 9 July 1996 (Washington, D.C.: GPO, 1996). THAAD "is to engage theater ballistic missiles at high altitudes and long ranges using hit-to-kill technology. High altitude intercepts reduce the probability that debris and chemical or biological agents from a ballistic missile warhead will reach the ground in damaging amounts. Long-range intercepts provide protection to wide areas, dispersed assets, and population centers. THAAD is to be deployed with the PAC-3 system and will consist of missiles, mobile launchers, ground based radars, a tactical operations center, and support equipment" (United States General Accounting Office, *Information on Theater High Altitude Area Defense (THAAD) System,* Report NSIAD-94–107BR, January 1994 [Washington, D.C.: GPO, 1994]).

3. Malcolm O'Neill, statement, *Administration's Proposal to Seek Modification of the 1972 Anti-Ballistic Missile Treaty,* 103d Cong., 2d sess., Senate Hearings, Committee on Foreign Relations, 3 May 1994 (Washington, D.C.: GPO, 1994), 14.
4. See Matthew Bunn, *Foundation for the Future: The ABM Treaty and National Security* (Washington, D.C.: Arms Control Association, 1990), 130–42.
5. These questions relate to competing theoretical estimates of missile defense system performance capabilities, including radar search power, battle management, intercept effectiveness, and other measures of system proficiency.
6. Technical discussion of THAAD's theoretical capability as a strategic defense system has direct ramifications for policy in the areas of international treaty compliance, diplomacy, and arms control.
7. Steven A. Hildreth, "The ABM Treaty and Theater Missile Defense: Proposed Changes and Potential Implications," CRS Issue Brief no. 94–374 F, 2 May 1994 (Washington, D.C.: GPO, 1994), 24.
8. Bunn, *Foundation,* 7.
9. See United States Strategic Defense Initiative Organization, *1989 Report to Congress on the Strategic Defense Initiative,* 13 March 1989 (Washington, D.C.: GPO, 1989); idem, "SDI: Goals and Technical Objectives," reprinted in *The Search for Security in Space,* ed. Kenneth N. Luongo and W. Thomas Wander (Ithaca, N.Y.: Cornell University Press, 1989), 144–59.
10. The treaty prohibits BMD testing "in an ABM mode," clarified in 1978 to mean "with the flight trajectory characteristics of strategic ballistic missiles . . . over the portions of the flight trajectory involved in testing" (quoted in Bunn, *Foundation,* 106).
11. In an op-ed piece for the *New York Times,* Holum claimed that "[a] theater missile defense system could protect against such threats [as North Korea]—but not against a Russian strategic missile attack. Missiles in a theater defense might theoretically shoot down one or another incoming long-range missile, but the multiple missiles in a theater could not shoot down significant numbers of enemy missiles during a real attack. The theater missile defense would not and could not be used to protect the United States" (John D. Holum, "Don't Put Allies at Risk," *New York Times,* 25 October 1994, A21).
12. Malcolm O'Neill, statement, *Hearings on National Defense Authorization Act for FY94: H.R. 2401 and Oversight of Previously Authorized Programs, Procurement of Aircraft, Missiles, Weapons, and Tracked Combat Vehicles, Ammunition, and Other Procurement,* 102d Cong., 2d sess., House Hearings, Committee on Government Operations, 10 June 1993 (Washington, D.C.: GPO, 1993).
13. O'Neill, Statement, *Administration's Proposal,* 14–15, emphasis added.
14. Quoted in John Mintz, "Arms Race with an Attitude?" *Washington Post,* 18 October 1994, C4.
15. John Rhinelander, statement, *Administration's Proposal to Seek Modification of the 1972 Anti-Ballistic Missile Treaty,* 103d Cong., 2d sess., Senate Hearings, Committee on Foreign Relations, 3 May 1994 (Washington, D.C.: GPO, 1994), 97; See also Jack Mendelsohn, "U.S. Takes Nuclear Gamble: Broader Policy for Weapon Use Could

Backfire," *Defense News,* 4–10 April 1994, 24; Jack Mendelsohn and John B. Rhinelander, "Shooting Down the ABM Treaty," *Arms Control Today* 24 (1994): 8–10.

16. "New Weapons Take Aim at Anti-Missile Treaty," *San Francisco Chronicle,* 25 June 1994, A24.
17. Theodore A. Postol, letter to the editor, *Washington Post,* 17 November 1994, A22.
18. George Lewis, statement, *Administration's Proposal to Seek Modification of the 1972 Anti-Ballistic Missile Treaty,* 103d Cong., 2d sess., Committee on Foreign Relations, Senate Hearings, 3 May 1994 (Washington, D.C.: GPO, 1994), 17.
19. Lisbeth Gronlund, George N. Lewis, Theodore Postol, and David Wright, "Highly Capable Theater Missile Defenses and the ABM Treaty," *Arms Control Today* 24 (1994): 3–8.
20. Gronlund et al. argued that the Clinton administration's proposed TMD demarcation threshold was untenable because THAAD-like TMDs exceed the maximum capabilities allowed for missile defense systems under the ABM Treaty: "[A]s our analysis indicates, it is far from obvious that it will be possible to deploy highly capable ATBMs without seriously undermining the ABM Treaty. It also casts serious doubts on statements by Administration officials that the proposed changes are only minor 'clarifications' to the Treaty'" (Gronlund et al., "Highly Capable Theater Missile Defenses," 8).
21. Arms Control Association director Jack Mendelsohn offered the duck/duckling, strategic/tactical metaphor as a way of highlighting the argumentative contortions that the Clinton administration went through to contend that on paper the THAAD system would have no strategic capability (John Mendelsohn, statement, "An ACA Panel: The Future of Arms Control," *Arms Control Today* 25 [1995]: 713).
22. The briefing was related to a paper Keith Payne had prepared for General O'Neill on the topic of "Proliferation, Potential TMD Roles, Demarcation and the ABM Treaty," and the briefing took place in BMDO front offices (Laura Lee, interview with the author, Rosslyn, Va., 12 July 1996).
23. The footprint research was covered under two ongoing BMDO contracts: (1) BMDO Super SETA Extension, SDI084–93-C-0018/J-700D-93014/Task Order 44, Work Area 1, and Task Order 65, Work Area 6; and (2) SETA for the BMDO Technology Readiness/Strategic Relations Deputate, HQ0006–95–0006/06S5000834, Task Order 10 (J. E. Lowder, "BMDO Request for Information Regarding SPARTA's 'Footprint Analysis,'" memo to Peter Van Name, Sparta, Inc., Ref: JEL95–085 [Rev. 3], 5 October 1995, released 7 November 1996 under FOIA request 96-F-1714, filed by the author). The estimated time spent "in support of the footprint analyses and briefings is less than 100 hours with a cost to the Government of less than $8,500" (Lowder, "BMDO Request for Information," 2).
24. See Joseph Cirincione and Frank von Hippel, eds., *The Last 15 Minutes: Ballistic Missile Defense in Perspective* (Washington, D.C.: Coalition to Reduce Nuclear Dangers), 1996.
25. Michael Isikoff, "Carving Out an SDI Niche: Local Defense Companies Scramble to Hitch Their Wagons to Star Wars," *Washington Post,* 13 October 1986, F1.

26. From the perspective of insiders, the "BMD community" includes those security-clearance holders directly and indirectly connected to the intelligence, planning, and procurement infrastructure of BMDO (Lee, interview with the author).
27. Lee, interview with the author; Dennis Kane, telephone conversation with the author, 22 July 1996.
28. "Hughes Establishes Organization for Air Defense/Anti-Missile Efforts," *Defense Daily,* 5 July 1991, 29; "Hughes Assumes Lead in Team Competing for THAAD Contract," *Defense Daily,* 4 November 1994, 182.
29. Quoted in Isikoff, "Carving Out an SDI Niche," F1. This experience was not unique to Sparta, Inc. Burgeoning SDI budgets sprouted an entire cottage industry of missile defense contracting in the mid-1980s (see Tim Carrington, "Star Wars Plan Spurs Defense Firms to Vie for Billions in Orders," *Wall Street Journal,* 21 May 1985, 1; William Hartung and Rosy Nimroody, "Cutting Up the Star Wars Pie," *Nation,* 14 September 1995, 200–202; Michael Weisskopf, "Conflict in 'Star Wars' Work Possible," *Washington Post,* 30 April 1985, A5).
30. Laura T. Lee, Martin T. Durbin, Rick B. Adelson, Sean K. Collins, and Wai H. Lee, *The Abuse of 'Footprints' for Theater Missile Defense and the ABM Treaty (U),* study by Sparta, Inc., under contract to the Ballistic Missile Defense Organization, September 1994, cleared for public release to Theodore Postol, February 1995, copy on file with the author.
31. Lee, interview with the author.
32. Ibid.
33. Lee et al., "Abuse of 'Footprints,'" viewgraph MCL94:104a/08.
34. The time for the track and intercept calculation was assumed to be 5 seconds, while the interceptor flight time (using the shortest possible path) was assumed to be 25.6 seconds. Added together, these estimates made up Sparta's total calculated battlespace time of 30.6 seconds (Lee, interview with the author).
35. Lee et al., "Abuse of 'Footprints,'" viewgraph MCL94:104a/08.
36. Ibid.
37. Lee et al., "Abuse of 'Footprints,'" viewgraph MCL94:104a/10.
38. Lee et al., "Abuse of 'Footprints,'" viewgraph MCL94:104a/11.
39. Lowder, "BMDO Request for Information," 2.
40. Lee, interview with the author.
41. Lowder, "BMDO Request for Information," 2.
42. Michael Nacht, interview with the author, Washington, D.C., 17 July 1996.
43. Ibid.
44. Nacht suggested that members of the following ACDA groups would have taken an interest in the Sparta briefing: Director's Office, Bureau of Strategic and Eurasian Affairs, Special Weapons Division, SCC Commission, Intelligence, General Counsel's Office, and Operations (Ibid.).
45. In an editorial by Sidney Graybeal and Keith Payne, the following reference was made to the Sparta study: "[Gronlund et al.'s] analysis is based on a calculation of the area that a THAAD-like system could theoretically defend—called a 'defensive footprint.'

However, *more recent and complete* computer simulations of a THAAD-like defensive footprint demonstrate that such a system has little or no significant capability against any realistic strategic missile attack" (Sidney Graybeal and Keith Payne, "ABM Debate Holds Cold War Echoes," *Defense News,* 9 January 1995, 20, emphasis added).

46. Postol sent copies of this letter to Ashton B. Carter (assistant secretary of defense for international security policy), John Harvey (deputy assistant secretary of defense for forces policy), John D. Holum (director, Arms Control and Disarmament Agency), Spurgeon Keeny (Arms Control Association), Laura Lee (Sparta, Inc.), Sidney Graybeal (Science Applications International, Inc.), Keith Payne (National Institute for Public Policy), Richard L. Garwin (Committee on International Security and Arms Control, National Academy of Sciences), John Holdren (chair, Committee on International Security and Arms Control, National Academy of Sciences), and Alexei Arbatov (chair, Subcommittee on International Security and Arms Control, Committee on State Defense, Russian State Duma).
47. Theodore A. Postol, letter to Malcolm R. O'Neill, 31 March 1995, 1, copy on file with the author.
48. Ibid.
49. Lisbeth Gronlund, George Lewis, Theodore Postol, and David Wright, "Documentation of Technical Errors in the BMDO Study 'The Abuse of "Footprints" for Theater Missile Defenses and the ABM Treaty,'" Main Appendix of letter by Theodore Postol to Malcolm O'Neill, 31 March 1995, 1–11.
50. As George Lewis and Theodore Postol explained in a separate technical paper attached as an appendix to Postol's letter to O'Neill, "[i]n order to find the optimal defended footprint, a balance must be struck between three related quantities: the detection range; the size of the area of the sky that can be searched by the radar; and the size of the defended footprint, which is in part determined by the distance versus time flyout characteristics of the interceptor. . . . Thus, the defended footprint is the result of a tradeoff between the search area detection range that can be achieved by the radar, and the size of the defended footprint, which depends on the time available for the interceptor flyout" (George Lewis and Theodore Postol, "Search Solid Angle Requirements and Detection Ranges for Theater and Strategic Ballistic Missiles," Appendix A of letter by Theodore Postol to Malcolm O'Neill, 31 March 1995, 1, copy on file with the author).
51. Lewis and Postol, "Search Solid Angle," 2.
52. Gronlund et al., "Documentation of Technical Errors," 3.
53. See Richard Garwin, FILE Memorandum, 10 October 1995, 2–3, copy on file with the author.
54. Gronlund et al., "Documentation of Technical Errors," 7.
55. O'Neill, statement, *Administration's Proposal,* 115.
56. Ibid.
57. Malcolm O'Neill, statement, *Department of Defense Authorization for Appropriations for Fiscal Year 1994 and the Future Years Defense Program,* 103d Cong., 1st sess., Senate Hearings, Committee on Armed Services, 10 June 1993 (Washington, D.C.: GPO), 345. As analyst Stanley Orman corroborates, "[c]ued from space-based sensors, the

THAAD system would have the ability to intercept attacking missiles at long range and high altitude. In addition to providing area defense, the system denies the adversary the use of deliberate or inadvertent countermeasures, which make the identification of warheads more difficult. The enlarged battlespace enables the defense to adopt shoot-look-shoot tactics, and to intercept warheads well away from their intended target" (Stanley Orman, "A Space Community Held Hostage," *Space News,* 30 October–5 November 1995, 14–18).

58. Postal, letter to O'Neill, 31 March 1995.
59. "The approximate launch point of an attacking missile is assumed to be provided by DSP satellites, which reduces the angular search required" (Gronlund et al., "Highly Capable Theater Missile Defenses," 6).
60. "The search area in each case depends not only on the anticipated detection range, but also on the size of the footprint to be defended, and is determined iteratively" (Ibid.).
61. In BMDO parlance, Gronlund et al.'s iterative radar search optimization mechanism is called "goaltending," and goaltending was not an assumption included in the Sparta study. As Lee explained, "[w]hat we were trying to do here, is perform it the way they [Gronlund et al.] did it, but we used the characteristics they told us about, using the concepts and the way that we employ radar, which is not goaltending. . . . Initially, they didn't tell us anything about goaltending" (Lee, interview with the author, 3).
62. Theodore A. Postol, letter to Malcolm O'Neill, 8 September 1995.
63. Postol, letter to O'Neill, 8 September 1995, 1. As Postol further explained in a letter to John R. Harvey, deputy assistant secretary of defense for forces policy, "I first spoke with Ms. Lee on the morning of 19 January 1995. During that conversation, all Ms. Lee was willing to tell me about her study was its name. She asked me for information about our interceptor model and we gave her that information on the afternoon of the same day. I again spoke with Ms. Lee on 20 January and she complained that she could not obtain reference three in our *Arms Control Today* article and suggested that perhaps this was because the reference did not exist. I asked her why she had not requested the reference five months earlier, before she published her study of our work, and she could offer no explanation. I sent her a copy of reference three that day and I have not talked to her since" (Theodore A. Postol, letter to John R. Harvey, 2 May 1995, 1, copy on file with the author).
64. Malcolm O'Neill, letter to Theodore Postol, 3 May 1995, 1, copy on file with the author.
65. Ibid.
66. Ibid.
67. Ibid.
68. See, e.g., Steven Epstein, *Impure Science* (Berkeley: University of California Press, 1996); Karin Garrety, "Social Worlds, Actor-Networks and Controversy: The Case of Cholesterol, Dietary Fat and Heart Disease," *Social Studies of Science* 27 [1997]: 727–73.
69. Sheila Jasanoff, "Contested Boundaries in Policy-relevant Science," *Social Studies of Science* 17 (1987): 198; see also Dorothy Nelkin, "The Political Impact of Technical Expertise," *Social Studies of Science* 5 (1975): 35–54.

70. See Lowder, "BMDO Request for Information," 3.
71. In addition to his position at IBM, Garwin is also an adjunct professor of physics at Columbia University, as well as a member of the National Academy of Sciences and the National Academy of Engineering; he sits on the Advisory Committee to the Arms Control and Disarmament Agency, and is also a Philip D. Reed senior fellow for science and technology at the Council on Foreign Relations. After completing his Ph.D. in physics under Enrico Fermi at the University of Chicago in 1949, Garwin went on to make fundamental contributions to the U.S. government air defense program in the 1950s. With the advent of long-range missile technology during the peak of the cold war, Garwin shifted his focus to missile defense, and he has made numerous public interventions on this subject since that time (see "The Garwin Archive—1990s," housed at the Federation of American Scientists Website, online at http://www.fas.org/rlg/90–96.htm).
72. Garwin faxed his findings to J. M. Cornwall (UCLA), R. Gottemoeller (Center for International Security and Arms Control [CISAC]), R. G. Henderson (JASONs), J. P. Holdren (CISAC), J. Husbands (CISAC), S. M. Keeny (Arms Control Association [ACA]), M. O'Neill (BMDO), J. E. Pike (Federation of American Scientists [FAS]), T. A. Postol (MIT), and R. Sherman (ACDA).
73. Lee et al., "Abuse of 'Footprints,'" viewgraph MCL94:104a/04.
74. Garwin, FILE Memorandum, 4. Garwin's findings on this point received support from Ted Gold, an independent defense analyst with the McLean, Virginia, consulting outfit Hicks and Associates. When asked about the TMD footprint controversy, Gold remarked that "THAAD certainly does have some footprint against strategic missiles" (Ted Gold, interview with the author, McLean, Va., 18 July 1996).
75. Garwin, FILE Memorandum, 2, emphasis added.
76. Ibid., 4.
77. See Gronlund et al., "Highly Capable Theater Missile Defenses," 6.
78. Garwin, FILE Memorandum, 1.
79. Lowder, "BMDO Request for Information," 2.
80. Laura Lee, e-mail to the author, 22 August 1999.
81. Richard L. Garwin, "Theater Missile Defense, National ABM Systems, and the Future of Deterrence," in *Post-Cold War Conflict Deterrence,* ed. National Research Council (Washington, D.C.: National Academy Press, 1997), 182–200.
82. Ibid., 192–93, emphasis in original.
83. Ibid., 193.
84. Lee, interview with the author.
85. Ibid.
86. O'Neill, letter to Postol, 1.
87. On the distinction between "constitutive" and "contingent" forums for scientific controversy, see Harry M. Collins and Trevor J. Pinch, "The Construction of the Paranormal: Northing Unscientific Is Happening," in *On the Margins of Science: The Social Construction of Rejected Knowledge,* ed. Roy Wallis (Keele, England: University of Keele Press, 1979), 237–70.

88. Alexei Arbatov, letter to Malcolm O'Neill, received 20 July 1995, 1, copy on file with the author.
89. Ibid., 1.
90. See, e.g., Graybeal and Payne, "ABM Debate."
91. As Mendelsohn explains, "[1]arge numbers of high-performance missile defense systems, even if disingenuously defined as 'theater' systems to circumvent ABM Treaty constraints, are likely to be perceived as the first step toward preparing a 'base' for national defense. In turn, that perception will chill efforts to shrink the strategic arsenals of Russia and the United States. In the worst-case scenario, large numbers of highly capable TMD deployments might actually reinvigorate the strategic arms race" (Jack Mendelsohn, "A Tenth Inning for Star Wars," *Bulletin of the Atomic Scientists* 52 [1996]: 28). When combined with space-based sensors, Pike writes that TMD will work "well enough to provoke offsetting buildups of offensive missiles by potential adversaries—a prescription for a new and spiraling arms race of offenses versus the defenses such as those now envisioned" (John Pike, "Theater Missile Defense Programs: Status and Prospects," *Arms Control Today* 24 [1994]: 14). For additional analysis, see Gordon R. Mitchell, "The TMD Footprint Controversy," Space Policy Project E-print, Internet, Federation of American Scientists Website, 12 September 1996, online at: http://www.fas.org/spp/eprint/gm1.htm.
92. Yuri K. Nazarkin and Rodney W. Jones, "Moscow's START II Ratification: Problems and Prospects," *Arms Control Today* 25 (1995): 8–14.
93. Quoted in Jack Mendelsohn, statement, "Arms Control at the Clinton-Yeltsin and Clinton-Jiang Summit Meetings," Program on Nuclear Policy Panel Discussion, 20 October 1995 (Washington: Atlantic Council, 1995), copy on file with the author.
94. Pavel Povdig, statement, *Congressional Record,* 22 December 1995, S19210.
95. Quoted in "Russian Analyst Warns of U.S. 'Back Door' Strategic Defense," *Missile Defense Report* 2 (26 April 1996): 1–2.
96. Strobe Talbott, "America and Russia in a Changing World," speech at 50th anniversary of Harriman Institute, 29 October 1996, Internet, Federation of American Scientists Website, online at http://www.fas.org.
97. Ibid.
98. Richard L. Garwin, "Keeping Enemy Missiles at Bay," *New York Times,* 28 July 1998, A15.
99. United States Department of Defense, "THAAD Test Flight Does Not Achieve Intercept Target," Department of Defense Press Release, 29 March 1999, Internet, Coalition to Reduce Nuclear Danger Website, online at http: www.clw.org/coalition/thaad99a.htm.
100. Bradley Graham, "Anti-missile Weapon Fails," *Pittsburgh Post-Gazette,* 30 March 1999, A-15.
101. Larry Welch, task leader, *Report of the Panel on Reducing Risk in Ballistic Missile Defense Flight Test Programs,* 27 February 1998, Internet. Federation of American Scientists Website, Online at http://www.fas.org/spp/starwars/program/welch/welch-1.htm.
102. See, e.g., Jesse Helms, statement, *Congressional Record,* 6 February 1996, S917–S919.

103. See Mark Genrich, "Suing the President; Sen. Kyl Leads Effort for Congress," *Phoenix Gazette,* 24 July 1996, B9.
104. The widely heralded 1998 Welch Commission Report echoed similar findings of a less publicized GAO report released two years earlier. Specifically, GAO found that "the THAAD program has already experienced significant cost, schedule, and technical performance problems" (United States General Accounting Office, *Issues Concerning Acquisition*). Reflecting on technical problems in the THAAD program, GAO concluded: "Our work has repeatedly shown that when production of weapon systems began on the basis of schedule or other considerations rather than on the basis of technical maturity, major design changes were often needed to correct problems. The design changes frequently led to additional testing and costly retrofit of units already produced." (Ibid., 10).
105. Welch, task leader, *Report of the Panel on Reducing Risk,* 7, emphasis in original.
106. Ibid., 24. Postol has argued that the poor technical research conducted by SDIO and BMDO has created a spiraling negative cycle, in which chronic failure feeds a brain drain effect that further degrades the technical quality of the organization: "And if you talk with these people at BMDO, it's pathetic. They have people there they call chief scientists. . . . These people they have now are jokes. They're jokes. And they're loaded with all these kinds of characters: chief scientists, principal scientists. Why do we have these jokes? Because anybody who's serious does not want to take this job and keep it" (Theodore A. Postol, interview with the author, Cambridge, Mass., 10 May 1995, full transcript on file with the author).
107. Chandra Mukerji, *A Fragile Power: Scientists and the State* (Princeton, N.J.: Princeton University Press, 1990), 3–21.
108. See Herbert N. Foerstel, *Secret Science* (Westport, Conn.: Praeger, 1993); Postol, interview with the author.
109. Isikoff, "Carving Out an SDI Niche," F1.
110. Quoted in David E. Sanger, "Pentagon and Critics Dispute Roles of Space Arms Designers," *New York Times,* 5 November 1985, C1.
111. "Hughes to Focus Air Defense on THAAD," *Defense Daily,* 24 January 1992, 117.
112. "Hughes Assumes Lead," 182.
113. The relevance of such concerns would seem to grow in light of Sparta's checkered history as a responsible steward of money flowing from BMDO coffers. Sparta was the same firm that triggered a GAO audit by spending $560,000 of DOD money for employees' conferences in Jamaica, Hawaii, Mexico, and the Grand Cayman Islands in 1994 (see John Mintz, "Sasser Cites 'Abuses' by Defense Firms," *Washington Post,* 4 March 1994, B2).
114. *HLI Lordship Indus., Inc. v Committee for Purchase from the Blind,* 615 F. Supp. 970, 978 (E.D.Va. 1985).
115. See Jerry W. Markham, "The Federal Advisory Committee Act," *University of Pittsburgh Law Review* 35 (1974): 557; Sheila Jasanoff, "Peer Review in the Regulatory Process," *Science, Technology, and Human Values* 10 (1985): 20–32.

116. See U.S. Congress, *Department of Defense/Strategic Defense Initiative Organization Compliance with Federal Advisory Committee Act,* 100th Cong., 2d sess., Senate Hearings, Committee on Governmental Affairs, 19 April 1988 (Washington, D.C.: GPO, 1998). During these hearings, Senator Levin noted that "SDIO's compliance has been dismal and requires a major overhaul" and that "SDIO has repeatedly violated FACA by using advisory committees without chartering them as required by law" (Ibid.). (Carl Levin, statement, *Department of Defense/Strategic Defense Initiative Organization Compliance with Federal Advisory Committee Act,* 100th Cong., 2d sess., Senate Hearings, Committee on Governmental Affairs, 19 April 1988 [Washington, D.C.: GPO, 1998], 6). An investigation by the Senate Committee on Governmental Affairs specifically found five unchartered committees advising SDIO in violation of FACA: "kinetic energy, lethality, national test bed, sensor steering group, and Eastport" (see Markham, "Federal Advisory Committee Act," 557). Sparta, Inc. was involved in the latter group, submitting a proposal in 1987 to provide SDI modeling and simulation support to the Eastport panel (Andre M. van Tilbourg, Memorandum, in *Department of Defense/Strategic Defense Initiative Organization Compliance with Federal Advisory Committee Act,* 100th Cong., 2d sess., Senate Hearings, Committee on Governmental Affairs, 19 April 1988 [Washington, D.C.: GPO, 1998], 188).
117. In considering amendments to FACA in 1992, Congress indicated that "technical and standard setting" groups fall under the scope of the act (see U.S. Congress, *Department of Defense/Strategic,* 6). There is a strong legal argument that the "analysis validation team" (AVT) formed at BMDO (Lee, interview with the author) to analyze footprint methodology met FACA's classification as a "technical and standard setting" group, and therefore fell within the law's prescribed openness requirements. If BMDO's AVT's was in fact working as an "advisory committee" during the Sparta-MIT footprint controversy, this committee appeared to contravene FACA guidelines on several counts.

    First, AVT appeared to violate section 4(b), by providing advice to the executive branch as an advisory committee without securing a charter. Second, AVT appeared to violate section 5(c)(1), by operating without a plan to assure balance of viewpoints in committee membership. Third, by failing to publicly release documents related to the Sparta footprint study, AVT appeared to violate section 7(e), calling for openness and transparency in meetings. Fourth, by holding meetings without providing advance notice in the *Federal Register,* AVT appeared to violate section 11(c)(6) (see Gordon R. Mitchell, "Sparta's TMD Footprint Advice to BMDO: Overstepping the Boundaries of FACA?," Issue brief for Senator David Pryor and Senator Carl Levin, 24 July 1996, subsequently published as Space Policy Project E-print, Internet, Federation of American Scientists Website, 8 August 1996, online at http://www.fas.org/spp/eprint/gm2.htm). AVT's disregard for section 5(c)(1) appears to have offended most seriously the spirit of FACA. BMDO's objective in commissioning the Lee et al. footprint study from Sparta, to "[u]nderstand why the conclusions they [Gronlund et al.] reached were at odds with the BMD community" (Lee et al., *Abuse of 'Footprints'*), and discussions leading up to the decision to commission the study intimate that the motivation for this advisory action was to insulate established internal BMDO footprint methodology from outside criticism.

Explaining the series of events that prompted BMDO to commission the Sparta study, Lee stated, "[a]pparently, Keith Payne's been doing some piece about the policies of the ABM Treaty. And when he was doing his work for BMDO, he briefed the BMDO staff about his conclusions, and in there, he talked about Gronlund's footprints. And they said 'Don't put any of their footprints in your paper,' you know, 'You should check them with *our people* [Sparta]. Our analysts don't get the same footprints, so when you put them in your paper, that looks strange, because it looks like a BMDO product, and you get a different footprint" (Lee, interview with the author, emphasis added). Additionally, the dearth of contact between the Sparta team and the authors of the *Arms Control Today* article prior to completion of the Sparta study is in direct tension with FACA's legislative history, which indicates that in pursuing balanced inquiry, "consideration shall be given to including members representing interests that would be directly affected by the committee's work and to obtaining technical and other relevant expertise" (see U.S. Congress, Department of *Defense/Strategic,* 17). As Postol explained, "existence [of the Sparta study] was only brought to my attention in early January 1995, months after its publication, when it was referred to [in an editorial by Keith Payne and Sidney Graybeal]. . . . No attempt was made by the authors of the study or by BMDO contract monitors to contact us to check the accuracy of statements made about our work in the BMDO study. There was no attempt by the BMDO-sponsored authors or the BMDO staff to query us about discrepancies between our results and those found by the BMDO-funded analysts" (Postol, letter to O'Neill, 31 March 1995).

118. Jasanoff, "Peer Review," 24.
119. Black programs represent the highest level of classification in the U.S. secrecy hierarchy (see Foerstel, *Secret Science*).
120. Laura Lee, interview with the author.
121. Malcolm O'Neill, "One on One," interview, *Defense News,* 3–9 April 1994, 30.
122. Lee, telephone conversation with the author, 17 July 1997.
123. Gordon R. Mitchell, "Another Strategic Deception Initiative," *Bulletin of the Atomic Scientists* 53 (1997): 22–23.
124. Garwin, "Theater Missile Defense," 193.

CHAPTER FIVE

Living with the Legacy of Strategic Deception in Missile Defense Advocacy

> After the people has set up a sanctuary of Zeus Syllanios and Athena Syllania, after the people has arranged itself by tribes and obes, after the people has set up a council of thirty including the kings, let them gather from season to season for the festival of Apellai between Babyka and Knakion; let the elders introduce proposals and decline to introduce proposals; but let the people have the final decision. But if the people make a crooked utterance, let the elders and the kings decline it.[1]
>
> —Spartan Great Rhetra, approximately 750 B.C.

Boilerplate histories of ancient Greece tend to play up contrasts between Athens and Sparta, counterposing the glory of progressive Athenian democracy against the dark shadow of Sparta (often portrayed as some kind of totalitarian boot camp).[2] Such accounts not only overstate the democratic openness of the Athenian polis; they also ignore the fact that Sparta's political institutions were far from tyrannical.[3] On this latter point, consider the Spartan Great Rhetra, what historian Guy Dickins calls a "treaty" between the ruling kings of Sparta and their people.[4] Enacted sometime in the seventh century B.C., the Great Rhetra gave Spartan citizens a voice in political affairs, and as historian A. H. M. Jones notes, "[t]he reports on popular assemblies at Sparta show that there was debate."[5] While the actual extent of political power enjoyed by the Spartan popular assembly is a matter of no small dispute in historical circles, there is little doubt that the existence of the Great Rhetra proves that Sparta was not a totalitarian dictatorship, but rather a sophisticated and complex society, a military state with at least the trappings of democracy.[6] These unique aspects of Spartan society have attracted the attention of many classical historians.[7]

Defense planners have also shown interest in Sparta's history as a heuristic resource for policy analysis. Struggling to develop reliable models that would explain Soviet defense spending patterns during the cold war, economists and Sovietologists such as Rush Greenslade, Henry Rowen, and Robert Gates proposed development of intelligence models based on an analogy between Sparta and the Soviet Union.[8] Acting on these recommendations in 1989, the U.S. Office of the Secretary of Defense commissioned the RAND Corporation to produce a remarkable study titled "Soviet Defense Spending: The Spartan Analogy."[9] In this report, RAND analyst Alvin H. Bernstein suggested that Sparta's "secretive oligarchy of elder statesmen" was "somewhat analogous to the Politburo during Brezhnev's last years."[10] Bernstein noted other likenesses: "The militarization of Spartan society evolved from the need to control a large, ethnically identifiable and unassimilable under-class. This need finds an analogue in the Soviet Union's need to control its unassimilated nationalities."[11]

One prominent undercurrent running through Bernstein's report was the notion that the flip side of the Sparta-Soviet Union analogy was a parallel between Athens and the United States. An analysis of the "restrictive institutions" that are part and parcel of a military garrison state "will be more important in understanding the workings of both the Soviet and Spartan economies than they are for the more secure and less politically constrained market economies of the West or of ancient Athens," Bernstein stipulated.[12] With these comments, Bernstein reflected a common tendency noted by classicist Michael Whitby: "Contemporary democrats have an inbuilt preference for Athens as a democratic power, and there is something very satisfying about the notion that Sparta, her great rival and destroyer, should herself be fatally flawed, with the mainspring of her economic system containing the worm of her future destruction."[13]

With more than a decade having passed since the completion of Bernstein's report, it might be appropriate to revisit his findings in light of recent developments. The Soviet Union has dissolved, the Warsaw Pact has unraveled, and Russia's defense spending has shrunk to a fraction of cold war levels. The fall of the Berlin Wall appears to have revealed the prescience of Bernstein's Sparta-U.S.S.R. analogy. But what about the flip side of the analogy? After its apparent cold war victory, is the United States proudly carrying on the grand tradition of Athenian democracy? It hardly seems so. In fact, the unquestioned political dominance of military imperatives in post–cold war America seems downright Spartan in character. In his 1998 book, *Fortress America*, journalist

William Greider writes that the United States has drifted into an odd type of command economy, where free market logic basically prevails, except in the largest economic sector, government defense spending.

> Fortress America remains mobilized to fight the big one but justifies itself now with vague threat scenarios that envision fighting two wars at once, twin regional conflicts that will be smaller in scale but simultaneous. Instead of a robust debate over new priorities or skeptical questioning of these fanciful premises, the political elites in both parties have settled into denial and drift—a status quo that argues only over smaller matters, like which new weapon system to fund or where they will be built. Defense spending, as one strategic analyst put it, has become "the new third rail of American politics." Most politicians are afraid to touch it.[14]

In *Citadel Culture,* historian O. K. Werckmeister offers a similar, sobering assessment of public life in the West today. Werckmeister's thesis is that citizens of democratic industrial societies such as the United States live in a culture of the citadel, a military-dominated culture where "the officeholders of democracy are being protected by elaborate security apparatuses as soon as they venture out into the public sphere."[15] Although the machinery of democratic decision making remains in place, notions of popular sovereignty are quickly trumped by military secrecy. "We are compulsively conscious of the all-pervasive military conditions of our existence, but we cannot gauge how they function technically and how they are guided politically. The democratic institutions that allow us to empower, change and control our governments have made sure that this entire operational system of weaponry is withdrawn from our political initiatives," explains Werckmeister.[16]

Instruments such as the so-called black budget and special access programs afford American defense officials latitude to pursue far-flung military projects that are completely insulated from popular objections raised through democratic channels.[17] For example, Lt. Col. Robert M. Bowman (USAF, Ret.), director of military space programs (including the precursor to SDI) in the Ford and Carter administrations, describes the strange case of the black-budgeted Precision Guided Re-Entry Vehicle (PGRV) program. This program was initiated to yield an "anti-ABM" weapon designed to enhance the first-strike effectiveness of U.S. nuclear forces.[18] In 1974, PGRV's existence "was purposefully leaked, the program was canceled, and then it continued in what we call the 'deep black.' The cancellation was just a ploy to fool both the

Russians, the American public, and Congress," Bowman recounts.[19] Huge vistas of opportunity opened up for similar ploys at the dawn of the Star Wars era, when Reagan kicked off his administration by boosting black budget weapons appropriations by 1400 percent during his first year in office.[20]

By 1985, political momentum for Star Wars had built to the point that it was possible for the Pentagon to subvert democratic control by openly defying Congress. In 1984 and 1985, Congress passed legislation requiring Pentagon officials to pursue arms control with the Kremlin as a precondition to further antisatellite (ASAT) testing.[21] As Representative George Brown (D-CA) noted, the Reagan administration subsequently made "a mockery of the law," by spurning such negotiations and going ahead with a Star Wars test of the Miniature Homing Vehicle ASAT weapon in September 1985. So blatant was this exercise of imperial executive power that on the floor of the House, Brown suggested drawing up articles of impeachment against Reagan.[22] Three years later, SDIO joined a purely civilian research project on nuclear-powered rockets, in order to design launch vehicles that could "carry particle beams, lasers, and homing rockets for missile defense purposes" into space.[23] Knowing that "public discussion of the dangers of nuclear-powered rockets would create difficult political and public relations problems,"[24] Pentagon officials hid this program (called TIMBERWIND) in the black budget.[25] Preston J. Truman, director of Downwinders, Inc., videotaped air force officials lying blatantly to citizens at a 1991 environmental impact hearing in Utah on the rocket program. As Truman recounts, they "outright lied to the public during hearings and to regional media stating that the proposed nuclear rocket was 100% civilian and had no military applications and that there was no relationship whatsoever to the SDI program, while knowing full well this was not the case."[26] Most recently, the House Appropriations Committee discovered that after it pulled the plug on the MEADS missile defense program, BMDO turned around and secretly shifted $2 million from another program to continue funding for MEADS in 1999.[27]

The case of PGRV's recess into the "deep black" budget, the massive hikes in black budget appropriations, Reagan's illegal ASAT test, the TIMBERWIND and MEADS projects, and the numerous other instances of strategic deception in missile defense advocacy explored in this book suggest that in the context of missile defense, our nation's democratic guarantees have become mere speed bumps on the Pentagon's gold-plated road to weaponization of American society. If you think such a society sounds a lot more like Sparta than Athens, your assessment would parallel the findings of scholars who have analyzed the significance of the Great Rhetra as a foundational document in Spartan history. Although the Great Rhetra instilled what historian A. Andrewes calls an ostensibly democratic "habit of free discussion" in Sparta; "[i]t was of course a pretence."[28] The last line of the Great Rhetra, "[b]ut if the people make a crooked utterance, let the elders and the kings decline it," ensured a way for the Spartan kings to overrule any popular sentiment questioning the necessity of incessant military buildup and warmaking.

There is obvious gravity in the proposition that as a guarantor of popular sovereignty, the U.S. Constitution has been eroded to the point that as a working document it now resembles the Spartan Great Rhetra. Taking such a warning sign as a point of departure, this chapter reflects on the lessons learned from the case studies explored in this book. Configuring the selected episodes of missile defense advocacy as interconnected events, it becomes possible to note links that tie the cases together, as well as discontinuities that represent evolutionary wrinkles in the history of BMD promotion campaigns. Insight gleaned from this panoramic perspective not only helps elucidate the overall dynamics of strategic deception in missile defense advocacy; it also brings historical experience to bear on current BMD discussions in the wake of Congress's May 1999 decision to mandate deployment of NMD "as soon as is technologically possible."

What are some general characteristics of historical attempts to bolster the case for missile defense through Spartanesque schemes of strategic deception? How do such deception campaigns affect ongoing public discussions about proposals for new systems costing billions of dollars? What can be done to counter the corrosive effects of strategic deception on American democracy? This chapter explores such questions in a three-part discussion. Part one focuses on the continuities and divergence in missile defense advocacy over time, reflecting on the common features and new twists in strategic deception's historical legacy. Part two applies lessons of this legacy to contemporary missile defense discussions, looking at ways in which past experience enables and constrains decision making on pending proposals for NMD. The chapter ends with a constructive turn in part three, where specific alternatives and suggestions for change are considered.

Common Threads and New Twists in the Legacy of Strategic Deception

Just as Newt Gingrich stressed in 1997 that contemporary missile defense debates should be informed by the experience of successful British air defense in World War II,[29] a strong argument can be made that the insights flowing from recent episodes of strategic deception in missile defense advocacy should be condensed and applied to current BMD discussions. In the pages that follow, I derive some of these insights by distilling perennial patterns of reasoning found in this book's case studies. These patterns could be used as resources to support future critiques of missile defense advocacy, and they could be used to inform analyses of similar scientific controversies outside the missile defense context. In what follows, I outline these perennial patterns of missile defense advocacy and review their characteristics.

The Paradox of Secret Scientific Objectivity

Chapter one explored the reasons why scientific truth claims resonate in the public sphere. Many suggest that science's rhetorical purchase in this regard is due to popular faith in the notion of scientific objectivity as a marker of unbiased and reliable knowledge. But what exactly is "objectivity"? As a rhetorical resource, objectivity confers status and power to truth claims; it signals to public audiences that a particular fact was produced out of a kind of "view from nowhere" (to use Nagel's positivist phrase). While this notion of "pure" objectivity still enjoys remarkable cachet as a rhetorical resource in public argument, positivism has been discarded to the dustbins of academic history in philosophy of science circles. A host of postpositivist studies of science have shown that in practice, reliable scientific knowledge is not snatched out of a "view from nowhere," but is instead worked up through a grinding process of intersubjective criticism.[30] On these accounts, scientific communities arrive at objective knowledge claims through a process of disclosure and discourse, where the checking function of dialogic interchange enables communities to thematize and examine idiosyncratic viewpoints critically, by making transparent and testing the background assumptions that undergird them. Because scientific truth is hitched to the notion of communication, truth claims require necessarily that their advocates submit such claims to the arena of intersubjective argumentation before legitimately redeeming the full rhetorical force of scientific objectivity in public spheres of deliberation.

Missile defense advocates frequently skip this last step, using secrecy to back their public arguments with the sanction of scientific objectivity, without first submitting such claims to intersubjective inspection and criticism. For example, Edward Teller pulled off his X-ray laser bluff by saying that he would disclose the scientific basis of his outrageous scientific claims if it were not for the onerous classification rules that prevented him from releasing the information. Similarly, in defending the Patriot's Gulf War performance, Raytheon spokesperson Robert Stein claimed substantial argumentative weight from secret studies. In public debate, Stein presented the studies as scientifically sound, although their secret status precluded detailed peer review or discussion of underlying methodology or assumptions. Likewise, when Keith Payne and Sidney Graybeal attacked Gronlund et al.'s computer footprint methodology for calculating the theoretical capabilities of THAAD in 1994, they cited a secret (and non-peer reviewed) study commissioned by BMDO as the key counterevidence. Even after the initial layers of classification shrouding this study were peeled away, Laura Lee, Sparta's lead researcher, refused to disclose key charts and data in subsequent scientific debates. Lee's justification for witholding such evidence was not based on any national security rationale, but instead the fear that disclosing such charts would result in them "being criticized out of context."[31]

While these strategies of manufacturing what might be called "secret objectivity" have proven effective as political levers, used to force massive budget hikes for missile defense projects, the intense secrecy involved has also been what John Tirman calls a "kiss of death" for Pentagon scientists. According to Tirman, "[w]hereas in nonclassified work both the company and the employee generally benefit from a rich and continuous cross-fertilization of ideas, this is not and cannot be so to the same extent in the blinkered environment of classified work. Rather, one tends to become an expert in a narrowly defined area, while the balance of one's training suffers from disuse. Professionally, this is the kiss of death in fields in which entire technological revolutions take place on the order of every five years."[32] Arvin S. Quist of the Oak Ridge Gaseous Diffusion Plant notes that security classification can "keep others ignorant of information already discovered, results in needless duplication of effort and delays scientific and technological advances."[33] Scientists and military leaders tend to agree with Tirman and Quist's conclusions. As William Burr and Thomas Blanton of the National Security Archive explain, "[s]ince the final days of the MED [Manhatten Engineer District], scientists and military leaders have assumed that nuclear secrecy had the potential to retard technological progress by cutting off formal and informal communications between scientists."[34]

By definition, secrecy and objectivity are mutually exclusive categories, since communicative openness is a major determinant of the quality of scientific inquiry.[35] Where the conditions for such criticism are lacking, objective research is jeopardized and the creative energy in scientific inquiry is dissipated.[36] William Broad's book *Star Warriors* documents how the secrecy surrounding Teller's X-ray laser project closed in on Livermore researchers, prompting them to lose perspective and drift into an isolated culture of technical failure.[37] The testimony of many defectors who quit the project in disgust chronicles vividly how this debilitating process unfolded. In the case of Patriot, extensive secrecy surrounding the Patriot's deployment, operation, and evaluation interfered with official efforts to ascertain reliable "experimental data" on Patriot performance.[38] Furthermore, the recent conclusions of Jeremiah Sullivan's American Physical Society review board indicate that classification of data from the Gulf War still blocks the path of scientific inquiry. The APS review board's findings appear to support Theodore A. Postol's conclusion that the BMDO culture of secrecy "has a corrupting effect on every aspect of weapons development," and that "denial of failure leads to institutionalized failure."[39] During preparation of "The Abuse of 'Footprints' for Theater Missile Defense and the ABM Treaty," Lee et al.'s failure to communicate effectively with the Gronlund et al. research team undermined the integrity of their own investigation, resulting in the production of a flawed study that excluded key components of the MIT group's footprint model. These examples illustrate the pitfalls involved in attempts to pursue objective science in secret.

However, it would be a mistake to overstate the generalizability of these instances. History is littered with secretive projects that have been completed successfully by military scientists. In the period from 1939 to 1945, a British top-secret group of naval scientists developed revolutionary anti-U-boat explosives and invented radio-controlled planes for laying smoke screens over assault beaches.[40] During a similar time period, R. V. Jones and other physicists made extraordinary advances in air defense technology under the constraints of severe secrecy.[41] Project ULTRA involved a highly successful cryptography effort by World War II Allies that was so secret its existence was concealed until 1973.[42] Then, of course, there was the Office of Scientific Research and Development (OSRD) in the United States, an ultrasecret organization that successfully built an atomic bomb in the Manhattan Project. In his 1946 book, *Scientists Against Time,* historian James Phinney Baxter chronicled the dramatic efforts of Manhattan Project scientists to engineer the world's first nuclear weapon. Although Baxter's overall narrative generally celebrated the atomic scientists' efforts, his cautionary closing words served as a stark warning that the secrecy of the Manhattan Project should not become a habit in future military projects.

> It was our good fortune, in World War II, to have a better organization of science for war than our friends or foes had. There is no reason to believe, however, that OSRD is the pattern we should follow in organizing science for war purposes in the days to come. . . . What is needed is the reorganization of scientific work within the armed services in such a way as to draw in first-rate scientists and give them the freedom necessary to do their best work.[43]

The Manhattan Project, ULTRA, and other secret military projects show that scientific openness is not a necessary condition for technical success. With the blessings of "good fortune" (to use Baxter's phrase), such projects can succeed. However, philosopher Helen Longino and others have shown that robust intersubjective critique in the scientific process is necessary to counter "Teller-esque" attempts to politicize science and hold entire projects hostage to the idiosyncratic and personal commitments of a few researchers with ulterior motives.[44] Intersubjective dialogue and critique, on this account, "block" the unwarranted domination of scientific practice by any one subjective preference. "It is the possibility of intersubjective criticism, at any rate, that permits objectivity in spite of the context dependence of evidential reasoning," explains Longino.[45] Subjective preferences, once out in the open, can be evaluated and either discarded or embraced collectively within the ambit of an objective enterprise. "As long as background beliefs can be articulated and subjected to criticism from the scientific community, they can be defended, modified, or abandoned in response to such criticism," argues Longino.[46]

World War II scientists achieved success without the benefit of such robust criticism because their cloistered research communities pursuing wartime projects shared a commitment to quality research. However, the corrupting influences of the military-industrial complex have created different constellations of power and more complicated mixtures of motives in post–World War II projects such as missile defense. Historical experience suggests that if missile defense science "remains cut off from the best physics minds in the country, and largely exempt from meaningful congressional oversight," as I noted recently in a piece for the *Bulletin of the Atomic Scientists*, "there is every reason to expect that new projects will careen out of control and their handlers will fabricate evidence to keep them afloat, as happened with Edward Teller's far-flung X-Ray laser campaign in the early 1980s."[47] The powerful interlocking interests that have coalesced around missile defense projects have built up political and economic inertia behind BMD to the point that politicized decision making now proceeds in a fashion largely unhinged from the technical constraints of physical engineering. The United States seems caught in the throes of what former Center for Defense Information analyst Sanford Gottlieb calls a "defense addiction," a compulsion for defense spending fed by "the eagerness of members of Congress to hold onto military contracts and facilities for their constituents."[48]

The Dangerous Momentum of Runaway Deception

One clear feature of government cover-up schemes ranging from Nixon's Watergate to Clinton's "zippergate" is a tendency for initial deceptions to quickly mushroom into elaborate webs of follow-on falsehoods. Protection of an initial lie often requires secondary and tertiary layers of deception, implemented to insulate the original canard. Underneath each justification for secrecy there lies a series of other unstated justifications for more secrecy. "[S]ecret practices permit abuses to grow, with corrupting effects on those who are empowered to deceive and manipulate others undetected. The practices of deception tend to spread precisely because they are so tempting and because of the power they confer," Sissela Bok explains.[49] As Presidents Nixon and Clinton learned in impeachment hearings, the danger of such "runaway deception" is that it is easy to get caught up in such a process as it spirals out of control. This can lead to unanticipated and sometimes dire consequences (even war), because deception strategies can take on a life of their own, leaving the original intentions and designs of deceivers behind.

By focusing on specific instances of seemingly benign or even beneficial strategic deception, missile defense advocates project the misleading illusion of deceptive precision. This illusion suggests that deception is a discrete communicative tool that can be deployed adroitly to accomplish exact ends. For example, a 1987 Joint Chiefs of Staff Memorandum ordered numerous branches of the military to "update their deception programs" in order to mislead the Soviets about U.S. weapon capabilities.[50] Such precision is extremely difficult to achieve in practice, however, because of the dangers posed by runaway secrecy. To safeguard the integrity of secret programs, sprawling information-control bureaucracies tend to expand their resources and authority along a steadily escalating curve.

This spiraling tendency is magnified by the fact that institutionalized strategic deception programs have runaway momentum built into their bureaucratic structures. As Burr, Blanton, and Schwartz detail, the DOD alone pays "hundreds of millions" of dollars each year to support a secrecy infrastructure designed to protect classified information.[51] The federal employees on this secrecy payroll are charged with storing secrets, enforcing classification rules, and conducting security clearance investigations. The bureaucratic momentum toward secrecy created by such a workforce has led to dramatic instances of spiraling classification and deception in the missile defense context.

Runaway deception occurred on a basic level in appropriations hearings held during Teller's heyday from 1983 to 1988, and immediately after the Persian Gulf War in 1991, when U.S. citizens and their elected representatives were hoodwinked by inflated, distorted, and fabricated evidence of missile defense feasibility and performance. Deceptive schemes in these episodes were designed initially to mislead foreign publics in the Soviet Union and Israel, but they ended up tricking the U.S. Congress as well. Such deception enabled missile defense advocates to cash in on rich new contracts worth billions of dollars. Apart from the sheer waste and expense involved, the fact that such democratic decisions were made with fundamentally flawed data is a matter of grave concern.

A more specific case of runaway deception involved Lawrence Livermore National Laboratory (LLNL) scientist Hugh DeWitt. In 1991, DeWitt wrote an article for Stanford University's alumni magazine, criticizing the lab's blatantly politicized Star Wars promotion campaign in the mid-1980s.[52] Even though DeWitt based all of his claims on open source materials, lab officials informed DeWitt two months later that his article constituted a "Class A security infraction."[53] DeWitt found it hard to believe that lab officials had issued such a reprimand, given that he had not used classified information in his article. Yet, when DeWitt told journalist Robert Scheer about the incident two years later, security officials admonished him again, claiming that the number of allegedly classified items in DeWitt's original article (nine) was itself classified information, which DeWitt had disclosed inappropriately to Scheer! "Ten months after this, on 8 March 1994, DeWitt's security clearance was revoked and his office moved thirty feet (nine meters), from inside a fenced-in area to an unclassified area," explain Burr, Blanton, and Schwarz.[54] All of this took place without Defense Investigative Services (DIS) officials ever proving that DeWitt had breached classification rules. Ultimately, Department of Energy (DOE) Secretary Hazel O'Leary ruled that this overzealous enforcement of classification rules was unacceptable, and she restored DeWitt's security clearance after his appeal.[55]

The DeWitt example illustrates how politically motivated strategic deception can quickly expand from primary (classification) to secondary (classification of classification decisions) and tertiary (security clearance revocation) levels of information control. In attempts to censor Theodore Postol's criticisms of the Patriot's Gulf War accuracy, Pentagon officials employed a similar "Kafkaesque" spiral of secrecy. Reporters for the *New Scientist* explained how Postol's publication of an article in the journal *International Security* based solely on open source materials led to multiple layers of retaliation by DIS officials.

> On 13 March [1992], Postol was visited by an investigator from the Defense Investigative Service. The DIS officer wanted Postol to attend a classified meeting to discuss where he had obtained the information for his article. Postol refused, saying that if he did, he really would learn secret information about the Patriot, which would prevent him from talking about it. The investigator then informed him that he would have to stop discussing the article in public anyway, because the U.S. Army had decided that it contained secret data. If Postol refused, he would be in violation of his secrecy agreement with the government, and could lose his security clearance. Postol says he found this order incredible, and asked to have it in writing. More than a week later, on 19 March [1992], he was told that a letter was waiting for him at the Mitre Corporation, a nearby military contractor. In a Kafkaesque twist, the letter itself was classified, so Postol is refusing to read it.[56]

Ultimately, a Department of Defense review board upheld Postol's right to publish his article, and ruled that "no further action is warranted."[57] Although in this case Postol's arguments eventually found the light of day, other examples show how runaway deception can lead to unanticipated and adverse policy outcomes. For example, in 1967 runaway deception resulted in Middle Eastern war, when, the Soviet Union attempted to inflate deliberately the threat of Israeli military mobilizations against Syria. "[T]he evidence pointing to deliberate deception about the supposed Israeli threat to Syria is overwhelming," explains Handel.[58] While this Soviet strategy worked for a time as a pressure tactic against the United States, eventually the campaign to flood Arab newspapers with exaggerated estimates of Israeli troop deployments inadvertently triggered the Six Day War, by sparking Israeli pre-emption. "Above all, this example illustrates the difficulty of controlling a deception operation involving volatile and unstable allies and of knowing where it will lead. The 'threat to Syria' operation is the quintessential runaway deception operation. It also backfired on the instigator, creating results that were exactly opposite of those desired," writes Handel.[59]

This sort of boomerang effect brings to mind the experience of former secretary of energy James Herrington. When it came to Star Wars research, Herrington insisted that research disputes between Los Alamos and Livermore remain hidden behind closed doors,[60] but he found that the same policy of strict information control interfered with his ability to manage the environmental and health risks associated with the nuclear bomb plants. "The entrenched culture of secrecy affected everyone in the DOE bureaucracy, including Secretary Herrington, who found that the secrecy made it difficult for him to assess the condition of the plants that were his responsibility," explains Michael D'Antonio.[61] The DOE's official policy of disinformation boomeranged on the agency. Although it proved to be politically convenient in the missile defense area, systematic communicative distortion compromised safety initiatives in the area of nuclear cleanup. Similarly, journalists Ron Grossman and Charles Leroux noted that the hype surrounding Livermore's apparent Star Wars breakthroughs helped shield the nuclear research establishment from general scrutiny. In the resulting permissive research environment, scores of DOE scientists conducted shocking human radiation experiments against vulnerable subjects without their informed consent. These examples reveal that precision in secrecy can be difficult to maintain even when insiders desire to delimit the scope of information classification.

An additional example casting doubt on the possibility of deceptive precision took place during the Gulf War. The fiction of miraculous Patriot effectiveness could only have been maintained during the war with the help of a smooth-running Pentagon information-control machine that shut down critique with zealous enforcement of draconian censorship rules. In addition to facilitating the arguably beneficial perception of Patriot effectiveness, however, censorship also gave the Pentagon breathing room to mothball gruesome videos shot from cameras mounted on Apache helicopters. These video clips showed so-called turkey shoots of surrendering Iraqi soldiers, as well as U.S. bombardment of civilian targets in Iraq. Since exposure of such U.S. conduct might have revealed evidence of U.S. war crimes,[62] the fact that the Pentagon's censorship machine was able to suppress such evidence is alarming.

There is another aspect of runaway deception in the Patriot case, involving the Clinton administration's accidental buildup of political momentum behind missile defense, caused by its cover-up of the Patriot's Gulf War failure. Postol explains that by passing up on the opportunity to clear the air on the Patriot when he first came into office in 1992, President Clinton locked himself into the lines of argument advanced by George Bush in his Gulf War address to Raytheon employees in Andover, Massachusetts. The result of this decision to continue a deception campaign exaggerating Patriot effectiveness (that started under Bush) has been an inexorable march toward NMD deployment—deployment that the Clinton administration worked feverishly to block from 1992 to 1998.

> People get trapped in a series of lies, which is for example what I think is going on with the Clinton administration. . . . [Instead of clearing the air on Patriot when it came into office] what it did instead was cover it up. The official position of the U.S. government is that the Patriot basically worked. . . . The evidence is overwhelming that the success rate was zero. . . . Now they have this Republican Congress in place and they're pushing for more missile defense activities, even beyond what the Clinton administration is doing, and so there's a whole series of lies that were told that trapped them in a very serious situation, because we could well lose the ABM Treaty in the process of fooling around at the edges with this missile defense activity. That would have profound implications for the effort to control proliferation of nuclear weapons. Profound implications.[63]

Later discussion in this chapter explores the ramifications of this momentum in light of recent congressional developments, including passage of the 1999 National Missile Defense Act, which mandates deployment of NMD "as soon as is technologically possible."

Wrong Rx: Harmful Side Effects of Habitual Placebo Defense

Justifications for strategic deception in missile defense advocacy frequently trade on the logic of medical placebos. Thus, "Reagan victory school" adherents such as Robert McFarlane, James Fallows, and Jay Winik tout Star Wars as precious cold war snake oil that tricked the Soviets into spending their way to economic ruin. Similarly, Patriot missile backers argue that even if Patriot did not actually intercept any Scuds in the Gulf War, the perception of Patriot effectiveness soothed the frayed nerves of Israeli civilians calling for IDF intervention against Iraq. Previous discussion has shown that the causal claims backing each of these versions of revisionist history are open to question. On a broader scale, however, there are also general dilemmas involved when placebo-like logic is institutionalized in missile defense politics. "To be an unqualified success," explain David A. Charters and Maurice A. J. Tugwell, "deception by a democracy in peacetime must be used like a medicinal but addictive drug: little and under continuous supervision."[64]

Doctors are well aware that one basic danger involved in habitual placebo use is the risk that chronic use will result in "self-deception." "The doctor may come to think that the agent has potency when, in fact, it has none. The danger is real," one participant explained in a 1946 medical conference on placebos.[65] Drawing from the Hesketh Report, an "unpublished after-action history of deception operations," Handel notes the applicability of this warning to military contexts.[66] "Once deception has proved itself, most commanders will quickly come to accept it as an indispensable part of all operational planning . . . at this point . . . '[t]here is a tendency on the part of those who are constantly at grips with compelling realities to regard deception as a swift panacea to be invoked when other remedies have failed,'" he observes.[67] This danger is magnified by the fact that, as journalist Tim Weiner notes, the potential for large-scale self-deception is great since "it is the nature of deception programs as run by the Pentagon that not every participant in a program knows that he is part of a deception program."[68]

Hazards from the placebo strategy may soon come home to roost in future military conflicts, since distortions introduced into the public debate over missile defense capabilities may filter down to the level of tactical battlefield decision making. "[N]ational defense preparedness and the lives of U.S. soldiers in future conflicts," writes Hildreth, "may well depend on contingency plans that assume a certain level of Patriot effectiveness against SSBMs (short-range ballistic missile) attacks."[69] If these assumptions are based on deliberately inflated figures, the intelligence capabilities of battlefield commanders could be compromised. For example, deliberately manufactured misconceptions about Patriot's capabilities may one day unnecessarily jeopardize the lives of U.S. soldiers. According to Postol, "soldiers' lives could be unnecessarily endangered if they are deployed in future conflicts based on optimistic assessments of the Patriot's capabilities. They may depend on destroying nine out of ten enemy missiles, as the Army now claims, when the actual capabilities are closer to one in ten, if that."[70] Counting on Patriot effectiveness, field strategists may undervalue the ballistic missile threats faced by troops in certain target areas. Instead of providing the safest possible routes of travel for troops, flawed deployment patterns issued by misinformed platoon leaders may lead soldiers directly into the line of enemy attack.[71] "American soldiers could be unnecessarily endangered if strategy and tactics in future conflicts are based on these incorrect assessments of Patriot's capabilities," argues Representative John Conyers (D-MI).[72]

NMD advocates point to the possibility of "rogue state blackmail" as an emerging threat justifying rapid pursuit of missile defense.[73] This "blackmail" scenario envisages the U.S. embroiled in a diplomatic or military dispute where an "enemy" (e.g. North Korea, Iran or Iraq—armed with weapons of mass destruction [WMD]) attempts to exact concessions from the U.S., by threatening the U.S. homeland with a long-range rocket attack. Despite the questionable assumptions upon which this analysis is based, let us accept the scenario for the sake of argument.[74]

In such a tragic situation, it is suggested that NMD would preserve U.S. "freedom of movement" by giving U.S. leaders room to call the "rogue state's" bluff. To do this, however, American officials would need to have supreme confidence in the NMD system. As BMDO Director Lt. Gen. Robert Kadish testified to Congress recently, "We must ensure that the NMD system we will eventually deploy will work with a very high level of confidence."[75] If such confidence was based on faulty or doctored feasibility data, the stage could be set for a miscalculation of tragic proportions. Misplaced faith in an illusory missile defense shield could embolden the president to take diplomatic risks that would recklessly expose thousands (perhaps millions) of civilians to nuclear, chemical or biological attack. One shudders at the prospect that discrepancy between actual and advertised effectiveness of a U.S. NMD system would be brought to light by a real ICBM launch on an American city. Yet this is precisely the scenario that seems possible if citizens embrace the Pentagon's suggestion that missile defense can be a potent and coercive tool to counter "rogue state blackmail." In such situations, rather than pursuing diplomatic alternatives in a stalemated conflict, shielded nations would dig in and dare a "state of concern" to follow through on its promise to launch an ICBM, hoping that NMD's ace-in-the-hole leverage would force the adversary to back down. But what if the "rogue state" wouldn't fold? Bluffing poker players who are forced to show their cards after engaging in a bluff often lose the hand. Presidents bluffing about the capability of their missile defense systems could lose entire cities. As John Pike of the Federation of the American Scientists notes, "If one nuclear weapon gets through, you have more dead Americans than every other war put together."[76] This is not just a concern for Americans, since in such a tragic scenario, Fylingdales, U.K. and Thule, Greenland (planned locations for NMD early warning radars) "would become prime targets in the event of a nuclear war," says Stephen I. Schwartz, publisher of the *Bulletin of the Atomic Scientists.*[77]

Thomas Jefferson described the placebo as a "pious fraud" in an 1807 letter to a physician friend.[78] The ostensibly fraudulent nature of the placebo stems from the fact that the effectiveness of placebo treatment is tied to manipulation of the psychological perceptions of patients. If the fund of trust underwriting patients' faith in medicine is depleted by revelations of widespread placebo usage, ethicists have warned that a crisis in confidence for the entire medical establishment might ensue.[79] When installed as a routine doctrine in wartime missile defense advocacy, strategic deception may likewise pay diminishing returns. If the United States becomes known around the world as a military power that inflates its claims of missile defense accuracy for the purpose of achieving political and psychological objectives, this reputation is likely to impede efforts to replicate the experience of successful deception in the Gulf War. In overlooking this pitfall, defenders of strategic deception ignore the generative function of rhetorical practice. The employment of certain rhetorical strategies, in addition to having local persuasive effects, also shapes the expectations and interpretations governing subsequent discourse.[80] "[T]he lessons learned from the Second World War, the existence of ULTRA, and the double-cross systems, the influence of which on the course of the war is widely recognized," Handel observes, "might make it difficult to play the same trick twice."[81]

Recent foreign policy crises have yielded palpable evidence that a lack of U.S. BMD credibility has constrained American rhetorical options and prevented repeat of the successful Gulf War deception campaign. In the spring of 1996, tensions between Israel and Lebanon flared up and escalated to military conflict. The IDF and Hezbollah military units traded short-range missile attacks, causing significant human death and structural damage. Katyusha rocket attacks on Israeli citizens sowed terror among the domestic population, terror which brought public pressure on the Israeli government to step up retaliatory attacks against Hezbollah targets in Lebanon.

Since the United States desired to head off full-scale war between Israel and Lebanon, American officials intervened in late 1996 and brokered a cease-fire. With the fragile cease-fire showing signs of eroding, U.S. secretary of defense William Perry attempted to shore it up by meeting with Israeli Prime Minister Shimon Peres. This meeting resulted in the announcement of a new U.S.-Israeli defense cooperation pact on 28 April 1996.[82] Secretary Perry announced that as part of the pact, the United States would ship Nautilus and Phalanx short-range rocket defense systems to Israel as soon as possible. U.S. defense officials suggested that these defense systems could protect Israeli towns such as Kiryat Shmona against Hezbollah's Katyusha rockets.

Perry's announcement was expected to fortify the Israeli-Lebanon cease-fire, by reassuring Israeli citizens that they would be protected by missile defense, thus persuading them that there was no need to press for preemptory military moves against Lebanon. The American strategy in this case bore a resemblance to that of the Gulf War, in which U.S. Patriot missile protection was offered to pacify Israeli civilians and dissuade the Israel Defense Force (IDF) from attacking Iraq. However, where the response by Israeli citizens in the Gulf War to U.S. missile defense was almost uniformly positive, this time the Israeli public greeted American promises of missile defense protection with pronounced skepticism.

In a piece published days after the Perry visit, commentator Avino'am Bar-Yosef, of the Israeli newspaper *Ma'ariv*, played up the "gimmick aspect" of the new Perry-Peres defense pact. The Israeli journalist specifically took aim at the credibility of Perry's assertion that U.S. missile defense would offer genuine protection, writing that "the effectiveness of the antikatyusha system is like using a fly repellent to reassure the residents of the northern border."[83] Bar-Yosef did not stop there. He went on to wonder whether the U.S. "interim solution" to "reassure the residents along Israel's northern border" was really a ploy to "promote U.S. arms sales" and "prevent unemployment in the [Israeli] military industries during an election year."[84] Bar-Yosef was not the only commentator to voice such concerns; *Ma'ariv* followed inking of the Perry-Peres security deal with a skeptical editorial, remarking that the agreement was nothing more than an "election-eve sales promotion."[85] These comments reflected general Israeli distrust of U.S. post–Gulf War missile defense assurances. In light of the Gulf War experience, the Patriot was a "joke" (according to Haim Asa, member of the Israeli Patriot team), "a myth" (in the view of Israeli general Dan Shomron), and "meaningless" (from the perspective of former Israeli defense minister Moshe Arens).[86] These assessments were consistent with the overall appraisal of Israeli expert Reuven Pedatzur: "There is no defense against ballistic missiles; it's all smoke and mirrors."[87]

It remains to be seen whether this skepticism will color Israel's perception of the Thermal High Energy Laser (THEL), a defense system built specifically by TRW for Israel to shield against Katyusha rocket attacks launched along Israel's northern border. In May 2000, TRW conducted an apparently successful THEL test at the White Sands National Missile Range in New Mexico, where a laser was focused on a stationary ground target.[88] According to Lt. Gen. John Costello, commander of the United States Army Space and Missile Defense Command, the highly-touted THEL system "could in fact revolutionize warfare," by protecting troops from rockets, mortars and other artillery.[89] Such ebullience may be premature, however, since the May 2000 THEL test had a design similar to the September 1985 Mid-Infrared Advanced Chemical Laser (MIRACL) experiment, which one scientist derided as an example of "strap-down chicken tests, where you strap the chicken down, blow it apart with a shotgun, and say shotguns kill chickens. But that's quite different from trying to kill a chicken in a dense forest while it's running away from you."[90] In fact, a recent report from DOD's High Energy Laser Executive Review Panel noted that when deployed to defend against moving rockets and mortar, "it would be very difficult [for systems like THEL] to acquire targets, control the HEL beam, and destroy targets."[91] Further, the review panel found that instead of being used for defense, THEL's most promising military applications may involve offensive missions, where it could generate lasers designed to burn up aircraft and surface craft.[92]

The U.S. strategy of using missile defense as a psychological tool was put to the test in another recent diplomatic crisis involving Korea. Tensions on the Korean peninsula rose to dangerous levels in January 1994 when North Korea and the United States faced off in an acrimonious dispute over Pyongyang's nuclear program. Attempting to calm South Korean fears in light of the North's promise to turn "Seoul into a sea of fire," the United States deployed Patriot missile defense batteries at U.S. bases located in the ROK. However, as in the Israeli example, this deployment was greeted with skepticism. Reacting to the shipment of Patriots, one senior South Korean official said, "[o]ur fear is that it could give a *false sense of security.*"[93] Writing in the Seoul newspaper *Choson Ilbo,* military commentator Chi Man-won expressed similar doubt that the Patriot would be up to the task.

> What did the United States do when North Korea threatened South Korea by mentioning "speculation on turning Seoul into a sea of fire?" It brought Patriot missiles, which it had wanted to sell to the ROK, into South Korea,

> and sold its anti-artillery radar to us while William Perry was personally involved in the arrangements. It is impossible for these weapons to cope with gun shells randomly fired at Seoul at the rate of 10,000 shells per minute.[94]

Later, the true motivation behind the American decision to deploy the Patriot in this case was revealed—Pentagon and Raytheon officials hoped to use the operation as an opening sales pitch in their campaign to peddle Patriots to the South Korean government. However, shortly after the crisis subsided, the South Korean government "decided not to purchase Patriotic [*sic*] missiles, considering their rate of performance."[95] ROK defense minister Chon Yong-taek declared that missile defense is "not an effective measure to counter Pyongyang's missile threats."[96] This was not the answer American officials wanted to hear. "When the ROK Government hesitated about buying the Patriot in view of its inefficiency and high price," noted journalists O Yong-hwan and Choe Sang-yon, "the [U.S.] Defense Department put pressure on the ROK through various channels."[97] Such pressure included a visit by U.S. secretary of commerce Mickey Kantor, who "called for the purchase" of the Patriot in a July 1996 meeting with Pak Chae-yun, ROK minister of trade, industry, and energy.[98] Defense Secretary William Cohen visited the ROK in June 1997 and "strongly demanded the ROK purchase U.S.-made Patriot missiles."[99] According to the South Korean newspaper *Chungang Ilbo,* these visits gave South Koreans the impression that "the U.S. government is waging an all-out campaign to sell its missiles to the ROK."[100] Journalist Kim Song-ho blasted this manipulation in a 1997 editorial for the South Korean newspaper *Hangyore 21.*

> The United States is hell bent [*hyorani toetta*] on selling out-moded [*hanmulgan*] Patriot missile [*sic*] to our country on the pretext of the ROK-U.S. joint defense system. . . . The attempt is a long-range and secret plan that the United States has consistently pushed since 1994. The U.S. pressure has continued ceaselessly. It is the general view that the U.S. pressure on the purchase of Patriot missiles has been prompted by the request of U.S. military industry firms. The U.S. Defense Department, which enjoys the three-legged race relations with the U.S. military industry, performed the most powerful and effective role as weapons salesmen.[101]

Kim Song-ho's comments reflect a serious limitation of the placebo approach to U.S. missile defense diplomacy—there is a tendency for such diplomacy to work as a crutch for U.S. officials seeking to avoid tough and perhaps embarrassing questions about American-made BMD systems. Rather than admit that their technological fixes may not be as miraculous as advertised, officials are tempted to gloss over such uncertainty with deception and coercion. Ethicist Jay Katz finds a similar phenomenon at work in medical circles, where doctors sometimes use placebos as conversation stoppers, in order to pre-empt challenging questions from patients, thereby reinforcing their power and authority in the clinical setting.

> Finally, it must be recognized that traditional justifications for placebos, faith, hope and reassurance have served physicians' needs to disregard uncertainty and profess certainty instead. Their use perpetuates the silence that has haunted doctor-patient interactions throughout history. It inhibits questions of recommendations by and with physicians. Placebo pills, unrealistic faith, hope, and reassurance are powerful conversation stoppers. Thus, their administration serves doctors' interests in silent compliance in order to avoid the embarrassment of having to reveal so much about the uncertainties of medicine to lay persons and the problems of discussing such difficult matters with patients.[102]

The skepticism about claims of missile defense effectiveness expressed by U.S. military allies in these two post–Gulf War scenarios was likely due in part to the dissipation of strategic deception's placebo-like qualities. Ironically, the generative rhetorical effects of Gulf War strategic deception appear to have undermined the trust of world audiences in the United States as a reliable source of information on its own missile defense capabilities. While the perception of Patriot effectiveness in the Gulf War may have been a useful fiction, it may very well have been a one-trick pony. Distrust of American missile defense assurances in recent crises indicates that wartime strategic deception appears to have entailed a serious hidden cost: the erosion of U.S. military credibility.

However, the Korean case signals an additional cause for concern. When the ROK government declined American offers of Patriot support, a broad array of U.S. officials (including the secretary of commerce!) pressured South Korea to buy the BMD system anyway, using veiled diplomatic threats as leverage. This tactic explains one reason why U.S. missile defense sales have shot up worldwide (even in the face of the Patriot's dismal Gulf War performance): American officials and trade representatives have simply bullied other countries into buying missile defense technology. Such a tactic represents a major threat to the security of the United States and its allies, because allied perception that U.S. defense policy has become captive to defense contractors such as Raytheon makes allies feel as if U.S. officials are more interested in exploiting their military weakness for profit than working with them to protect their people in a serious and cost-effective way. Editors of South Korean newspaper *Tong-a Ilbo* spelled out these concerns in a 1997 article: "To add to matters, the Raytheon Company of the United States, which manufactures Patriot missiles, is reportedly lobbying a great deal with the U.S. Congress and Government for missile sales. It would be nonsensical for the U.S. administration to pressure an ally in response to commercially-motivated push from an arms manufacturer."[103] The editors of *Chungang Ilbo,* another Seoul-based newspaper, protested even more stridently: "[W]e would like for the United States to do away with the idea of making a sucker out of the ROK [*hangugul ponguro samkettanun palsang*] by taking advantage of our weakness."[104]

Deception Goes up in Smoke: The "Tobacco Institute Strategy"

In public discussions on the X-ray laser and Patriot missile defense systems, Pentagon and DOE officials worked hard to shut down and close off avenues of controversy, since such disagreement bred uncertainty that interfered with efforts to stabilize scientific facts proving the technical feasibility of BMD. For example, in the Star Wars era, LLNL director Roger Batzel silenced top Livermore missile defense scientists such as Roy Woodruff and Lowell Morgan, in the process preempting public controversy over Edward Teller's wild claims of X-ray laser brightness. Similarly, DIS officials attempted to prevent controversy over the Patriot's Gulf War performance by declaring Postol's *International Security* article "secret," because its publication challenged all the Pentagon's scientific evidence on Patriot missile accuracy. In the terminology of sociologists Tristram Engelhardt and Arthur Caplan, these attempts constituted efforts to "end controversy by force."[105]

The forced termination of scientific controversy can prove to be an efficacious political tool, allowing those with authority to multiply their power in public debate by leveraging arguments with the weight of scientific consensus. Interestingly, this process also works in reverse; when science does not appear to be breaking in a way favorable to those with deep pockets, it is possible for well-moneyed interests to buy controversy by funding scientific counterresearch. David Helvarg, columnist for *The Nation,* dubs such deliberate attempts to foment scientific disagreement "tobacco institute strategies." Helvarg developed this moniker in recognition of the Tobacco Institute, a now-defunct trade group that kept debate about cigarette smoking's health harms going for years by churning out endless scientific studies of dubious merit.[106] In addition to discussing the tobacco research example as a case where prolonged scientific controversy worked to defer the necessity of definitive judgment by policy-making authorities, political scientist David Dickson documents how this same phenomenon has prolonged disputes over the regulation of formaldehyde and the environmental effects of acid rain.[107] These examples show that conflict among scientific experts reduces the experts' potential impact as authoritative arbiters of policy disputes.[108]

Other similar "debates about closure"[109] have unfolded over the issue of climate change. On the one hand, a large group of scientists (the Intergovernmental Panel on Climate Change, or IPCC) recently argued that a definitive and robust scientific consensus has emerged in support of the theory that greenhouse gas emissions from human sources are causing rapid global warming. Essentially declaring the global warming debate over, this group of scientists said in 1995 that the emergence of a scientific consensus marked the end of an intense period of controversy on climate science and that immediate and aggressive action to cut greenhouse gas emissions was warranted.[110] Following the so-called tobacco institute strategy, a consortium of coal and oil corporations launched a counterinitiative to challenge the findings of the IPCC study. By hiring a number of Washington, D.C.-based think tanks to produce dissenting scientific evidence, these threatened corporate interests engineered a "stealth greenhouse spin campaign" that Helvarg says has "managed to convince the public that there is still significant 'debate' on global warming."[111] History professor Gale Christianson recently noted that in this context, the "tobacco institute strategy" of prolonging scientific controversy has turned a "scientific discovery [global warming] into an ideological, spiritual, or political debate."[112]

Just as today fossil fuel industry spokespersons point to ongoing "scientific debate" surrounding the IPCC report to leverage claims that action to curb warming would be premature, in the mid-1990s BMDO director Malcolm O'Neill stimulated "healthy scholarly debate" over THAAD's theoretical capability, to shield the system from political criticism. O'Neill reasoned that headlong pursuit of the system would not flout ABM Treaty limits on strategic defenses, as long as the jury was still out on key scientific questions relating to THAAD's theoretical capabilities. BMDO's rhetorical approach here featured a two-track strategy to generate sham debate as a tool to defer closure of scientific controversy. After commissioning a secret study from a private SDI boutique (Sparta, Inc.), BMDO officials arranged for Sparta scientists to amplify their findings, without rebuttal, to high-ranking Russian and U.S. diplomats, in government-sponsored official briefings. The Sparta scientists used these briefings as a platform to discredit arguments advanced by Gronlund et al. regarding the technical soundness of the Clinton administration's ABM Treaty negotiating strategy. When the Sparta study drew intense criticism after its public release, however, BMDO Director O'Neill pulled a bait-and-switch. Rather than defending the Sparta study or answering Gronlund et al.'s extensive technical rebuttals, O'Neill simply rose above the fray and declared the controversy a "healthy scholarly debate." However, O'Neill ignored charges that Sparta researchers had systematically subverted the possibility of such debate by withholding evidence and failing to communicate with Gronlund et al. Here was one manifestation of a common strategy of institutional argument noted by Erik Doxtader: "[I]nstitutions use the contingency of public discussion to tactically disarm the force of criticism," he explains, and in this process, "bureaucratic rule-makers presuppose forms of deliberation that have not yet occurred."[113]

Argumentation in scientific controversies can yield heterogeneous layers of rhetorical performance addressing different targeted audiences. The complex textual layering that results poses serious problems to methods of critical analysis that presuppose essentialized models of closure debates, but it also opens up opportunities for critique that take the interplay between tracks of argumentation as key points of departure. When significant cleavages between tracks are identified, possible vistas of inquiry open up. For example, in the recent Florida verdict against cigarette industry representatives, "jurors found that cigarette makers committed fraud, concealment and negligence by privately acknowledging the health risks of smoking and the addictiveness of nicotine while publicly playing down those concerns."[114] At trial, victorious attorney Stanley Rosenblatt decried the Tobacco Institute's "diversionary tactics to keep alive a nonexistent scientific controversy about smoking and health."[115]

In reviewing the TMD footprint controversy, independent expert Richard Garwin noted similar dynamics. Although Garwin's primary motivation for reviewing the controversy between Lee et al. and Gronlund et al. was to examine the technical aspects of TMD footprint modeling, his most hard-hitting conclusions dealt with communicative aspects of the controversy. Garwin first unpacked the burdens Lee et al. assumed by taking on the job of a scientific "review," then moved on to examine the extent to which Lee et al. lived up to such burdens in the course of their study. His findings raised deep questions about the seriousness of Sparta's commitment to legitimate scientific peer review, and aroused grim reservations about BMDO's credibility in the eyes of world audiences.

Although resort to the "tobacco institute strategy" in this case may have been politically expedient in the short term for BMDO and Sparta, the mistrust and suspicion that surfaced after public release of the Lee et al. study was diplomatically damaging. Recent evidence suggests that Russian suspicion about the potential capability of proposed U.S. BMD systems continues to hamper arms control diplomacy designed to achieve reciprocal "deep cuts" of offensive strategic missiles.[116] As Arbatov observed in 1999, Russian concerns about the U.S. TMD program have "also deadlocked the U.S.-Russian talks at the Standing Consultative Commission on the delineation of strategic and tactical defenses."[117] In spring 2000, Russia upped the ante in arms control negotiations by ratifying the Comprehensive Test Ban Treaty (CTBT) and START II accords, then promising to pull out of these agreements if the U.S. built a missile defense system in violation of the ABM Treaty. "The prevailing system of arms control agreements is a complex and quite fragile structure. Once one of its key elements has been weakened, the entire system is destabilized. The collapse of the ABM Treaty would, therefore, undermine the entirety of disarmament agreements concluded over the last 30 years," said Russian Foreign Minister Igor Ivanov, in remarks at the Nonproliferation Treaty (NPT) review conference in New York.[118]

National Missile Defense "As Soon as Technologically Possible"

On 20 May 1999, members of the U.S. House of Representatives were debating the 1999 Missile Defense Act, which calls for mandatory deployment of a National Missile Defense (NMD) system "as soon as technologically possible." During the debate, Representative Lloyd Doggett (D-TX) rose to speak: "Mr. Speaker, how appropriate the timing of this debate. As we speak, folks are lined up around the block across America to see the new Star Wars movie [*The Phantom Menace*]. Just like the original movie, this bill puts a tractor beam in the Capitol dome and aims it right at the wallets of the American taxpayer to support this defective system."[119] Doggett's pith did little to stem the tide of pro-missile defense sentiment that engulfed both chambers of

Congress and led to springtime passage of the 1999 Missile Defense Act by substantial, veto-proof majorities. Missile defense stalwarts reveled after their congressional victory, linking approval of the new bill with Reagan's 1983 Star Wars address as comparable milestones in the annals of great missile defense moments.

Following approval of this legislation, journalists Bill Sweetman and Nick Cook pronounced that "the debate about how to implement such a defence will begin in earnest."[120] The current era represents a watershed period for U.S. missile defense, comparable in significance to other decision points in recent history, such as 1972 (leading up to the ABM Treaty), 1983 (the period after Reagan's Star Wars address), and 1991 (the aftermath of the Persian Gulf War). Since the Missile Defense Act of 1999 stipulates that technological feasibility will be the key factor controlling upcoming deployment decisions, future missile defense debates are likely to center on questions of scientific fact. Testing methodology, extrapolation of experimental results, and interpretation of data are topics likely to receive substantial play in near-term discussions. As such, the historical record of missile defense advocacy contained in this book merits consideration as a reference point for understanding the debates to come. The following pages elucidate the political context surrounding passage of the 1999 Missile Defense Act and explore how discussions leading up to the looming deployment decision might unfold. This exploration pays particular attention to historical ties between contemporary missile defense debates and the case studies examined in previous chapters.

Political Origins of the 1999 Missile Defense Act

President Bill Clinton attacked the concept of NMD relentlessly in his successful 1996 re-election campaign against missile defense stalwart Bob Dole. Lambasting Dole's proposal for NMD, Clinton argued that missile defense "is the wrong way to defend America," and that it would be folly to build "a costly missile defense system that could be obsolete tomorrow."[121] When he assumed office after his victory over Dole, Clinton could have followed up on this line of argument by ordering a full investigation of the Patriot's Gulf War performance, redeeming his campaign promise to restore integrity and balance to the U.S. missile defense program. Such a course might have allowed Clinton to correct inaccurate official assertions about missile defense feasibility still left over from Bush's inflated assessments of Patriot accuracy.

Instead of choosing this path, Clinton, the centrist "New Democrat," opted to pursue a strategy of "triangulation," charting a middle course between BMD skeptics and GOP missile defense diehards. Specifically, Clinton approved major spending hikes for theater missile defense systems such as THAAD, Patriot, and the navy "Upper Tier" TMD system. While missile defense partisans cheered Clinton's about-face, they still questioned his genuine commitment to the issue, decrying what they saw as a slothful testing schedule for THAAD. Specifically, missile defense lobbyists and pro-BMD think tank analysts such as Frank Gaffney and Baker Spring hammered Clinton on his failure to pursue more ambitious NMD programs.

This partisan pressure built up initially in 1994, when Republicans seized control of Congress in midterm elections.[122] In the face of strident GOP calls for a more comprehensive missile defense program, Clinton found it increasingly difficult to defend his vision of limited theater defense. His predicament paralleled the situation of Johnson administration officials such as Robert McNamara, who found themselves painted into a similar political corner in the mid-1960s. Although McNamara and others in the Johnson cabinet were convinced that large-scale BMD deployment would do more harm than good, they discovered that it had become impossible to resist the pro-missile defense sentiment swelling in Congress and among the Joint Chiefs of Staff. Thus, on 18 September 1967, McNamara relented in a famous speech to the Press Club in San Francisco, where he called for deployment of Sentinel, an ABM system designed primarily to protect against Chinese rocket attack. Charting the path of a similar political retreat in 1996, the Clinton administration developed an "NMD Deployment Readiness Program," referred to as the "3 plus 3" plan.[123] This plan called for an initial three-year evaluation period, designed to assess NMD's prospects, paving the way for a possible follow-on three-year deployment phase that would be triggered by good initial assessment results.

While Clinton's "3 plus 3" plan bought him time, GOP missile defense advocates continued to step up their pressure. In 1998, Senator Thad Cochran (R-MS) introduced legislation mandating NMD deployment "as soon as technologically feasible."[124] This proposal drew a Clinton veto threat, and it ultimately died on the floor of the Senate when forty-one senators blocked debate on the motion. However, circumstances changed dramatically in July 1998, when a bipartisan commission led by former secretary of defense Donald Rumsfeld released a report on ballistic missile threats. This report found that countries such as Iran and North Korea could develop and field an intercontinental ballistic missile in just five years.[125] Following release of the Rumsfeld Commission Report, journalist Jacob Heilbrunn noted that "[o]vernight, the debate changed."[126] "For conservatives," Heilbrunn explained, "the report was heaven-sent."[127]

With missile defense advocates using the Rumsfeld Commission's findings to bolster their case for quick deployment of an NMD system, Clinton found that he needed to "triangulate" his missile defense policy once again, yielding to GOP pressure for a more aggressive approach. On 20 January 1999, Secretary of Defense William Cohen announced that the administration would spend $6.6 billion in new money for research and development of a national missile defense system, and that the decision to deploy such a system would be made as early as June 2000.[128] Clinton "finally caved," said World Policy Institute analyst William Hartung; by co-opting the GOP's call for NMD, he looked "to give Gore some political cover" in the 2000 presidential election.[129] Commenting on Clinton's late conversion as a NMD believer, John Isaacs of the Council for a Livable World wrote that "he filched the issue just as he had welfare reform, deficit reduction, crime, and most recently, military spending."[130] "Apparently, in a White House with its eyes firmly fixed on the 2000 election," noted journalist Mark Thompson, "the idea of co-opting a Republican hobbyhorse, especially one likely to win congressional approval, was just too delicious."[131]

With Clinton having started the NMD ball rolling, the Republican-controlled Congress followed suit with its own proposal mandating NMD deployment "as soon as technologically possible." Drafted by Senator Cochran and Senator Daniel Inouye (R-HA), the Missile Defense Act of 1999 passed the Senate on 17 March 1999 by a 97–3 margin. A companion piece of legislation passed the House of Representatives 317–105, in short order. "As a declaration of national policy," journalist William Neikirk wrote, the bill "could provide powerful momentum for funding and development."[132] The Missile Defense Act of 1999 calls for deployment of NMD "as soon as technologically possible," thus setting up a mandatory near-term deployment decision that could come as soon as the summer of 2000.

The debates leading up to this eventual deployment decision will be conducted with billions of dollars and huge national security issues hanging in the balance. Additionally, a positive deployment decision will likely shape missile defense policy for decades to come, since according to Joseph Cirincione of the Carnegie Endowment for International Peace, "once factories start 'bending metal,' weapons systems acquire a serious constituency of sub-contractors, chambers of commerce, labor unions, and workers' families, not to mention congressional hearts and minds."[133] As Senator Richard Durbin (D-IL) notes, the Missile Defense Act of 1999 "is going to be the reference point for the budget debate in the future."[134] The potential cost of any eventual NMD system has been estimated from $28 to $60 billion.[135] Given the act's emphasis on technical feasibility as a key assessment criterion, scientific evidence is likely to figure prominently in upcoming NMD debates. The following pages explore the likely role of scientific proof in the upcoming NMD deployment decision, using findings from this book's earlier case studies as a backdrop to contextualize the discussion.

Scientific Evidence and Crucial Experiments

Previous discussion highlighted the 1998 Rumsfeld Commission report as a key factor driving 1999 spending hikes and congressional votes for ambitious NMD systems. After the report came out, conservative firebrand Phyllis Schlafly noted that it "provided Congress with enough talking points to win the argument [on missile defense] both in the strategic arena and in the 20-second soundbite television debates."[136] John Holum, acting undersecretary of state for arms control and international security affairs, said that the Rumsfeld Commission had "a profound impact" on Clinton administration missile defense policy.[137] Since the Rumsfeld Commission's limited research objectives dealt only with the issue of ballistic missile threat assessment (not missile defense as an appropriate response to such threats), it is somewhat curious that missile defense advocates were able to use the report's conclusions so convincingly as evidence proving the need for aggressive NMD development.

This is particularly noteworthy, given that, as Isaacs pointed out, the Rumsfeld Commission "did make it clear that it was not advocating the deployment of a ballistic missile defense."[138] For example, noted commission member Richard Garwin stipulated clearly that the commission's work should not be taken as definitive evidence in favor of NMD. In a letter to Congress on 2 September 1998 Garwin explained that "[t]he Rumsfeld Commission Report may have inadvertently given rise to the impression that, if the missile threat is near-term than otherwise assumed, a missile defense system is more urgently indicated than before."[139] In contradistinction to this general perception, Garwin made clear in a 1998 piece for the *Bulletin of the Atomic Scientists* that the Rumsfeld report "should—and must—be regarded as neutral regarding missile defenses."[140] In a *New York Times* op-ed piece, Garwin wrote that he was "alarmed that some have interpreted [the Rumsfeld Commission's] findings as providing support for a new national missile defense system."[141]

Specifically, Garwin's concerns stemmed from what many commentators see as a disconnect between the possible threats isolated in the Rumsfeld Commission report and the capability of NMD systems to counter such threats. As Garwin stated on the *McNeil-Lehrer NewsHour,* "[i]f these countries really wanted to hurt us, they would use shorter-range missiles from ships, nuclear weapons blowing up in harbors, purchased cruise missiles if they like, small airplanes that could fly out of shipping containers on a ship."[142] In an open letter to Congress, Garwin further explained that the proposed NMD system "would have zero capability against the much more realistic and important threat from North Korea, Iran, or Iraq—short-range cruise or ballistic missiles fired from merchant ships near U.S. shores, a nuclear weapon detonated in a harbor, or biological warfare agent (BW) disseminated in the United States or from a ship in harbor."[143] Additionally, Garwin noted that the proposed NMD system would not be effective against the threat that missile defense advocates hyped the most, a nuclear-tipped ICBM fired from North Korea to California, Alaska, or Hawaii. It would be simple and cheap, Garwin explained, for North Korea to outfit their missiles with common-sense "countermeasures" designed to confuse and frustrate missile defense radar and tracking instruments.[144]

While it remains to be seen how prominently such arguments regarding NMD countermeasures will play in future discussions about missile defense feasibility, if the historical track record is any guide, it is likely that such concerns will be swept under the rug. In October 1983, the so-called Fletcher panel delivered a report on the feasibility of the X-ray laser as a possible basis for Reagan's SDI system. As William Broad documented, the seventh volume of the report, on possible Soviet countermeasures, was "left out of the overall antimissile assessment . . . to keep bad news from the public."[145]

The historical record suggests that rather than emphasizing countermeasures and other "real world" factors in determining BMD feasibility, missile defense advocates prefer to play up the significance of recent, apparently successful "crucial experiments" as litmus tests for BMD feasibility. For example, following the 22 December 1983 Romano experiment at Livermore, designed to test X-ray laser brightness, Edward Teller labeled the test a "clear-cut scientific breakthrough," and wrote that the results of the test were so breathtaking that "we are now entering the engineering phase of X-Ray lasers."[146] Similarly, after the fourth Homing Overlay Experiment over the Kwajalein Atoll on 10 June 1984 the official DOD line was that the test had demonstrated the possibility of "hitting a bullet with a bullet." Missile defense advocates such as William F. Buckley claimed that HOE validated the "axiomatic geometry of missile defense."[147] In a similar vein, the Patriot's apparent success in the 1991 Persian Gulf War was treated by missile defense advocates as a positive litmus test for general BMD feasibility. Shortly after the war, Patrick Buchanan announced, "the United States has shown it can attack and kill ballistic missiles. . . . The [SDI] debate is over."[148] Missile defense officials delivered more buoyant rhetoric following a string of apparently successful THAAD tests in 1999. After a THAAD interceptor collided with a target missile during a test on 10 June 1999 at the Kwajalein Missile Range, Air Force Lt. Gen. Robert T. Kadish said, "I guess today is probably one of the watershed events in the technological history of our country, because today . . . we basically hit a bullet with a bullet in outer space."[149]

In these episodes, Teller, Buckley, Buchanan and Kadish invoked a rhetoric of crucial experiments. Such experiments are thought to be definitive trials offering valid "up or down" judgment on particular scientific theories under consideration. In the Romano experiment, the Homing Overlay Experiment, the desert laboratory of the 1991 Persian Gulf War, and the Kwajalein Missile Range, commentators suggested that the theoretical soundness of missile defense was put to the test. In each case, the apparent success of the experimental apparatus in delivering predicted results led advocates to conclude that such results confirmed the theory that indeed, missile defense could work. In hindsight, such claims have failed to hold water. The Romano experiment yielded nowhere near the X-ray laser brightness that Teller touted in 1983. The fourth HOE experiment in 1984 turned out to be a hoax. No quality evidence exists to prove that the Patriot successfully neutralized a substantial number of Iraqi Scud missiles during the 1991 Persian Gulf War. Despite recent test successes, doubts about the credibility of THAAD's testing program persist.

Such disclosures not only cast real doubt on the credibility of "crucial experiment" claims advanced by missile defense advocates; they also provide support for a classic theory in philosophy of science, the Duhem-Quine thesis. This thesis states that there is always a significant gap between experimental data and confirmation or refutation of a particular scientific theory.[150] As philosopher of science Allan Franklin explains it, "if an experiment seems to refute a theory it, in fact, refutes the conjunction of both the theory and background knowledge and one doesn't know where to place the blame."[151] Donald MacKenzie explains that the crucial experiment notion is seriously flawed, given the multitude of ways in which the connection between evidence and theory can be challenged.

> Recent sociology of science, following sympathetic tendencies in the history and philosophy of science, has shown that no experiment, or set of experiments however large, can on its own compel resolution of a point of controversy, or more generally, acceptance of a particular fact. A sufficiently determined critic can always find a reason to dispute any alleged "result." If the point at issue is, say, the validity of a particular theoretical claim, those who wish to contest an experimental proof or disproof of the claim can always point to the multitude of auxiliary hypotheses (for example about the operation of instruments) involved in drawing deductions from the given theoretical statement to a particular experimental situation or situations. One of these auxiliary hypotheses may be faulty, critics can argue, rather than the theoretical claims being tested. Further, the validity of the experimental procedures can be attacked in many ways.[152]

In *Inventing Accuracy,* MacKenzie shows how these generalizations play out dramatically in the context of experiments conducted to prove ballistic missile accuracy.[153] The extraordinarily complicated and hypothetical world of nuclear war contains so many variables, it is extremely difficult to design experiments in peacetime that accurately simulate the conditions of an actual ICBM launch. As a result, MacKenzie concludes that there is an inherent level of uncertainty built into experiments on missile accuracy. Whether or not such uncertainty is perceived and factored into military and political decision making is another matter. Drawing on Collins's theory of the "certainty trough,"[154] MacKenzie shows that some defense industry insiders "directly involved in knowledge production" are likely to fall victim to a form of the experimenter's regress, where the desire to see positive experimental results becomes so strong that it structures perceptions in such a way as to filter out contrary information.[155]

In the upcoming round of debates on the question of whether NMD is "technologically possible," these issues are likely to take on great significance. In a general sense, any NMD test conducted prior to a possible deployment decision is likely to be classified as a crucial experiment, given the paucity of tests on the system that will be possible. The danger is that in the process of overstating the import of a few successful tests, missile defense advocates may distort the deployment decision by invoking the now-familiar rhetoric of crucial experiments. The problem with such rhetoric is pointed out by General Henry H. Shelton, chairman of the Joint Chiefs of Staff: "The Chiefs question putting additional billions of taxpayers' dollars into fielding a system now that does not work or has not proven itself—and we do not think that one round hitting one missile is proof positive that we should start fielding."[156]

The previous discussion has provided several arguments supporting General Shelton's conclusion. First, the legacy of strategic deception in missile defense advocacy warrants healthy skepticism of any official claims regarding experimental validation of missile defense feasibility. When such claims are based on a small number of missile defense tests, the historical record suggests that the possibility of rigging, design manipulation, and embellishment of test results should not be discounted. Second, early NMD tests contain assumptions and auxiliary hypotheses that may not hold in actual "real world" NMD use. For example, the NMD testing apparatus includes what Garwin calls "cooperating missiles," that is, target missiles shot up like clay doves, "launched like puppy dogs, wagging their tails, and wanting to be slapped with hit-to-kill interceptors."[157] As Garwin explains, "none of these tests will simulate real world conditions," because they contain assumptions that significantly favor the defense (e.g., few countermeasures, stable target missile trajectory).[158] Casting doubt on the validity of the apparently successful and much ballyhooed THAAD tests, Philip Coyle, the Pentagon's top testing official, said the THAAD intercepts were "tightly scripted" experiments that were not "operationally realistic."[159] Indeed, empirical evidence suggests that successful test range performance does not necessarily translate into effectiveness in the field. The Patriot system had a perfect seventeen-for-seventeen intercept record on the test range prior to the Persian Gulf War, but these tests did little to simulate the challenging Gulf War environment, in which the Patriot ultimately failed to deliver.

Perhaps these concerns motivated Rep. John Hostettler (R-IN) to attach an amendment to the FY 1999 Defense Authorization Bill that would "allow the Pentagon to proceed with the production of the National Missile Defense system *regardless of whether initial operational testing and evaluation of the system has been completed*," as Michael Sirak of *Inside Missile Defense* describes.[160] Tom Collina, director of the Arms Control and International Security program at the Union of Concerned Scientists, called the Hostettler amendment a "buy before you fly" clause enabling the Pentagon "to begin production of a national missile defense system 'without regard' to whether it will even work. . . . [T]he move suggests that House Republicans fear their missile defense system won't pass any of its upcoming tests, and, as odd as it seems, want to 'front load' deployment before the system has a chance to fail. This is a blatant display of politics overriding the realities of science."[161]

Collina's warnings hearkened back to similar arguments offered by Los Alamos scientists in a 1985 brief questioning the wisdom of Livermore's decision to rush the schedule of the Goldstone X-ray laser experiment. The Los Alamos group recommended that Goldstone should be delayed because of problems with the beryllium detector, a key component of the experimental apparatus. Ultimately, Edward Teller failed to heed the Los Alamos group's advice, and Goldstone went off as originally planned, although the experiment was a near total failure. The Goldstone example serves as corroborating evidence for Garwin's thesis that allowing political imperatives to speed up the experimental timetable for missile defense tests invites technical mishaps. "In fact by taking such special risks," explains Garwin, "you are more likely to delay employment than speed it up. As the Welch Report stated . . . 'The virtually universal experience of the study group members has been that high technical risk is not likely to accelerate field capabilities. It is far more likely to cause program slips, increased costs and even program failure.'"[162]

These comments suggest that the persistent strain of politicized science that infected earlier missile defense projects is still present in contemporary initiatives. Is history repeating itself? Is the U.S. Congress poised to make yet another multi-billion-dollar decision using inflated or doctored missile defense data? The more that recent episodes of missile defense advocacy begin to resemble past campaigns of strategic deception, the more appropriate such questions become.

Consider the NMD testing program. To date, very few tests have been conducted on NMD's capability to intercept incoming missiles. Out of the nineteen Integrated Flight Tests (IFTs) that have been scheduled, only five have been completed, and two of these (IFT-1A and IFT-2) did not involve intercept attempts. The IFT-3 test in October 1999 produced an apparently successful intercept, but subsequent disclosures revealed that a faulty star map caused the exo-atmospheric kill vehicle (EKV) to drift off course and home in on a decoy until the very last second, when the interceptor inexplicably veered into the target missile. In that test, according to Collina, "they got lucky."[163] An internal Pentagon assessment of the test, supplied to the online publication *CounterPunch*, noted how the test conditions in IFT-3 differed in important ways from likely attack scenarios, casting further doubt on the test's validity as a realistic measure of NMD effectiveness.[164]

The IFT-4 test, carried out in January 2000, resulted in a missed intercept when infrared seekers on the Raytheon-built EKV failed. The IFT-5 test could not evaluate the ability of the overall NMD architecture to work in harmony, because some components of NMD were not ready in time to be included in the experiment. Lockheed Martin is not expected to be finished with its EKV production booster until 2001, which would be tried for the first time in IFT-7.

Some have explained the decidedly mixed results of early NMD tests by pointing out that an artificially compressed test schedule has resulted in a politically driven "rush to failure."[165] Coyle cautions that the current testing timetable puts NMD on an "extremely high-risk schedule."[166] "They're moving too fast, heading for a deployment decision based on primitive, simple developmental tests," says Joseph Cirincione of the Carnegie Endowment for International Peace.[167]

The "rush to failure" approach codified in the current NMD timetable is connected to a related, but independent, "hush to failure" pitfall that may undermine the credibility of NMD testing data more fundamentally. A premature decision to build NMD before it is proven to be feasible would lock in at least $60 billion in future U.S. missile defense spending (according to a new estimate by the U.S. Congressional Budget Office).[168] Presumably, disbursement of these funds to private vendors would be contingent on such contractors demonstrating a credible track record of engineering success. However, the history of strategic deception in the U.S. BMD program shows that military officials and industry leaders take extraordinary measures to keep Pentagon dollars flowing, even when poor test results suggest that continued funding of beleaguered missile defense systems would be wasteful and imprudent.

Current NMD deployment plans call for no significant institutional changes in the overall leadership structure or bureaucratic make-up of the U.S. BMD enterprise. It is reasonable to expect this enterprise to suppress unfavorable NMD test results, to silence whistleblowers, and to use the classification system strategically to protect the funding windfall of a positive NMD deployment decision.[169] The result of this process could be deployment of a deeply flawed NMD system that is technically bankrupt, yet has an illusory veneer of effectiveness. Such a "hush to failure" outcome would not only pose grave security risks, it would also rival Reagan's SDI as the most lavish and wasteful "big science" program ever funded by a government.

The case of Nira Schwartz provides one disturbing example of this "hush to failure" phenomenon at work. As a former senior engineer at TRW, Schwartz worked from 1995-1996 on computer software enabling NMD interceptors to discriminate between target missiles and decoys. In documents filed in a Los Angeles federal district court, Schwartz alleges that TRW "knowingly made false test plans, test procedures, test reports and presentations to the United States Government," then retaliated by firing her when she refused to cover up such fraudulent activity.[170] Schwartz's charges concerned TRW's suppression of unfavorable data in the Kalman Filter Extractor (KFE) discrimination technology program, a computer engineering effort designed to develop software for the exo-atmospheric kill vehicle (EKV) component of National Missile Defense. Schwartz's charges were corroborated in an affidavit filed by retired TRW senior staff engineer Roy Danchick, as well as an official Pentagon investigation conducted by the U.S. Defense Criminal Investigative Services (DCIS), an investigative arm of the DOD Office of the Inspector General.[171]

During its investigation of Dr. Schwartz's initial charges, DCIS turned up other evidence of suspicious data manipulation by TRW that took place subsequent to her dismissal. One such finding involved handling of test data in IFT-1A, the very first NMD flight test carried out in July 1997.[172] IFT-1A was a "fly-by" experiment, designed to validate the capability of the EKV on-board sensor to "see" the target missile while in flight. At a DOD news briefing shortly after IFT-1A, Brig. Gen. Joseph Cosumano explained that "the purpose of this fly-by test was to take a first look at our ability to discriminate between the RV and other objects that were decoys or replicas"[173] Since the on-board EKV sensor did not perform "real-time" discrimination in IFT-1A, evaluation of the system's ability to pick out the warhead from other decoy objects hurtling through space had to be done on the ground, where scientists combed through the test flight data after the experiment was completed.[174] At the same news briefing, Cosumano called the test "a success," indicated that "[n]ominally and mechanically, everything worked exactly as scripted," then gushed, "I'm elated. This sets the cornerstone, I believe, for a good start for the program that is really three months old."[175]

Cosumano's exhilaration here is puzzling, in light of the fact that at the time he shared such excitement with the Pentagon press corps, BMDO had not yet even *received* TRW's analysis of the IFT-1A flight test data (which would have enabled him to evaluate fully the NMD system's performance in the aspect of the experiment he said was most important—target discrimination).[176] Cosumano sounded like someone who had just purchased a lottery ticket, then popped the cork on celebratory champagne, long before the winning numbers were announced.

Later, Samuel W. Reed, Jr., a Special Agent with DCIS, filed findings that took the luster off Cosumano's ebullient assessment of IFT-IA. In his analysis of TRW's presentation of the test data to BMDO, Reed found that "TRW heavily censored the IFT-1A signature data, deleting approximately the first 20 seconds and the last 11 seconds." This deletion permitted TRW "to go from poor results in its 45-Day Report, to excellent results in its 60-Day Report Addendum."[177] DCIS Group Manager Robert A. Young reached similar conclusions in his investigation of the IFT-1A experiment.

> TRW artificially chopped the amplitude of the sensor signal output which results in arbitrarily forcing the signal mean to a predetermined value. This resulted from the sensor output noise level being too high for the baseline discrimination algorithm to process. . . This allowed an artificial improvement in the Probability of Assigned Target (PAT). The PAT should be solely dependent on the actual measurement of data from the signature(s), versus creating a signature(s) with gaps of missing data and filling in this data artificially to increase the value of the calculated PAT. *This GPA* [Gap Filling Algorithm] *process is done outside the scope of statistical scientific methods. . . .* This only leads us to believe that the PAT percentage figure reported in both the 45 and 60 day deports, for the IFT-1A flight test, were *invalid* and *well below* the contract requirements.[178]

Reed punctuated the seriousness of DCIS's finding with the following conclusion: "There is no crime in producing a failed algorithm during a Research and Development Project. . . . The crime is in producing a failed algorithm and knowingly covering up its failure."[179] There is disagreement in official circles as to whether TRW's actions in this matter broke laws or just flouted norms of good scientific practice.[180] However, the strenuous attempts of TRW officials to squelch internal criticism and punish dissenting employees recall earlier episodes of controversy containment ordered by Pentagon and industry leaders to protect the reputations of the Star Wars and Patriot missile defense systems. It seems like business as usual in the corrupt BMD enterprise.

The disappointing flight test data gathered from IFT-1A presented a serious problem to NMD program managers, since the EKV's ability to discriminate between warheads and decoys in space is a crucial index of the system's potential to perform effectively in challenging combat environments, where decoys would likely be in abundance. One strategy employed by officials in response to this problem involved simplifying test conditions in subsequent experiments. Initially, the number of decoys for follow-on tests was reduced from nine to three. Then test conditions were simplified further, when plans were made to disguise the warhead reentry RV in these experiments by using only one decoy.[181] To help ensure success in the testing program, BMDO Director Kadish then declared that the original benchmark of two successful intercepts necessary to trigger a positive NMD deployment decision would be revised downward, so that only one intercept was required "to proceed with the award of the construction contracts."[182]

Responding to these moves, Postol wrote a remarkable letter to John Podesta, White House Chief of Staff.[183] In this letter, Postol synthesized and extended many of the charges made by Nira Schwartz regarding fraud and cover-up in the NMD testing program. Specifically, Postol criticized TRW's "tampering with both the data and analysis" from the IFT-1A experiment, as well as BMDO's reconfiguration of follow-on flight tests to "hide the program-stopping facts revealed in the IFT-1A."[184] Assessing TRW / BMDO's strategy of polishing official test reports by simply discarding bad data, Postol argued that " . . . the procedures followed by the BMDO were like rolling a pair of dice and throwing away all outcomes that did not give snake eyes, and then fraudulently making a claim that they have scientific evidence to show that they could reliably predict when a roll of the dice will be a snake eyes."[185]

Postol's analysis was based entirely on sources labeled "UNCLASSIFIED" by the U.S. government. As *New York Times* reporter William J. Broad described a few days after Postol released his findings, "the letter has circulated widely throughout the administration, Congress and the world, with hundreds, if not thousands, of electronic copies speeding across the Internet."[186] Despite such wide circulation, the Pentagon shot back by declaring that Postol's entire letter was classified, even though there was no realistic way to put the proverbial toothpaste back in the tube. Air Force Lt. Col. Richard Lehner, spokesperson for BMDO, explained that this extraordinary classification ruling was made not because Postol had inappropriately leaked secret information, but because the Pentagon's own declassification officers had erred upstream, mistakenly releasing "three or four charts" in material that Postol subsequently referenced.[187] This exchange stirred memories of a similar struggle over classification that occurred nearly a decade before, when BMDO attempted to classify Postol's entire *International Security* article regarding performance of the Patriot missile in the 1991 Persian Gulf War.

Just as the debate over Patriot's Gulf War accuracy eventually made its way into congressional deliberations, Postol's most recent charges also colored arguments on Capitol Hill over the proper level of NMD funding to be included in the FY 2001 Defense Authorization Bill. In floor debate on 18 May 2000, Rep. Dennis Kucinich (D-OH) said "[t]his time we know about the scandal before we vote on the money," then proposed an amendment that would have slashed funding for NMD by $2.2 billion.

> When you have the credibility of the Pentagon and of defense contractors being called into question by a prominent scientist at the Massachusetts Institute of Technology, when this report says they are covering up flaws in data, this makes it a national security matter, because if this system cannot work, then we are telling the American people to pay $2.2 billion in the hope that somehow a system will work when there is data that has been according to this scientist . . . phonied up. . . . It's a shame to have the bill clouded up with deception by the Pentagon and by defense contractors.[188]

Although Rep. Kucinich's amendment did not pass, elsewhere on Capitol Hill Postol's letter stirred more questions about the NMD program. During an 18 May 2000 Pentagon news briefing, DOD spokesperson Kenneth H. Bacon fended off press queries based on Postol's charges, saying, "He [Postol] focused primarily on one test, so-called the Integrated Flight Test 1A. . . . Dr. Postol's letter focuses primarily on this first test that involved the Boeing TRW interceptor. Now we are using interceptors, and have for several tests, they've been using interceptors made by Raytheon."[189] Bacon's rejoinder, which smacked of a shell game, glossed over the broader and more fundamental point of Postol's letter. Although Postol did focus much of his technical analysis on data manipulation in the IFT-1A experiment, he used this analysis as a point of departure for a more sweeping indictment of the testing methodology employed to test the new Raytheon interceptor in IFT-2, IFT-3, and IFT-4. In these follow-on tests, Postol argued that BMDO radically simplified test conditions in order to maximize the chances for good experimental results, by reducing the number of decoys from nine to three, then from three to one. This backtracking in the NMD testing program, coupled with announcement of the curious downward revision in the number of successful intercepts needed to trigger an NMD deployment decision (from two to one), brings to mind the image of an over-matched high jumper who keeps lowering the bar to salvage some modest success. In the NMD program, "[s]o far the bar has been set very, very low," comments Lisbeth Gronlund, of the Union of Concerned Scientists, "[m]uch, much lower than it should be, which leads me, at least, to conclude that this is really not about assessing the technology."[190] If the missile defense testing program is not about assessing the technology, what is its real purpose? According to Robert Scheer, missile defense in the new millennium is about finding "a new justification for subsidizing defense contractors that admits its true purpose to create profit and jobs for the military-industrial complex."[191]

FUTURE PATHS

In 1968, twenty-seven notable scientists from around the globe gathered for a Pugwash Conference, called for the purpose of discussing the "implications of anti-ballistic missile systems." In the conference proceedings, European physicists C. F. Barnaby and A. Boserup summarized the main points raised by participants. Much of their summary dealt with the standard issues usually highlighted in missile defense discussions, such as the effectiveness of proposed systems, how such systems would affect international relations, and whether widespread missile defense would enhance or jeopardize strategic stability.[192] However, near the end of their report, Barnaby and Boserup raised a different issue, suggesting that ABM deployment would "increase the role of the military in decision-making and increase public suspicion in both blocs."[193] With this warning, they hinted that missile defense proposals transcend questions of military strategy and laboratory science, by implicating much broader issues regarding decision-making power in democratic societies. Barnaby and Boserup's concerns in this regard turned out to be quite prescient. The historical record of missile defense advocacy is shot through with examples where military officials have abused their considerable powers to control and manipulate democratic decision making. The following list, culled from the case studies in this book, details some of these techniques.

- Classification power used to cover up rigged scientific experiments and censor criticism.
- Power over personnel decisions used to punish and deter in-house whistle-blowers.

- Security clearance review power used to intimidate and silence oppositional voices.
- Black budget spending power used to circumvent congressional appropriations authority.

As classicist A. Jardé notes, in public discussions, Spartans "took great care to keep the figures both of levies and of losses [from war] secret."[194] Earlier discussion in this chapter has explored the possibility that missile defense advocates' frequent abuse of military power to commit strategic deception shows that the United States may have evolved into a kind of modern-day Sparta.[195] The strength of this analogy is bolstered by the fact that Spartans (like Americans today) were expected to tolerate official deception as a matter of course. As historian Anton Powell explains, "Spartan official deceit included not only lying to the helots as to whether they would be rewarded or killed [in battle] . . . but also lying to their own citizens about the outcome of battles involving Spartan forces. This last form of deceit, recorded twice, suggests that Spartans *were expected not to object strongly to being so misled,* for the unpleasant truth would predictably emerge before long."[196]

In questioning a possible historical link between contemporary America and ancient Sparta, skeptics might point out that U.S. heritage is still tied to Athens, since the Athenian government was democratic, yet it lied, too. However, Powell's comments on this point speak volumes about the unique status of state-sponsored and coordinated strategic deception as a threat to democracy, when such deception is adopted as a routine instrument of foreign and domestic policy.

> At Athens there was no shortage of public lies, as students of the orators will testify. But the Athenian law against deceiving the demos reminds us of the danger that lay in conspiring—with several potential witnesses against one—to mislead the sovereign body. In contrast to Sparta's tight oligarchic structures, the vast membership of the ruling assembly at Athens made it difficult to mislead even foreigners by secret diplomacy. What distinguishes Sparta is not self-interested mendacity by ambitious individuals, but high-minded, often elaborate, official conspiracy.[197]

The parallel between Sparta and the contemporary "citadel culture"[198] of "Fortress America"[199] is more than just an intriguing historical footnote. Earlier discussion has highlighted profound consequences of democratic erosion brought about by strategic deception in missile defense advocacy. Over $100 billion has been spent on frequently dubious missile defense projects since 1962.[200] The current NMD budget commits the United States to new missile defense outlays of approximately $5 million per day over the next six years. Excessive secrecy has compromised the technical integrity of scientific research at U.S. missile defense laboratories. Runaway deception has boomeranged on defense officials attempting to minimize the environmental hazards of nuclear weapons cleanup. Repeated lies have drained the worldwide credibility of U.S. missile defense assurances, undermined domestic confidence in government, and injected fear and distrust into international relations. Some day Americans may be forced to pay the ultimate price for this legacy of official deceit, warns Garwin: "Time after time our national security choices have been misdirected by false argument, concealed assumptions, and hidden agendas. Some of the best options have been ruthlessly suppressed. We have all so far paid for that with our wealth and our well-being. If we don't restore integrity to our government, we may well pay for it with our lives."[201]

Can this tragic outcome be avoided? In Brazilian educator Paulo Freire's progressive pedagogy, the first step toward changing unjust, exploitive, or dangerous conditions in the world is to imagine alternative worlds worth seeking.[202] The following pages sketch the features of some possible alternative worlds in this regard. The complex and layered nature of strategic deception schemes makes the process of challenging strategic deception resemble the peeling of an onion: once one layer of secrecy is turned back, previously obscured dimensions of deception come into view. In keeping with this phenomenon, the following discussion moves through three dimensions of change (surface-level, institutional, and cultural) that address progressively deeper and more fundamental aspects of the problem. This discussion is offered as a heuristic resource that might be used to counter the current drift in missile defense politics.

Surface-Level Change

The prospect of a momentous NMD deployment decision looms in the immediate future. Previous decisions of a similar nature in 1983 and 1991 were made on the basis of systematically distorted scientific evidence. To increase the likelihood that the upcoming deployment decision will be made with sound information, short-term steps could be taken to generate high-quality data on the key question of whether missile defense is "technologically possible." For example, the American Physical Society has recommended additional study of the Patriot's Gulf War performance that would include a comparison between Postol et al.'s video analysis and the Pentagon's classified evidence.[203] So far, the Pentagon has thwarted the possibility of such a review by shielding its classified evidence from public inspection. This strategy has unnecessarily impeded efforts to reach a clear and unambiguous assessment of the first-ever use of a missile defense system under battlefield conditions. Although the Patriot differs from proposed NMD systems in important technical respects, an analysis of the Patriot's Gulf War performance would nevertheless yield pertinent insight regarding the challenges involved in extrapolating the results of NMD field tests to actual wartime environments.

In congressional debates on the 1999 Missile Defense Act, the Rumsfeld Commission report on ballistic missile threats was a pivotal piece of evidence for many elected representatives voting for the act, including swing vote Bob Kerrey (D-NE). However, in numerous publications and letters prior to the vote, Rumsfeld panelist Richard Garwin warned that assessing foreign missile threats and deciding the appropriateness of missile defense to counter such threats are entirely different tasks. "If the Rumsfeld commission spent another six months looking at the capability of the proposed national missile defense system against the threats that we have identified," Garwin explained, "I believe that we would have come to a similarly uniform conclusion—that it would not be technically effective."[204] Thus, before the Rumsfeld Commission findings are used again to support a possible NMD deployment decision, it would seem prudent to reconvene the commission in order to study the "technological possibility" of effective missile defense. In a 1998 letter to Congress, Garwin made this suggestion explicitly: "One way to rectify this unfortunate misapprehension [of the Rumsfeld Commission's findings] is to reconvene the Rumsfeld Commission with a related relevant mandate or to create a second Commission."[205] A second Rumsfeld Commission study of missile defense feasibility could help insulate Congress from the manipulative effects of any possible "crucial experiment" rhetoric flowing from fresh and successful NMD tests. As previous discussion has shown, because such tests do not ordinarily incorporate countermeasures and other "real world" complications faced by defensive systems, they have only limited utility as proof that missile defense is "technologically possible."

Collina recently echoed Garwin's call for an expert panel to assess missile defense feasibility, although Collina stipulated that for such a review to be useful, it must be conducted by an independent group of scientists who consider the factor of countermeasures directly in their evaluation.[206] Given the demonstrated propensity of the Pentagon to mothball critical BMD reports on countermeasures (e.g., the 1983 Fletcher report), special care should be taken to ensure that expert opinions on countermeasures are solicited and incorporated into the methodology of any expert panel.[207] This is a crucial component of any NMD deployment decision, since as Garwin stated in a 1999 Senate hearing, "[u]ltimately, the U.S. NMD system will succeed or fail based on its ability to deal with countermeasures."[208]

Institutional Change: Time for BMDO's Own "Openness Initiative"?

As members of Congress learned recently from a futile attempt to cancel the resilient MEADS missile defense program, changes made at the surface level of missile defense politics can be easily circumvented in deeper layers of the secret BMDO bureaucracy. According to Gottlieb, "basic change in the defense industry is not possible so long as the sole buyer of weapons (the Pentagon) and the sellers (defense firms) necessarily only deal with each other."[209] Similarly, it would be folly to expect that completion of a few independent studies on missile defense effectiveness would be enough to restore overall integrity and accountability to the missile defense enterprise. For this task, more fundamental and systemic changes are necessary. Such changes require basic shifts in the way that BMDO's bureaucratic culture treats the notions of secrecy and public accountability. Perhaps the best place to start would be to revisit classification policy. Currently, the Pentagon's classification policy runs largely on autopilot, since the vast majority of classified material is "born secret," regardless of content. The phrase "born secret" is a term of art in intelligence circles that refers to information automatically classified, without review.[210] This automatic classification procedure removes accountability for classification decisions and invites abuse. BMDO officials need not explain why a particular piece of information is classified, they only have to explain that it is. As Burr, Blanton, and Schwartz explain, "by its very nature, the idea of 'born secret' promotes overclassification."[211]

One simple step that could be taken to counteract this trend would be to shift the presumption of secrecy, so that BMDO material would be what I call presumptively public, not born secret.[212] From this starting point of presumed nonclassification, the burden of proof would rest with those desiring to keep information under wraps. While such a measure would go far to reduce frivolous and politically motivated classification, truly sensitive information could still be classified with a defensible, written rationale. A National Academy of Sciences review board, which studied DOE's classification policies in 1995, concluded that such a remedial shift in the burden of proof would be essential for the department to carry out its post–cold war mission.

> DOE should be guided by the presumption that information should not be classified unless there is an identifiable reason why the release of the information could damage national security or a reason for concluding that the costs of release outweigh the benefits. The burden of proof should be on those who argue for classification, not on those who propose declassification.[213]

A similar shift in burden of proof for BMDO employees might prompt valuable reflection about the classification of missile defense information in the post–cold war milieu. The pressure to provide reasons for each measure of information control could desediment the organization's ossified cold war secrecy norms. Prompted to rethink the institutional logic that led to the establishment of the many "born secret" categories of information in the first place, missile defense officials might develop approaches to information control that are more appropriate in the post–cold war environment.[214] Additionally, such a change in burden of proof would provide a mechanism to increase post-hoc accountability, whereby citizens could monitor BMDO decisions about classification. A bill proposed recently by Senator Daniel Patrick Moynihan (D-NY) would require that an official "who makes the decision to classify information shall identify himself or herself and provide a written justification for the action."[215] Stripped of the easy "born secret" rationales for secrecy, under this proposal classification officers would have to justify their decisions, and their rationales could be more readily scrutinized and evaluated as to their appropriateness.[216]

Classification policy is just one pillar of BMDO's prevailing culture of secrecy and strategic deception. As former secretary of energy Hazel O'Leary demonstrated in her "Openness Initiative" (1993–97), there are a broad array of institutional practices that need to be changed in order to introduce accountability into a massive, defense-related government agency. Incentive structures for personnel decisions need to be adjusted to reward responsible initiatives taken to increase transparency and challenge unnecessary secrecy and fraud. Better communication with contractors must be achieved in order to enhance competition, increase efficiency, and reduce costs. Relations with external oversight bodies, members of the media, and the general public need to be improved to reduce fear and increase trust and cooperation.[217]

Even if BMDO leaders make ostensible commitments to these principles, such commitments may still not be sufficient to ensure accountability in the missile defense enterprise. Ultrasecret government agencies have a poor track record of following through on bold new promises of openness. For example, soon after they were confirmed, CIA directors Robert M. Gates and James Woolsey each promised to open records—"warts and all"—on several controversial covert operations.[218] However, as George C. Herring, history professor at the University of Kentucky and member of a CIA panel that helps the agency declassify its data, recently concluded, Gates and Woolsey's promises were "a brilliant public relations snow job," and the prospect of greater CIA openness was a "carefully nurtured myth."[219] Given the propensity of missile defense researchers to fabricate data, rig experiments, and then directly deceive Congress in scientific reports, Congress should mandate external peer review to monitor the reliability of BMDO experimentation, even if BMDO unilaterally promises more openness. In short, "there should be an independent oversight of the overall NMD testing program," Garwin explains.[220]

Such oversight need not entail the creation of vast new surveillance agencies, since legislation already exists authorizing close congressional monitoring of BMDO activities. The majority of research products generated by BMDO's peripheral supporting network are classified secret, designed for the eyes of BMDO officials only. However, when private research advice is furnished to BMDO through the channel of an explicitly established or de facto "advisory committee," it is a statutory requirement under the Federal Advisory Committee Act (FACA) that such advice must be proffered through a public forum, and that the research products from such collaborative efforts must be publicly available.[221]

FACA can be a powerful tool for contesting excessive secrecy. In the missile defense context, previous FACA challenges have resulted in the termination of several unchartered advisory committees that were operating in violation of the law.[222] However, termination of illegal committee contact is only one possible outcome that can result from a successful FACA challenge. Because FACA does not prescribe specific remedies for statutory violations, courts have ruled that creative remedial measures may be tailored to fit particular situations.[223] Congressional investigators and litigants are thus free to use evidence of FACA noncompliance to leverage their own preferred strategies for demanding transparency and accountability in the government's solicitation of private advice.

More energetic congressional oversight can have a significant effect. "Indeed, some of the worst effects of nuclear secrecy could have been mitigated if Congress had pursued more aggressively its oversight responsibilities and ensured that the AEC [Atomic Energy Commission] conducted tests and constructed and operated fissile materials production and storage facilities in ways that minimized risks to the public. The JCAE [Joint Committee on Atomic Energy] could have protected public interests by prodding the AEC and DOD [Department of Defense] to bear in mind more fully the environmental and health effects of nuclear weapons. In this way, Congress could have strengthened the constitutional system of checks and balances, instead of allowing them to atrophy," explain Burr, Blanton, and Schwartz.[224]

Cultural Change

While greater institutional openness and heightened congressional oversight would represent positive steps toward improved accountability in the missile defense business, these procedural changes alone may not be sufficient to counter the deceptive inertia built up in the BMDO bureaucracy. Perhaps consideration of another highly controversial public policy issue will shed light on this dilemma. In the 1970s, the U.S. Supreme Court came close to banning capital punishment on the grounds that discriminatory application of the death penalty violated the Eighth and Fourteenth Amendments to the U.S. Constitution.[225] However, the Court stopped short of a flat ban on capital punishment, ruling instead in the 1976 case of *Gregg v. Georgia* that states could use the death penalty, as long as they implemented a set of procedural safeguards designed to ensure racial justice in capital cases.[226] Assessing the results of widespread experimentation with such procedural safeguards almost fifteen years after the *Gregg* decision, sociologists Adalberto Aguirre Jr. and David V. Baker concluded that "racial discrimination continues to influence the imposition of the death penalty," and that "[t]he proposed safeguards that surround the imposition of the death penalty amount to no safeguards at all."[227] Reaching a similar conclusion in the 1987 case of *McCleskey v. Kemp*, dissenting Supreme Court justices William Brennan and Thurgood Marshall argued that "the effort to eliminate arbitrariness in the infliction of that ultimate sanction is so plainly doomed to failure that it—and the death penalty—must be abandoned altogether."[228] Systemic and unconscious racism was so persistent and ubiquitous throughout the capital punishment machinery, Brennan and Marshall reasoned, that at best, procedural protections against discrimination would provide only the illusion of fairness in death penalty cases.

While on first glance capital punishment and missile defense appear to have little in common, the futility of procedural safeguards to ensure justice in death penalty jurisprudence points to an important insight that may connect the two topics. When bureaucratic "machinery" is irreparably broken, no amount of procedural reform may restore its integrity. Just as jurists and commentators suggested that the death penalty "machinery" is inherently stained by systemic racism, perhaps the U.S. missile defense bureaucracy suffers comparable intrinsic defects that invite chronic failure and breakdown. The predilection for fraud, waste, and deception built into BMDO's bureaucratic structures may be so endemic that no amount of procedural reform could right the course of the U.S. missile defense program and enable citizens to reap the potential benefits of missile defense protection. If this is indeed the case (and the case studies explored in this book suggest strongly that it is), then citizens and policy makers worldwide need to think long and hard about what may be the only sound policy option: termination of the flawed BMD program.

As the most potentially destabilizing and provocative element of the U.S. BMD program, space-based weapons are most eligible for outright cancellation. Although official U.S. declarations currently portray the proposed U.S. NMD system as a limited ground-based defense, designed to intercept only a handful of missiles, a premature deployment decision could leverage the efforts of advocates for spaceborne weapons to shoehorn their projects into expansive missile defense budgets.

For example, Gen. Joseph Ashy, Commander-in-Chief of the U.S. Space Command, has linked explicitly the hit-to-kill (HTK) mission of BMD with plans to weaponize outer space. As journalist Karl Grossman observes, "The PR spin is that the U.S. military push into space is about 'missile defense' or defense of U.S. space satellites. But the volumes of material coming out of the military are concerned mainly with offense—using space to establish military domination over the world below."[229]

There is the frightening potential of spaceborne weapons to be used for offensive military missions. According to US Senator Charles Robb (D-VA),

"During the Reagan years advocates of the Strategic Defense Initiative ran an effective television spot featuring children being saved from nuclear attack by a shield represented by a rainbow. If we weaponize space, we will face a very different image—hundreds of weapons-laden satellites orbiting directly over our homes and our families 24 hours a day, ready to fire within seconds. If fired, they would destroy thousands of ground, air, and space targets within minutes."[230]

The parade of well-respected scientists who have defected from the space-based BMD program on grounds of scientific principle punctuates former defense industry insider Carol Rosin's point that "most of the people working on these projects, of space-based weapons related projects, I think they don't want to be doing it."[231] However, Rosin notes that the restrictive bonds of military-industrial-complex contracts prevent retractors from expressing their reservations, even though many researchers employed by the Pentagon would prefer to be working on nonweaponized projects.

> The boys in the industry can't speak out, because they're afraid they'd lose their jobs. . . . They would rather see alternatives researched and developed in space that eliminate space-based weapons, but allow them to go on and expand the military space program in a nonweapons way, because the military can really be of service, and do great things like they already are doing. We can expand their civil and commercial space program, which is so lacking now because the largest research and development program in recorded history is the space-based weapons program. Most people don't even know it.[232]

Rosin's assessment suggests that to the extent that calls for straightforward bans on space-based weaponry are issued without accompanying plans for researching and developing commercial, nonweapons programs, such claims are bound to lose traction as measures of constructive change. The tremendous economic, political, and social momentum already gathered behind missile defense research and engineering is so powerful that it is probably unrealistic to demand that such work halt dead in its tracks. Instead, Rosin recommends an "aikido" strategy of resistance,[233] where the prowess of military researchers is acknowledged, affirmed, and embraced, then rechanneled into projects that promise to transcend the endemic and chronic flaws of space-based missile defense. Rosin identifies the commercial space program as a project that holds potential in this regard as a new outlet for

Pentagon researchers to continue their technical work in a context that breaks out of the debilitating cycle of deceptive, zero-sum planning found in missile defense circles. "This space program, military-civilian-commercial, is the key to getting us out of the deception, and into the openness of the vastness of space itself, a whole new way of thinking and communicating," Rosin explains.[234]

Unfortunately, the prevailing cultural climate does not lend itself to generation of the necessary political pressure required to deflect the substantial bureaucratic, economic, and political inertia gathered behind weaponized space projects. Widespread citizen alienation and withdrawal from the political process limits grassroots efforts to demand government accountability. As Burr, Blanton, and Schwartz explain, the fallout of Secretary O'Leary's Openness Initiative appears to have ironically exacerbated the cynicism and alienation that made previously hidden DOE abuses possible in the first place.

> Secretary O'Leary's decisions to release information about once-secret AEC and DOE activities enriched public debate and restored greater accountability to the political system. Unfortunately, revelations about human radiation experiments and intentional radiation releases may also have contributed to already high levels of public mistrust toward, and cynicism about, the government. . . . Scandals create the impression that the government is hiding even more serious misdeeds. Thus, even when secrecy serves legitimate public interests, the public continues to believe, and not without reason, that it is excessive.[235]

There is a danger that this prevalent crisis in government trust is creating a spiraling dynamic, whereby citizen alienation spurs political withdrawal, and political withdrawal creates a vacuum inviting more dubious government policies. According to Steven Aftergood of the Federation of American Scientists, "[m]ore and more people just do not believe that government is telling the truth. This is a profoundly disturbing development that threatens our political system at its roots. A healthy society does not have to agree on every subject or even have a consensus on those subjects, but it does require some minimal degree of public confidence in government. Without it, the system is on a path toward breakdown."[236]

How can a vital political culture be invented, and such a breakdown be averted? Full exploration of this daunting question could easily fill the pages of another book. However, three points here deserve at least passing mention. First, the stranglehold of big money over American electoral politics and mass media sucks the breath out of grassroots efforts by citizens to assert democratic control over the political process. Speaking about this problem in the context of attempts to counter ill-conceived missile defense initiatives, Bowman argues that such attempts "won't make real progress until we first all get together and get campaign finance reform, electoral reform, and media reform, so we can restore our democracy."[237] Second, attention to this broader picture should not obscure the fact that small acts, although insignificant when measured in the overall context of a comprehensive solution, can still achieve important effects. "Citizens need to challenge government secrecy daily by asking questions of their elected officials, of the administration in power, of the press, of scholars and public interest watchdogs. Each question loosens a grain of sand in the bricks of secrecy. The principal benefit of the resulting greater openness is good government through accountability to the people," suggests Scott Armstrong, founder of the National Security Archive.[238]

Third, while citizen questions can be important tools of transformation, perhaps the most crucial agents of transformation are the scores missile defense scientists who have lingering reservations about the U.S. BMD program, yet feel constrained to act on such reservations because they perceive a lack of alternative options. Scientists currently building U.S. missile defense systems possess world-class skills—missile defense engineering demands a dizzying array of technical talent in divergent fields, from chemistry to computer science to physics. These same skills could be put to use in peaceful and cooperative space ventures that might forge the foundation of a new and strong world security framework. For centuries, dueling nations have weaponized the Earth's land, oceans, and atmosphere. Before individual scientists uncritically buy into the project of grafting old war games onto outer space, they should pause to consider whether their talents might be better utilized in peaceful space projects designed to break out of the secretive and deceptive framework of cold war-style military confrontation. Those individual scientists who share these concerns, yet fear that their isolated acts of resistance might amount only to spitting into the wind, should take stock in Buckminster Fuller's hopeful aphorism: "With a handful of people, we can make it by the skin of our teeth."

Conclusion

The practice of rhetoric involves the invention of persuasive arguments in contingent moments of public decision. The turbulent currents of complexity and uncertainty rushing through the post–cold war milieu call out for recovery of rhetorical deliberation as a mode of collective learning and decision making. Collective deliberation was foreclosed systematically during the cold war, when the bitter bipolar conflict between nuclear superpowers prompted systematic distortion of discussion about the role of military science in society. In the vacuum of public discourse, events were interpreted through a black-and-white filter that framed issues in zero-sum terms and reified a self-referential logic that redoubled the technological momentum of an "internal arms race" pitting American defense planners against themselves. In the context of offensive weapons, this closed logic yielded dubious constructs such as the "missile gap" and the "window of vulnerability," illusory notions that primed the American public's thirst for crash episodes of nuclear bomb production.[239] Eventually, citizens grew weary of the looming threat of nuclear annihilation, and calls for nuclear arms control and disarmament became more numerous and strident as years of cold war hostility piled up.

In 1983, President Ronald Reagan changed fundamentally the dynamics of the debate with his advocacy of the Star Wars missile defense system. The historical record indicates that in a strictly instrumental sense, this advocacy campaign turned out to be a smashing success. Using Reagan's argument, missile defense advocates successfully lobbied for approximately $70.7 billion in congressionally appropriated research funds related to SDI. In the process, the momentum of peace and antinuclear movements was blunted. However, the glitter of such instrumental successes is tarnished by a very basic blemish. Reagan's argument for SDI was based on a lie—the argument for "leakproof" missile defense held water only when buoyed by systematically distorted scientific evidence. During the Star Wars era, missile defense advocates could "steal the language" of peace movement activists only by peddling scientific snake oil. In public arguments, such data appeared credible only because strict secrecy guidelines insulated the data from rigorous scientific peer review and criticism. When whistle-blower leaks and chronic technical failure in the SDI program began to shatter the illusion of "perfect defense," missile defense advocates retreated to argue that even if imperfect, BMD could still force the Soviets to the arms control table and serve as a useful bargaining chip.

Retrospective historical accounts suggest that SDI largely failed in this capacity as well, provoking superpower hostility and suspicion that postponed the end of the cold war by at least two years.

Although startling disclosures of Reagan-era missile defense fraud have cast real doubt on the integrity of the entire missile defense enterprise, the key elements of such an enterprise have not only persisted but flourished in the post–cold war milieu. During the Gulf War, for example, missile defense advocates resorted to many of the same tactics of strategic deception honed during Star Wars promotion campaigns. Evidence of such deception can also be found in more recent episodes of missile defense advocacy involving contemporary BMD systems such as THAAD and NMD. Although missile defense officials tend to evince a strong commitment to the notions of scientific peer review and democratic accountability in press conferences and public appearances, these normative guarantees often turn out to be mere window dressing. When missile defense advocates successfully "stole the language" of peace movements during Reagan's Star Wars heyday, such thievery also locked in the authority of a secretive and antidemocratic missile defense technocracy. Rhetorical criticism of these episodes may help facilitate recovery of deliberation as a category of critique that identifies departures from prevailing communicative norms, as well as instances where costs and risks of secrecy are shifted to vulnerable populations. Controversy over missile defense represents a site where many of the pertinent issues about the role and significance of military science in democracy have played out in public argument. These pertinent issues include the authority of science in society, the role of the public in national security decision making, the power of the military as a patron of scientific enterprise, the meaning of scientific consensus, and the very definition of legitimate scientific inquiry. The wide array of important and overlapping concerns implicated here makes recovery of deliberative practice in missile defense discussions imperative.

This chapter has explored several possible strategies for recuperating deliberative rhetoric as a form of lived human interaction that has real potential to help us make better collective judgments about important matters of war and peace. On a surface level, government agencies and citizens might press for more reliable data on the effectiveness of proposed missile defense systems, by insisting that such data be forged through communicative interchanges that square more fully with the tradition of scientific inquiry as an open process of communal dialogue. Those contemplating action on a deeper plane of political engagement might consider calling for more fundamental changes in the institutional structures of the missile defense bureaucracy, in hopes that such pressure could help promote more open and accountable institutional norms within government agencies and corporations involved in the missile defense business. Since there is always the possibility that these strategies for change will simply be overwhelmed by the huge military, political, and economic inertia currently gathered behind the missile defense enterprise, it is also important to consider the prospect of more radical cultural change as a potential remedy.

One way to conceive of such a cultural sea change is to envision fundamental alterations in the interlocking constellation of interests that make up what President Dwight D. Eisenhower called the "military-industrial-complex." In this book, I have made a few token gestures in this direction, but my efforts have been necessarily preliminary in nature. Campaign finance reform, individual citizen and social movement protest, and exuberant whisteblower activism each have potential as catalysts of cultural change. But in the case of BMD, there may be another approach that has not yet been considered in detail.

The rhetorical fulcrum leveraging Ronald Reagan's advocacy of SDI was a simple, yet powerful equation: BMD = disarmament. This equation was codified in secret planning documents, voiced in Star Wars promotion campaigns, and accepted by a considerable segment of the American population who identified with the appeal of defensive arms as a hopeful strategy to "make all nuclear weapons obsolete" (as actor Paul Newman speculated in the 1966 Alfred Hitchcock film *Torn Curtain*, and Reagan echoed in his 1983 Star Wars address).

One of the clear lessons of the cold war is that "purely defensive" weapons can be highly provocative and destabilizing when introduced into a climate of hostility and suspicion. Yet, defensive arms may have more utility as tools of peace if they are shared openly and introduced into a world where nations have already abolished nuclear weapons as tools of statecraft. This point was made by the unlikeliest of pro-BMD sources, nuclear abolition advocate Jonathan Schell, who wrote in 1984 that "if defenses were arrayed against the kind of force that could be put together in violation of an abolition agreement, they could be crucial."[240] On this logic, missile defense could work as an insurance policy that would underwrite the credibility of a worldwide agreement to abolish nuclear arms.

> President Reagan recently offered a vision of a world protected from nuclear destruction by defensive weapons, many of which would be based

> in space. The United States, he said, should develop these weapons and then share them with the Soviet Union. With both countries protected from nuclear attack, he went on, both would be able to scrap their now useless nuclear arsenals and achieve full nuclear disarmament. Only the order of events in his proposal was wrong. If we seek first to defend ourselves, and not to abolish nuclear weapons until after we have made the effort, *we will never abolish them*, because of the underlying, technically irreversible superiority of the offensive in the nuclear world. But if we abolish nuclear weapons first and then build the defenses, as a hedge against cheating, we can succeed. *Abolition prepares the way for defense.*[241]

The Heritage Foundation/High Frontier rhetorical strategy for using the equation BMD = disarmament to boost political support for Star Wars during the 1980s relied on a "public argumentation campaign" designed to "steal the language and cause" of the nuclear freeze movement.[242] Schell's counter-recipe, "abolition prepares the way for defense," represents a dramatic reversal of this logic, by using disarmament as a tool to transform missile defense. This might sound like a chicken-or-egg quibble. However, it is important to note Schell's stipulation that his equation only runs one way. A secret missile defense program, rushed into existence before the emergence of a nuclear abolition regime, and designed to protect only a handful of rich nations, is likely to sow seeds of international fear and distrust that could make disarmament impossible. On the other hand, nuclear abolition would likely transform the idea of missile defense in profound ways. In fact, abolition could turn out to be the most promising strategy for overcoming the myriad pitfalls of strategic deception in missile defense advocacy.

In the context of a nuclear abolition regime, it would seem appropriate to conduct scientific research on BMD out in the open, and to develop defenses that would protect all nations who rejected nuclear armaments. Since the eyes of the world would follow every test, calculation, and conclusion of missile defense scientists, there would be little place for secrecy, trickery, and waste in such a program. Scientists and engineers from around the globe could collaborate on peaceful space projects that would have as one of their primary objectives strengthening the nuclear abolition regime. Since research would be conducted in the open scientific community, scientists could throw off the shackles of military classification that constrain and corrupt the process of inquiry. While such notions may sound fanciful, it is worth noting that similar visions have received support from unexpected quarters. Consider the following quotation, which appeared in a 1960 paper written by one of the most die-hard cold warriors of our time:

> Toward the end of World War II and in the years following Hiroshima and Nagasaki, Niels Bohr suggested a method of dealing with the problems of nuclear arms. The suggestion was clear-cut and radical. Its central part was to abandon secrecy. He strongly advocated that we return to the free discussion of discoveries and ideas which were characteristic of scientific work before World War II. It is obvious that if freedom of information were fully established throughout the world *all arms-control problems would at once become more manageable.* It would be necessary to bring about the situation where the freedom to exchange information would be guaranteed by enforceable international law. Under such conditions it would become extremely difficult to keep the development of nuclear weapons secret, whether the development were to be pursued by testing or by other procedures. The production and deployment of weapons might become known at the same time. . . . I believe that Bohr's suggestion deserves serious consideration. It strikes at the root of our difficulties. It stresses the kind of openness which is natural in free countries and which has been the lifeblood of science.[243]

Readers familiar with the history of the hydrogen bomb might recognize these words from Edward Teller, who penned them years before he would begin work on a defensive system that he hoped would ameliorate the destructive potential of his most frightening invention, the H-bomb. Ultimately, Teller's vision of spaceborne X-ray laser defense never came to fruition, in part because the secrecy and corruption that pervaded his laboratory caused a parade of talented scientists to abandon the project in protest. In a nuclear abolition regime, those nations and groups seeking to develop advanced *offensive* weapons could find themselves in a similar predicament, since they would be forced to labor in isolation, cut off from what Teller calls "the lifeblood" of science. Alternatively, those scientists working on defensive weapons would enjoy the creative fruits of shared inquiry that are nurtured in open scientific communities. According to Schell, one surprising result of this contrast could be a reversal of the traditional view of nuclear strategy which holds that technical improvements in offense can always overwhelm defense. "As the years passed after the signing of the agreement the superiority of the defense would be likely to increase," Schell speculated, "because defensive weapons would continue to be openly developed, tested, and deployed, while offensive weapons could not be."[244]

The obvious practical obstacle standing in the way of such a plan is the fact that the nation currently possessing the most advanced BMD technology also has the world's largest offensive nuclear arsenal, and has shown little or no willingness to give it up. The task of convincing U.S. leaders to re-evaluate their stance on disarmament may be an uphill struggle, given the doggedness with which they cling to cold war concepts of nuclear deterrence and "strategic stability." "I have never seen a moment where the U.S. seemed so isolated from the mainstream of international opinion on the nuclear weapons issue," says William Hartung of the World Policy Institute.[245]

Cursory examination of the U.S. negotiating stance in talks with Russia over proposed ABM Treaty modifications reveals how far American officials are behind the curve of swelling momentum for nuclear disarmament. In ABM Treaty "Talking Points" presented by U.S. negotiators to their Russian counterparts earlier this year in Geneva (and leaked to the public in April 2000), the U.S. attempts to ease Russian fears regarding NMD with the alarming reminder that "under the terms of any possible future arms reduction agreements, large, diversified, viable arsenals of strategic offensive weapons" will exist on both sides, and that forces of this size "can easily penetrate a limited NMD system of the type that the United States is now developing."[246] In essence, the U.S. negotiating position tells Russians not to worry about NMD (because it is not reliable enough to work against their sophisticated weapons), but that just in case, Russian leaders might want to hang on to a huge nuclear arsenal (kept on "hair trigger" alert status), in order to make sure that they can overwhelm the planned U.S. missile defense shield.[247]

Republican presidential candidate George W. Bush recently made a redoubled pitch for missile defense, issuing a full-throated call for land, sea, and space-based defenses, and packaging such a prescription as part of what he called a "new approach to nuclear security that matches a new era."[248] Apparently dismissing cold war logic that led to massive buildups in offensive nuclear weapons, Bush coupled his appeal for robust missile defense with proposals for unilateral reductions in the U.S. nuclear stockpile, as well as adjustments in the U.S. nuclear force posture designed to increase nuclear safety. Bush's suggestion for unilateral nuclear cuts was greeted with approval by surprised members of the arms control community,[249] and his new approach seemed to represent a striking turnabout in Republican thinking on military policy. In fact, one could compare Bush's formula for combining nuclear build-down and missile defense build-up to Schell's strategy of using BMD as a tool of nuclear disarmament. However, such similarities only run skin-deep. Asked whether he would work for nuclear disarmament after his speech, Bush replied: "I will never reduce the levels of the nuclear stockpile of the United States to a position where it would jeopardize our safety and security."[250] Rather then following Schell's proposed path of "abolition prepares the way for defense," or Reagan's BMD equals disarmament formula, Bush seems to have chosen a third approach that uses admittedly imperfect missile defenses to leverage the U.S. nuclear deterrent. As several commentators pointed out after Bush's speech, this approach has limited potential as a comprehensive strategy for achieving world-wide nuclear safety, given that Bush's stated intention of hanging on indefinitely to the U.S. nuclear aresenal may prompt foreign leaders and citizens to see American missile defense as a tool for American nuclear hegemony. Such perceptions are likely to block the emergence of any abolition regime that offers a genuine escape from the cold war's nuclear grip.[251]

Current NMD plans call for Upgraded Early Warning Radar (UEWR) improvements to be made at U.S. military bases in Thule, Greenland; and Fylingdales, U.K. A parallel U.S. Air Force project, designed to enhance the long-term effectiveness of NMD, calls for construction of a Space-Based Infrared System (SBIRS) on U.K. soil at the Menwith Hill Signals Intelligence Centre near Harrogate, North Yorkshire.[252] Presumably, these early warning radar upgrades would consist of replacing existing computers, graphic displays, communication equipment, and radar receiver/exciters to perform the NMD mission (i.e. identification and precise tracking of a ballistic missile launched against the U.S.). Additionally, software programs would probably have to be re-written to enable the radars to identify different types of missiles and discriminate them from decoys or chaff.

Schell's proposal for defense-insured nuclear abolition regime raises an important question for European nations considering whether to support NMD by participating in U.S. plans to upgrade early warning radars housed on their soil. Will such radars gather data for the purpose of bolstering U.S. nuclear hegemony, or will they contribute to the more hopeful project of world-wide nuclear disarmament?

Europeans would do well to realize that they currently enjoy a significant measure of control over the answer to this question. Negotiations addressing the terms of U.S.-NATO radar co-operation have the potential to shape U.S. defense policy in significant ways. During the Cold War, the Heritage

Foundation's secret blueprint for SDI advocacy prescribed that proponents "should spend a great deal of time trying to get an 'offshore' constituency, particularly the governments of major U.S. allies. . ." Allied support for missile defense, the report reasoned, "will lend credibility to domestic U.S. proponents of BMD, while enlisting political support overseas that inevitably will begin to register on the entire U.S. national security bureaucracy."[253] In the contemporary political milieu, there is potential for this strategy to work again, but for a different cause. U.S. allies favoring nuclear abolition could use their positions on missile defense to shape the domestic U.S. debate on nuclear disarmament.

Specifically, NATO members might use their leverage as providers of crucial early warning data to stipulate that their participation in U.S. BMD programs would be contingent on a corresponding U.S. commitment to abolish its offensive nuclear arsenal and share missile defense technology and know-how with nations similarly committed. Additionally, European leaders could fortify the Outer Space Treaty of 1967, by making their approval of U.S. missile defense plans contingent on an American pledge to refrain from research and deployment of spaceborne weapons. Such a development could reverse the alarming trend toward space weaponization and clear the way for collaborative and peaceful ventures designed to explore outer space in the name of mutual interdependence.[254]

Given the durability of strategic deception as a staple of missile defense advocacy, it should be expected that we will see the patterns of argumentation analyzed in this book recurring and resurfacing in future contexts. "Whatever path is taken now, history indicates that in the coming decades proposals for strategic defense will re-emerge," writes MIT physicist Joseph Romm.[255] Like a gullible and forgetful diner who keeps ordering the mystery soup after a long series of unsavory surprises, will the United States keep writing blank checks for secret and exotic missile defense projects proposed by zealous advocates in the future, or will officials acknowledge a feasible emerging vision of cooperative space research, development, and exploration that breaks out of the restrictive zero-sum logic of the cold war?

NOTES

1. Raphael Sealey, trans., *A History of the Greek City States 700–338 B.C.* (Berkeley: University of California Press, 1976), 74–75.
2. For example, see Kevin Reilly, *The West and the World* (New York: Harper and Row, 1980), 69–77; Donald Kagan, Steven Ozment, and Frank M. Turner, *The Western Heritage to 1715* (New York: Macmillan, 1979), 51–66.

3. On the tendency to idealize Athens and demonize Sparta, see Elizabeth Rawson, *The Spartan Tradition in European Thought* (Oxford: Clarendon Press, 1969), 1–2.
4. See Guy Dickins, "The Growth of Spartan Policy," *Journal of Hellenic Studies* 32 (1912): 8.
5. A. H. M. Jones, *Sparta* (Oxford: Basil Blackwell, 1967), 20.
6. Conservative assessments of Spartan citizens' political power (see, e.g., Sealey, *History of the Greek City States,* 74–75) can be traced to Aristotle, who suggested in the *Politics* that Sparta's assembly could only ratify decisions by authorities, and listen (but not speak) during political arguments (see A. Andrewes, "The Government of Classical Sparta," in *Ancient Society and Institutions: Studies Presented to Victor Ehrenberg on His 75th Birthday* [New York: Barnes and Noble, 1967], 2–4). However, Andrewes suggests that "Aristotle wrote the relevant chapter of the *Politics* before he had studied the Rhetra, and indeed without having seriously considered the historical record of the first half of the fourth century" (Ibid., 7). For an account emphasizing the Rhetra's role as a tool of democratization, see L. H. Jeffery, "The Pact of the First Settlers at Cyrene," *Historia* 10 (1961): 139–47.
7. Anton Powell and Stephen Hodkinson's *The Shadow of Sparta* (London: Routledge, 1994) and Carol G. Thomas's *Paths from Ancient Greece* (Leiden, Netherlands: E.J. Brill, 1988) contain a wide range of essays on the so-called Spartan question.
8. This proposal was made explicitly at a 1984 RAND Conference (see Gregory G. Hildebrandt, ed., "Rand Conference on Models of the Soviet Economy, October 11–12, 1984" [Santa Monica, Calif.: RAND Corporation, 1985], 139–41).
9. See Alvin H. Bernstein, "Soviet Defense Spending: The Spartan Analogy," RAND Note N-2817-NA, prepared for the director of net assessment, Office of the Secretary of Defense (Santa Monica, Calif.: RAND Corporation, 1989).
10. Ibid., 12.
11. Ibid., v.
12. Ibid., 5.
13. Michael Whitby, "Two Shadows: Images of Spartans and Helots," in *The Shadow of Sparta,* ed. Anton Powell and Stephen Hodkinson (London: Routledge, 1994), 89.
14. William Greider, *Fortress America* (New York: Public Affairs Press, 1998), 10.
15. O. K. Werckmeister, *Citadel Culture* (Chicago: University of Chicago Press, 1991), 8.
16. Ibid., 5.
17. According to Aftergood, "approximately 15% of the Defense Department budget (about $35 billion per year) for weapons acquisition is classified. The way this money is spent is secret not only from the public, but even from the overwhelming majority of members of Congress" (Steven Aftergood, statement, *Government Secrecy After the Cold War,* 102d Cong., 2d sess., House Hearing, Committee on Government Operations, 18 March 1992 [Washington, D.C.: GPO, 1992], 179). As Tim Weiner notes, this translates into approximately $10 million in unaccountable Pentagon spending per day. Programs funded through the black budget are so-called special access programs, and knowledge of their existence is restricted to those with the highest levels of classification clearance (see Tim Weiner, *Blank Check: The Pentagon's Black Budget* [New York: Warner Books,

1990]). "Secrecy," according to Gregory Foster, a professor at the Industrial College of the Armed Forces, "is the most lasting, visible and destructive feature of the Cold War ethos. . . . Obsessive secrecy has had the unintended effects of disguising government abuse, obscuring accountability, and engendering public distrust, fear, alienation and apathy" (Gregory D. Foster, *In Search of a Post-Cold War Security Structure* [Washington, D.C.: National Defense University Institute for National Strategic Studies, 1994], 23).

18. This program was designed to outfit U.S. ICBMs with capability to maneuver around enemy ABM interceptors.
19. Robert M. Bowman, telephone interview with the author, 21 July 1999. As Bowman explains further, the "Precision Guided Re-Entry Vehicle Program was canceled by Congress, because they said 'We don't need a Precision Guided Re-Entry Vehicle with zero-zero accuracy; its only use is for first-strike, and that's not our national policy.' So the program was canceled. The next day, I became program manager of MARV (the Maneuvering Re-Entry Vehicle program). The mission of that program was to maneuver out of the way of an ABM system, and then maneuver back toward its target. It was exactly the same thing; it had a different rationale, but it was the same program."
20. Bowman, telephone interview with the author. "When Reagan was elected," Bowman explains, "the people around him recognized that the [black program] was a wonderful way to keep things out of the scrutiny of Congress, and to create weapons programs that Congress and the American people wouldn't support, but they can do it anyway in the black program, without financial accountability."
21. For details of the legislation, see Paul B. Stares, *Space and National Security* (Washington, D.C.: Brookings, 1987), 42–52.
22. On 4 September 1985 Representative Brown raised the possibility of impeachment hearings on the floor of the House: "These articles would recite the failure of the President to well and faithfully execute the law, as required by his oath of office" (George Brown, statement, ASAT Testing Certification, *Congressional Record,* 4 September 1985, H1610). Looking back on the episode, Bowman concludes that Reagan's decision to go ahead with the 1985 ASAT test was "clearly an impeachable offense" (Bowman, telephone interview with the author).
23. William Burr, Thomas S. Blanton, and Stephen I. Schwartz, "The Costs and Consequences of Nuclear Secrecy," in *Atomic Audit: The Costs and Consequences of U.S. Nuclear Weapons Since 1940,* ed. Stephen I. Schwartz (Washington, D.C.: Brookings Institute Press, 1998), 480. The Timberwind program used technology derived from a secret nuclear-powered aircraft program that ran from 1946–1961. Initially, Timberwind was cosponsored by the DOE and NASA (see John E. Pike, Bruce G. Blair, and Stephen I. Schwartz, "Defending Against the Bomb," in *Atomic Audit: The Costs and Consequences of U.S. Nuclear Weapons Since 1940,* ed. Stephen I. Schwartz [Washington, D.C.:" Brookings Institute Press, 1998], 292). For further discussion of Timberwind's historical origins, see Nancy B. Kingsbury, statement, *Development of Nuclear Thermal Propulsion Technology for Use in Space,* 102d Cong., 2d sess., House Hearings, Committee on Science, Space, and Technology, 1 October 1991 (Washington, D.C.: GPO, 1991), 4–11.

24. Burr, Blanton, and Schwartz, "Costs and Consequences," 480.
25. As Truman explains, "[m]ost of the development of the N-rocket project to date was carried out under the Timberwind program and was done so in secrecy and with complete classification. Much of the funding came from the Pentagon's black budget removed from public eyes" (Preston J. Truman, letter to Rep. Howard Wolpe, printed in House Hearings, *The Development of Nuclear Thermal Propulsion Technology for Use in Space,* 102d Cong., 2d sess. [Washington D.C.: GPO, 1 October 1992], 333–36).
26. Truman, letter to Howard Wolpe, 334.
27. A report of the House Appropriations Committee made the following dramatic findings: "Regarding OSD acquisition officials, in addition to the example involving LOSAT cited above, the Committee is little short of amazed when it comes to their actions on the Medium Altitude Air Defense (MEADS) program. This program was specifically terminated in the conference report accompanying the fiscal year 1999 Defense Appropriations Act. Internal DOD financial management documents issued this spring noted this action and correctly stated that: 'This item has been denied by the Congress and is *not subject to reprogramming*' [emphasis added]. Nonetheless, the Committee has since learned that officials in the OSD acquisition structure as well as in the Ballistic Missile Defense Organization, an OSD acquisition organization, directed the use of over $2 million of funds specifically provided for another program to continue MEADS-related activities, and actually announced the winner of the MEADS contract competition. All for a program explicitly terminated in the fiscal year 1999 appropriations process" (U.S. Congress, *Department of Defense Appropriations Bill, Fiscal Year 2000,* 109th Cong., 2d sess., House Report 106–244, Appropriations Committee, 22 July 1999 [Washington, D.C.: GPO, 1999], emphasis added). Members of the committee concluded with an additional warning by noting that this type of deliberate circumvention of congressional authority is "occurring with increasing frequency in both the budget requests submitted by the Department of Defense as well as in the execution of program funding once appropriations have been provided by the Congress" (Ibid.). For further commentary, including discussion of Defense Secretary Cohen's excuse that "about 99.9 percent of the time, we seem to be doing things right," see John Martin and David Ruppe, "Finding a Way to Pay: House Report Says Pentagon Ignored Spending Rules," Internet, ABC News Website, 23 July 1999, online at htttp://abcnews.go.com:80/sections/us/DailyNews/pentagon990723.html.
28. Andrewes, "Government of Classical Sparta," 15.
29. Gingrich made the following statement during his announcement of the daily agenda on the floor of the House of Representatives on 17 March 1997: "Just as Britain had to have the foresight to build radar in the 1930s to survive the Battle of Britain, the time to prepare to defend ourselves is not when the crisis occurs, not when we are blackmailed, but now. And every evidence, I think, of every independent observer is that the threat is real, it is already here and that we should be building today a national missile defense system capable of protecting the United States, capable of protecting Europe and Israel, and capable of protecting our allies in the Far East, if necessary, so that no

one who has a missile can think that with impunity they can blackmail the free countries of the world" (Newt Gingrich, statement, *Congressional Record,* 17 March 1997, H1029). Gingrich's appeal glossed over the historical fact that defending against enemy aircraft and enemy ballistic missiles are entirely different technical feats. Most planes can be tracked by radar, and their relatively large surface areas make them vulnerable to surface-to-air (SAM) missiles. Ballistic missiles, on the other hand, travel at supersonic speeds that make radar detection difficult and effective interception ("hitting a bullet with a bullet") even more challenging.

30. One group of science studies scholars attempts to stake out a middle ground between staunch empiricist and radical social constructivist approaches. This effort has in large part been motivated by the twin weaknesses of empiricism and social constructivism: an intolerance to ambiguity and indeterminacy in the case of the former, and an inability to explain scientific progress, in the case of the latter (see, e.g., Helen Longino, *Science as Social Knowledge: Values and Objectivity in Scientific Inquiry* [Princeton, N.J.: Princeton University Press, 1990], 81). Drawing openly but gingerly from empiricist philosophers such as Willard Quine and Karl Popper, "post-individualist" empiricists such as Helen Longino, Sandra Harding, Lynn Hankinson Nelson, Theodore Porter, and Marcello Pera have sought to retain but revise the notion of objectivity, refashioning the concept as a collective property of discursive interchange. As Harding explains, "[t]he notion of objectivity—like such ideas as science, rationality, democracy, and feminism—contains progressive as well as regressive tendencies. In each case, it is important to develop the progressive and block the regressive ones" (Sandra Harding, *Whose Science? Whose Knowledge?* [Ithaca, N.Y.: Cornell University Press, 1991], 161). Longino's formulation carries with it echoes of Popper's falsificationist epistemology, that is, the idea that "objectivity is based, in brief, upon mutual rational criticism, upon the critical approach, the critical tradition" (Karl Popper, *The Myth of Framework* [London: Routledge, 1994], 70). Nelson's insights receive substantial support from Quine's coherence theory of truth, the idea that the validity of theories is based not on their degree of correspondence to nature, but rather the coherence they exhibit when juxtaposed with other theories, data, and background, "auxiliary" hypotheses featured within the discursive community (Lynn Hankinson Nelson, *Who Knows: From Quine to a Feminist Empiricism* [Philadelphia: Temple University Press, 1990]; see also Willard Quine, *Word and Object* [Cambridge: MIT Press, 1960]; Lorraine Code, *rhetorical spaces* [New York: Routledge, 1995]; and Marcello Pera, *The Discourses of Science,* trans. Clarissa Botsford [Chicago: University of Chicago Press, 1994]).
31. See Garwin, FILE Memorandum, 10 October 1995, copy on file with the author.
32. John Tirman, ed., *The Militarization of High Technology* (Cambridge, Mass.: Ballinger, 1984), 159.
33. Arvin S. Quist, *Security Classification of Information. Vol. 1, Introduction, History, and Adverse Impacts* (Oak Ridge Gaseous Diffusion Plant, Tenn.: U.S. Department of Energy, 1989), 92.
34. Burr, Blanton and Schwartz, "Costs and Consequences," 478.

35. See Sissella Bok, *Secrets* (New York: Pantheon Press, 1982); William D. Carey, "The Secrecy Syndrome," *Bulletin of the Atomic Scientists* 38 (1982): 9–10; Rosemary Chalk, "Overview: AAAS Project on Openness and Secrecy in Science and Technology," *Science, Technology and Human Values* 10 (1985): 28–35; idem, "Security and Scientific Communication," *Bulletin of the Atomic Scientists* 39 (1983): 19–23; Daryl E. Chubin, "Open and Closed Science: Tradeoffs in a Democracy," *Science, Technology and Human Values* 10 (1985): 73–81; David Faure and Matthew McKinnon, "The New Prometheus: Will Scientific Inquiry Be Bound by the Chains of Governmental Regulation?" *Duquesne Law Review* 19 (1981): 651–730; Dorothy Nelkin, *Science as Intellectual Property: Who Controls Scientific Research?* (New York: Macmillan, 1984); Harold C. Relyea, "Shrouding the Endless Frontier—Scientific Communication and National Security; Considerations for a Policy Balance Sheet," *Government Information Quarterly* 1 (1984): 1–14; Edward Shils, *The Torment of Secrecy* (Glencoe, Ill.: Free Press, 1976); and Edward Teller, "The Feasibility of Arms Control" *Daedalus* 89 (1960): 749–99.
36. This formulation jibes with Longino's analysis that the degree of objectivity achieved in science is correlated linearly with the degree to which intersubjective scientific criticism entertains competing perspectives embedded in background beliefs. "The greater the number of different points of view included in a given community," she notes, "the more likely that its scientific practice will be objective, that is, that it will result in descriptions and explanations of natural processes that are more reliable in the sense of less characterized by idiosyncratic subjective preferences of community members than would otherwise be the case" (Longino, *Science as Social Knowledge,* 80).
37. See William Broad, *Star Warriors* (New York: Simon and Schuster, 1985), 206–20.
38. For example, a command decision to not equip Patriot missile batteries with optical recording equipment compromised severely the pool of hard data used to evaluate Patriot performance during the war. Secret postwar studies funded to ascertain the precise level of Patriot effectiveness were spotty in coverage and generally poor in technical quality. "On close examination, many data sources consist of only bits and pieces of information," Congressional Research Service analyst Steven Hildreth explained in testimony to Congress (Steven A. Hildreth, "Evaluation of U.S. Army Assessment of Patriot Antitactical Missile Effectiveness in the War against Iraq," report, *Performance of the Patriot Missile in the Gulf War,* 102d Cong., 2d sess., House Hearing, Committee on Government Operations, 7 April 1992 [Washington, D.C.: GPO, 1992], 28).
39. Quoted in James Glanz, "Patriots Missed, but Critics Hit Home," *Science* 284 (16 April 1999): 413–16.
40. See Gerald Pawle, *The Secret War: 1939–1945* (New York: William Sloane, 1957). For discussion of other successful weapons systems generated from Britain's secret military research program during 1939–45, see R. V. Jones, *The Wizard War* (New York: Coward, McCann and Geogheghan, 1978).
41. See Jones, *Wizard War.*

42. See Ewen Montagu, *Beyond Top Secret ULTRA* (New York: Coward, McCann and Geoghegan, 1978); Thomas Parrish, *The ULTRA Americans: The U.S. Role in Breaking the Nazi Codes* (New York: Stein and Day, 1986).
43. James Phinney Baxter, *Scientists Against Time* (Boston: Little, Brown and Company, 1946), 449.
44. The frank acknowledgment and critical testing of such value-laden background assumptions can permit science to progress without being compromised by a reign of idiosyncratic, subjective belief over collective, deliberative judgment. According to Pera, "[w]e shouldn't attempt to eliminate subjective wishes and social conventions from science; rather we should try and incorporate them into science without sacrificing its undeniable nature of rigorous and objective knowledge. My claim is that this is possible provided we transfer science *from the kingdom of demonstration to the domain of argumentation,* and conceive its constraints not as universal methodological rules but as historical Dialectical factors on which concrete interlocutors in concrete discussions rely (Pera, *Discourses of Science,* 47, emphasis in original). "Dynamic objectivity," Keller states in a similar light, "is thus a pursuit of knowledge that makes use of subjective experience (Piaget calls it consciousness of self) in the interests of a more effective objectivity" (Evelyn Fox Keller, *Secrets of Life: Essays on Language, Gender, and Science* [New York: Routledge, 1992], 117).
45. Longino, *Science as Social Knowledge,* 71.
46. Ibid., 74.
47. Gordon R. Mitchell, "Another Strategic Deception Initiative," *Bulletin of the Atomic Scientists* 53 (1997): 23.
48. Sanford Gottlieb, *Defense Addiction: Can America Kick the Habit?* (Boulder, Colo.: Westview Press, 1997), 9.
49. Sissela Bok, *A Strategy for Peace* (New York: Vintage Books. 1989), 41–42.
50. See Tim Weiner, "General Details Altered 'Star Wars' Test," *New York Times,* 27 August 1993, A19.
51. Burr, Blanton, and Schwartz, "Costs and Consequences," 457.
52. See Hugh DeWitt, "The Selling of a Wonder Weapon," *Stanford* (March 1991): 28–33.
53. See Burr, Blanton, and Schwartz, "Costs and Consequences," 467.
54. Ibid.
55. See Hugh E. DeWitt, letter to Secretary Hazel O'Leary, 4 April 1994, quoted in *Atomic Audit: The Costs and Consequences of U.S. Nuclear Weapons Since 1940,* ed. Stephen I. Schwartz (Washington, D.C.: Brookings Institute Press, 1998), 467; Kay Davidson, "Nuke Lab Scientist Regains Security Status," *San Francisco Examiner,* 6 May 1994, A2.
56. "Pentagon Homes in on Patriot Critic," *New Scientist* 133 (28 March 1992): 13.
57. See Nina J. Stewart, letter to Hon. John Conyers, reprinted in *Government Secrecy after the Cold War,* 102d Cong., 2d sess., House Hearing, Committee on Government Operations, 18 March 1992 (Washington, D.C.: GPO. 1992), 72–73.
58. Michael I. Handel, "A 'Runaway Deception': Soviet Disinformation and the Six-Day War, 1967," in *Deception Operations,* ed. David A. Charters and Maurice A. J. Tugwell (London: Brassey's, 1990), 167.

59. Ibid.
60. See Dan Morain, "Energy Secretary Warns Scientists Not to Disagree in Public," *Los Angeles Times,* 23 July 1988, 1.
61. Michael D'Antonio, *Atomic Harvest* (New York: Crown Press, 1993), 237.
62. As former attorney general Ramsey Clark observes, "[t]he Pentagon has hundreds of hours of such film including contemporaneous pictures of the aerial bombardment of civilians. This film is direct evidence of criminal actions, and the media and Congress know it but do nothing" (Ramsey Clark, *The Fire This Time: U.S. War Crimes in the Gulf* [New York: Thunder's Mouth Press, 1992], 140). See chapter three for further discussion.
63. Theodore A. Postol, interview with the author, Cambridge, Mass., 10 May 1995, full transcript on file with the author.
64. David A. Charters and Maurice A. J. Tugwell, eds., *Deception Operations* (London: Brassey's, 1990), 321.
65. Quoted in "Conferences on Therapy: The Use of Placebos in Therapy," *New York State Journal of Medicine* 46 (1976): 1726. For further discussion relating this phenomenon to the distinction between "pure" and "impure" placebos, see Howard Brody, *Placebos and the Philosophy of Medicine: Clinical, Conceptual, and Ethical Issues* (Chicago: University of Chicago Press, 1980), 102–3.
66. See Michael I. Handel, *War, Strategy, and Intelligence* (London: Frank Cass, 1989), 442.
67. Ibid., 403.
68. Tim Weiner, interview with Noah Adams, *All Things Considered,* National Public Radio, 18 August 1993, Internet, Online, Lexis/Nexis (an electronic information database with over 28,000 full-text sources online at http://www.lexis-nexis.com/lncc).
69. Steven A. Hildreth, "Evaluation of U.S. Army Assessment of Patriot Antitactical Missile Effectiveness in the War against Iraq," report, *Performance of the Patriot Missile in the Gulf War,* 102d Cong., 2d sess., House Hearing, Committee on Government Operations, 7 April 1992 (Washington, D.C.: GPO, 1992), 24.
70. Theodore A. Postol, "Reply to letter of Robert Stein," *International Security* 17 (1992): 229.
71. See Gordon R. Mitchell, "Placebo Defense: The Rhetoric of Patriot Missile Accuracy in the 1991 Persian Gulf War," *Quarterly Journal of Speech* 86 (2000): 137.
72. John Conyers, "The Patriot Myth: Caveat Emptor," *Arms Control Today* 22 (1992): 10.
73. According to Anthony Cordesman, a Mideast expert with the Center for Strategic and International Studies, "A lot of the value of [NMD] is deterrent. It keeps the United States from being held hostage" (quoted in John Diamond, "'Rogue States' Want Weapons for Leverage," *Pittsburgh Post-Gazette,* 21 February 2000, A1); see also Jeffrey Gedmin and Jon Kyl, "The Case for Missile Defense," *Washington Post,* 31 December 1999, A31.
74. According to Michael T. Klare, "The rogue-state concept is a product of a determined Pentagon effort to create a new foreign threat to justify military spending in the wake of the cold war. . . . The danger arising from this impulse to brand certain states as rogues and outlaws, and to threaten them with severe military punishment, is that the

policy can take on a life of its own (spurring countermoves and counterthreats until the White House is forced to back up its words with a show of force" (Michael T. Klare, "The New 'Rogue State' Doctrine," 260 [8 May 1995] *The Nation*, 625-28).

75. Lt. Gen. Robert Kadish, statement, *Military Procurement (Missile Defense Programs,* 106th Cong., 2d Sess., Armed Services Committee, House Hearings, 16 February 2000 (Washington, D.C.: GPO, 2000), Internet, Online, Lexis/Nexis.
76. Quoted in Diamond, "'Rogue States,'" A1.
77. Stephen I. Schwartz, telephone conversation with the author, 3 May 2000.
78. Quoted in W. B. Blanton, *Medicine in Virginia in the Eighteenth Century* (Richmond: Garrett and Massie, 1931), 199.
79. "[I]t is convenient to designate as 'placebo' effects all those psychological and psychophysiological benefits or detriments which quite directly involve the patient's expectations and depend directly upon the diminution or augmentation of the patient's apprehension by the symbolism of medication or the symbolic implications of the physician's behavior and attitudes" (quoted in R. C. Cabot, *Social Service and the Art of Healing* [New York: Moffat, Yard and Co., 1909]). Consistent with this hypothesis, some ethical theorists have posited that full public appreciation of the widespread use of deceptive placebos in medical treatment would result in a diminution of the effectiveness of all medical treatment, as public confidence in the medical establishment would decline (See, e.g., Sissela Bok, "The Ethics of Giving Placebos," *Scientific American* 231 [1974]: 17–23).
80. See Thomas B. Farrell, "Knowledge, Consensus, and Rhetorical Theory," *Quarterly Journal of Speech* 62 (1976): 1–14.
81. Handel, *War, Strategy, and Intelligence,* 394.
82. For a text of the pact, see Israel Government Press Office, *Text of Joint Peres-Clinton Settlement* 30 April, 1 May 1996, trans. Foreign Broadcast Information Service (NES-96–086).
83. Avino'am Bar-Yosef, "U.S. 'Not Crazy About' Peres Despite 'Show' During Visit," *Ma'ariv,* 2 May 1996, 3, trans. Foreign Broadcast Information Service (NES-96–087).
84. Ibid.
85. Quoted in Israel Government Press Office, *Text of Joint Peres-Clinton Settlement.*
86. Quoted in "Wonder Weapon," *Progressive* 58 (January 1994): 10–14.
87. Quoted in Scott Peterson, "High Hopes for Foolproof Missile Defense Fizzle," *Christian Science Monitor* (online edition), 30 July 1997, Internet, Christian Science Monitor Website, online at http://www.csm.org.
88. See James Glanz, "Israeli-American Laser Passes a Missile Defense Test, U.S. Says," *New York Times,* 4 May 2000, 11.
89. Quoted in Glanz, 11.
90. Quoted in Union of Concerned Scientists, *Empty Promise: The Growing Case Against Star Wars,* ed. John Tirman (Boston: Beacon Press, 1986), 18.
91. Quoted in Hunter Keeter, "Laser Master Plan Okays HEL Systems for Offensive and Defensive Applications," *Defense Daily,* 3 April 2000, Internet, online, Lexis/Nexis.

92. See Keeter, "Laser Master Plan;" Frank Wolfe, "Army Running Short of THEL Funds," *Defense Daily*, 3 April 2000, Internet, online, Lexis/Nexis.
93. Quoted in David E. Sanger, "North Korea Warns U.S. on Patriot Missiles," *New York Times*, 30 January 1994, 4, emphasis added.
94. Chi Man-won, "Missile Sovereignty," *Choson Ilbo*, 8 February 1996, 5, trans. Foreign Broadcast Information Service (EAS-96–028).
95. "Do We Have to Buy Patriots?" *Tong-a Ilbo*, 5 April 1997, 3, trans. Foreign Broadcast Information Service (TAC-97–095).
96. Quoted in "ROK Reaffirms 'No Plans' to Join U.S.-led TMD Plan," *Korea Herald*, 4 May 1999, trans. Foreign Broadcast Information Service (EAS-1999–0503).
97. See Paul Richter, "Defense Secretary Warns Seoul Not to Buy Russian Arms," *Los Angeles Times*, 6 April 1997, A1; O Yong-hwan and Choe Sang-yon, "Officials' 'Displeasure' with U.S. Pressure on Weapons Deal," *Chungang Ilbo*, 13 June 1997, 3, trans. Foreign Broadcast Information Service (TAC-97–164).
98. Kang Hyo-son and Pak Chong-hun, "Kantor Asks ROK to Purchase Theater Missile Defense System," *Choson Ilbo*, 2 July 1996, 2, trans. Foreign Broadcast Information Service (EAS-96–128).
99. Reporting on Cohen's 8 April 1997 visit to South Korea, Mary Jordan, of the *Washington Post*, wrote that "many South Koreans say he has come as a glorified missile salesman" (Mary Jordan, "Cohen Urges Seoul to Buy U.S. Missiles," *Washington Post*, 8 April 1997, A1; see also Pak Won-ung, "U.S. Applies 'Virtual' Pressure on ROK in Selling Weapons," Seoul KBS-1 Radio Network, 12 June 1997 broadcast, trans. Foreign Broadcast Information Service [TAC-97–163]).
100. "U.S. Pressure on Missile Purchase," *Chungang Ilbo*, 6 April 1997, 6, trans. Foreign Broadcast Information Service (TAC-97–096).
101. Kim Song-ho, "The Arrogant U.S. Missile Sales," *Hangyore 21* (24 April 1997): 42–44, trans. Foreign Broadcast Information Service (EAS-97–083).
102. Jay Katz, *The Silent World of Doctor and Patient* (New York: Free Press, 1984), 195.
103. "Do We Have to Buy," 3.
104. "U.S. Pressure."
105. In Engelhart and Caplan's seminal piece, "Patterns of Controversy and Closure," they outline five different mechanisms that are used to close controversies: "1) Closure through loss of interest; . . . 2) Closure through force; . . . 3) Closure through consensus; . . . 4) Closure through sound argument; . . . 5) Closure through negotiation" (Tristram H. Engelhardt and Arthur L. Caplan, *Scientific Controversies: Case Studies in the Resolution and Closure of Disputes in Science and Technology* [Cambridge: Cambridge University Press, 1987], 13–15).
106. David Helvarg, "The Greenhouse Spin," *Nation* 263 (16 December 1996): 21.
107. David Dickson, *The New Politics of Science* (Chicago: University of Chicago Press, 1988).
108. See Dorothy Nelkin, "The Political Impact of Technical Expertise," *Social Studies of Science* 5 (1975): 53–54; Simon Shackley and Brian Wynne, "Representing Uncertainty

in Global Climate Change Science and Policy: Boundary-ordering Devices and Authority," *Science, Technology, and Human Values* 21 (1996): 276.

109. Steven Epstein, *Impure Science* (Berkeley: University of California Press, 1996), 29.
110. IPCC 95, *Climate Change 1995: The Science of Climate Change* (Cambridge: Cambridge University Press, 1995).
111. Helvarg, "Greenhouse Spin," 21.
112. Gale E. Christianson, "Naysayers, Thriving in the Heat," *New York Times*, 8 July 1999, A25; see also Shackley and Wynne, "Representing Uncertainty."
113. Erik Doxtader, "Learning Public Deliberation through the Critique of Institutional Argument," *Argumentation and Advocacy* 31 (1995): 186. For a similar commentary recommending "counteranalysis" of the "production and use of scientific and technological knowledge" in the context of SDI, see Frans Birrer, "Counteranalysis: Toward Social and Normative Constraints on the Production and Use of Scientific and Technological Knowledge," in *Controversial Science: From Content to Contention*, ed. Thomas Brante, Steve Fuller, and William Lynch (New York: SUNY Press, 1993), 43, 47.
114. Milo Geyelin, "'Class' Trial Finds Tobacco Firms Liable," *Wall Street Journal*, 8 July 1999, A3.
115. Quoted in Barry Meier, "Tobacco Industry Loses First Phase of Broad Lawsuit," *New York Times*, 8 July 1999, A18.
116. See David Hoffman, "Moscow Proposes Extensive Arms Cuts; U.S., Russia Confer over Stalled Pacts," *Washington Post*, 20 August 1999, A29; Adam Tanner, "General Raps U.S. Anti-missile Effort; Says Changes Sought to ABM Pact Threaten Arms Talks," *Washington Times*, 21 August 1999, A5.
117. Alexei Arbatov, "Deep Cuts and De-alerting: A Russian Perspective," in *The Nuclear Turning Point: A Blueprint for Deep Cuts and De-alerting of Nuclear Weapons*, ed. Harold A. Feiveson (Washington, D.C.: Brookings Institution, 1999), 316.
118. Quoted in Barbara Crossette, "Russians Heat Up Dispute on ABMs," *New York Times*, 26 April 2000, 1.
119. Lloyd Doggett, statement, *Congressional Record*, 20 May 1999, H3428.
120. Bill Sweetman and Nick Cook, "Getting to Grips with Missile Defense," *Interavia* (Geneva) 54 (1999): 34.
121. Quoted in Mark Thompson, "Star Wars: The Sequel," *CNN Interactive*, Internet, CNN Website, 15 February 1999, online at http:www.cnn.com/ALLPOLITICS/time/1999/02/15/star.wars.html.
122. See Sweetman and Cook, "Getting to Grips," 32.
123. See Steven A. Hildreth, "National Missile Defense: The Current Debate," CRS Issue Brief #96–441 F, 7 June 1996 (Washington, D.C.: GPO, 1996).
124. See John Isaacs, "Missile Defense: It's Back," *Bulletin of the Atomic Scientists* 55 (1999): 23.
125. For commentary on the Rumsfeld Commission report, see Center for Strategic and Budgetary Assessments, "The Rumsfeld Commission Report: Where Do We Go From Here?" Center for Strategic and Budgetary Assessments Report, Internet, Center for

Strategic and Budgetary Assessments Website, 4 August 1998, online at http://www.csbahome.com/Publications/Rumsfeld.html; Center for Security Policy, "Critical Mass #2: Senator Lott, Rumsfeld Commission Add Fresh Impetus to Case for Beginning Deployment of Missile Defenses," Center for Security Policy Decision Brief no. 98-D 133, 15 July 1998, Internet, Center for Security Policy Website, online at http://www.security-policy.org/papers/1998/98-D133.html.

126. Jacob Heilbrunn, "Playing Defense," *New Republic* 219 (18 August 1998): 16–18.
127. Ibid., 16.
128. See Isaacs, "Missile Defense," 23.
129. Quoted in "The Return of Reagan," *Progressive* 63 (March 1999): 8.
130. Isaacs," Missile Defense," 23.
131. Thompson, "Star Wars."
132. William Neikirk, "Senate Overwhelmingly Ok's Bill for Anti-Missile Defense," *Chicago Tribune* (internet edition), Internet, Chicago Tribune Website, 18 March 1999, online at http://chicagotribune.com/news/nationworld/article/0,1051,SAV-9903180135,00.html.
133. Joseph Cirincione, "The Persistence of the Missile Defense Illusion," paper presented at the Conference on Nuclear Disarmament, Safe Disposal of Nuclear Materials or New Weapons Development? Como, Italy, 2–4 July 1998, Internet, Carnegie Endowment for International Peace Website, online at http://www.ceip.org/people/cirincio.htm.
134. Richard Durbin, quoted in Neikirk, "Senate Overwhelmingly Ok's Bill for Anti-Missile Defense," *Chicago Tribune* (internet edition), Internet, Chicago Tribune Website, 18 March 1999, online at http://chicaotribune.com/news/nationworld/article/0,1051,SAV-9903180135,00.html; see also "Senate Backs Missile Defense System," CNN Interactive, Internet, CNN Website, 17 March 1999, online at http://www.cnn.com/ALLPOLITICS/stories/1999/03/17/missile.defense/; "House OK's Missile Defense Plan," CNN Interactive, Internet, CNN Website, online at http://www.cnn.com/ALLPOLITICS/stories/1999/03/18/missile.defense/.
135. See William J. Broad, "After Many Misses, Pentagon Still Pursues Missile Defense," *New York Times*, 24 May 1999, A1; United States Congressional Budget Office, *Budgetary and Technical Implications of the Administration's Plan for National Missile Defense*, Congressional Budget Office paper (Washington, D.C.: GPO, 2000).
136. Phyllis Schlafly, "ABM Should Be Republicans' Unifying Issue," Eagle Forum Column, November 1998, Internet, Eagle Forum Website, online at http://www.eagleforum.org/column/1998/nov98/98–11–11.html.
137. Quoted in Tom Raum, "GOP, Democrats Agree on Defense," Associated Press news feed, 29 June 1999, Internet, posted at Council for a Livable World Website, online at http://www.clw.org/ef/bmdnews/
138. John Isaacs, "Rumbles from Rumsfeld," *Bulletin of the Atomic Scientists* 54 (1998): 17.
139. Richard L. Garwin, letter to Senator Carl Levin, 2 September 1998, copy on file with the author.
140. Richard L. Garwin, "The Rumsfeld Report—What We Did," *Bulletin of the Atomic Scientists* 54 (1998): 40.

141. Richard L. Garwin, "Keeping Enemy Missiles at Bay," *New York Times*, 28 July 1998, A15.
142. Garwin, quoted in "The Return of Reagan," 9; see also Rep. Peter DeFazio, statement, *Ballistic Missile Defense: The Emperor's Newest Clothes, American Defense Monitor*, Internet, Center for Defense Information Website, online at http://www.cdi.org/adm/939/index.html.
143. Richard L. Garwin, "Effectiveness of Proposed National Missile Defense Against ICBMs from North Korea" letter prepared for distribution to Congress and the Press, 17 March 1999, Garwin Archive, Internet, Federation of American Scientists Website, online at http://www.fas.org/rlg/930504-imsa.htm.
144. Specifically, Garwin explained that defeat of a "hit-to-kill" NMD system "is easily done by the use of an enclosing balloon made of aluminum-foil coated mylar that can be put together by anyone who buys this article of commerce and spends $20 on a hand-held tool for heat sealing the plastic to make a large balloon. Even a balloon ten meters (33 ft.) in diameter, inflated after the RV separates from the missile, would render it unlikely that an interceptor would actually strike the warhead rather than plunging harmlessly through the balloon" ("Effectiveness of Proposed"). Another countermeasure cited by Garwin was use of individually dispersing biological weapons "bomblets": "It is far more effective militarily for an ICBM payload of biological warfare agents to be arriving in the form of individual reentry vehicles (bomblets) spread over an area 10 or 20 kilometers in extent. . . . Given this undisputed increase in military effectiveness, any nation with the capability to make an ICBM and reentry vehicles would almost surely arrange to package the BW in the form of bomblets, released just as soon as the ICBM reached its final positions in a rack within a spinning final stage, the release of the bomblets would then allow them to spread during their 20-minute or more flight to reentry, with the initial rotation rate determining precisely the spread, and the pattern being that in which the bomblets were stored in the missile. This threat of BW bomblets released on ascent is to be expected whether or not a defense is deployed, but the proposed NMD would have *strictly zero capability* against these bomblets" (Garwin, "Effectiveness of Proposed," emphasis added). See also Garwin, "Keeping Enemy Missiles at Bay"; George N. Lewis, Theodore A. Postol, and John Pike, "Why National Missile Defense Won't Work," *Scientific American* 281 (1999): 36–41.
145. William Broad, *Teller's War* (New York: Simon and Schuster, 1992), 148–49.
146. Edward Teller, letter to George Keyworth, 22 December 1983, copy on file with the author.
147. William F. Buckley, "Intercepting the Truth about SDI Testing," *Buffalo News*, 14 October 1993, 3.
148. Patrick Buchanan, "Evidence of SDI's Potential," *Washington Times*, 30 January 1991, G2.
149. Quoted in John J. Miller, "The Rocket Boys: Missile Defense on its Way," *National Review* 51 (25 October 1999), 52.
150. See Pierre Duhem, *The Aim and Structure of Physical Theory* (Princeton, N.J.: Princeton University Press, 1954).

151. Allan Franklin, *Can That Be Right? Essays on Experiment, Evidence, and Science* (Dordrecht, Netherlands: Kluwer 1999), 168.
152. Donald MacKenzie, "From Kwajalein to Armageddon? Testing and the Social Construction of Missile Accuracy," in *The Uses of Experiment,* ed. David Gooding, Trevor Pinch, and Simon Shaffer (Cambridge: Cambridge University Press, 1989), 412.
153. See Donald MacKenzie, *Inventing Accuracy: A Historical Sociology of Nuclear Missile Guidance* (Cambridge: MIT Press, 1990).
154. Ibid., 371–72.
155. Ibid., 353.
156. Gen. Harry H. Shelton, interview with *Sea Power Magazine,* quoted in *Briefing Book on Ballistic Missile Defense,* Internet, Council for a Livable World Website, online at htttp://www.clw.org/ef/bmdbook/legis.html.
157. Richard L. Garwin, statement, *Ballistic Missile Defense,* 108th Cong., 2d sess., Senate Hearing, Foreign Relations Committee, 4 April 1999 (Washington, D.C.: GPO, 1990).
158. See Garwin, statement, *Ballistic Missile Defense.* As Sweetman and Cook explain, "the [NMD] tests planned for this year [1999] and next [2000] will use prototype EKVs and a surrogate booster (a modified Minuteman), and the Vandenberg-Kwajalein range cannot fully demonstrate realistic engagement geometries and closing velocities. The tests do not include decoys or multiple targets" (Sweetman and Cook, "Getting to Grips," 34).
159. Quoted in John Donnelly, "THAAD Intercepts Were Unrealistic, Top Tester Says," *Defense Week,* 23 August 1999, Internet, online, Lexis/Nexis. According to Donnelly, Coyle said there were three aspects of the THAAD tests that made them unrealistic simulations of actual combat situations: "The tests used a THAAD missile other than the one that would be bought; the targets were shorter range than the system might really face; and the test conditions were contrived" (Donnelly, "THAAD Intercepts"). Furthermore, the Army Space and Missile Defense Command (SMDC) raised a number of "serious issues" about Lockheed Martin's recently drafted proposal for moving THAAD into the engineering phase of development. Echoing Coyle's concerns about previous tests, SMDC questioned the "reliability of planned engineering tests" proposed by the manufacturer (Kerry Gildea, "SMDC Rejects Lockheed Martin Proposal for THAAD EMD Phase," *Defense Daily,* 21 April 2000, Internet, online, Lexis/Nexis).
160. Michael C. Sirak, "House Waives Initial Testing Requirements on NMD Production," *Inside Missile Defense,* 16 June 1999, 1, emphasis added.
161. Quoted in Sirak, "House Waives," 1.
162. Garwin, statement, *Ballistic Missile Defense.*
163. Quoted in James Glanz, "Military Experts Debate the Success of Warhead Missile Interceptor Test," *New Orleans Times-Picayune,* 14 January 2000, 3A; see also "Pentagon Admits Problems with Missile Defense Test," *St. Louis Post-Dispatch,* 15 January 2000, 21.
164. See "How the Pentagon Fixed the Star Wars Test," *CounterPunch,* 9 January 2000, Internet, CounterPunch Website, online at http://www.counterpunch.org/starwars.html.

165. This phrase was used in a "stinging 40-page report" delivered to Congress by the so-called Welch Commission in November 1999 (see Bradley Graham, "Panel Faults Antimissile Program on Many Issues," *Washington Post*, 14 November 1999, A1).
166. Phillip Coyle, *Missile Defense and Related Programs Director, Operational Test & Evaluation (DOT&E) FY'99 Annual Report*, February 2000, Internet, Director, Operational Test & Evaluation (DOT&E) Website, online at http://www.dote.osd.mil/reports/FY99/index.html.
167. Quoted in Bradley Graham, "Missile Shield Still Drawing Friends, Fire; Verdict on Deployment Due in Political Climate," *Washington Post*, 17 January 2000, Internet, online, Lexis/Nexis.
168. U.S. CBO, *Budgetary and Technical Implications.*
169. See Gordon R. Mitchell, "The National Missile Defense Fallacy," *Pittsburgh Post-Gazette*, 29 April 2000, A17.
170. Nira Schwartz, Third Amended Complaint for Violation of False Claims Act 31, USC 3729, Civil Action CV96-3065, United States District Court, Central District of California, 21 September 1999; see also William J. Broad, "Ex-Employee Says Contractor Faked Results of Missile Tests," *New York Times*, 7 March 2000, A1.
171. See Roy Danchick, Declaration filed under Civil Action CV96-3065, United States District Court, Central District of California, 18 May 1999; Samuel W. Reed, *Defense Investigative Service Report* (DCIS/DOD-IG), 1999. These documents (and others on the case) are archived at the Federation of American Scientists Website, online at http://www.fas.org/spp.starwars/program/news00/000203-trw.htm.
172. IFT-1 was scrubbed on 17 January 1997, when a data-link malfunction prevented the Payload Launch Vehicle (PLV) carrying the EKV from launching.
173. Brig. Gen. Joseph Cosumano, statement, Department of Defense News Briefing (Subject: National Missile Defense Flight Test), Office of the Assistant Secretary of Defense, 2 July 1997, Internet, Federation of American Scientists Website, online at http://www.fas.org/spp/starwars/program/news97/t07181997_t702bmdo.html.
174. Keith Englander, the NMD joint program office's deputy for systems integration, confirmed that in IFT-1A and IFT-2, the early NMD "fly-by" tests, the EKV sensor did not perform on-board discrimination, but rather gathered sensor data enabling technicians on the ground to use such data in discrimination simulations later on. "Although the EKV models did not perform on-board discrimination during the sensor fly-by tests, Englander said ground simulations afterwards using the data collected by the EKVs showed both models would have successfully discriminated the mock warhead from the decoys. "We did not exorcise the algorithms on the kill vehicle," Englander said, "We exorcised the algorithms on the ground . . . after we got the data telemetered down. We ran the scenes and the information through processors with the algorithms"(quoted in Michael Sirak, "BMDO Begins 'Orderly Phaseout' of Boeing Backup NMD Kill Vehicle," *Inside Defense*, 19 May 2000, online at http://www.InsideDefense.com).
175. Cosumano, statement.
176. See Nira Schwartz, letter to Dennis C. Egan, 29 July 1998, copy on file with the author.

177. Samuel W. Reed, Jr., letter to Keith Englander, 1 February 1999, copy on file with the author; see also Robert A. Young, letter to Keith Englander, 25 March 1998, copy on file with the author.
178. Robert A. Young, letter to Keith Englander, 10 April 1998, emphasis in original, copy on file with the author.
179. Reed, letter to Englander, 1 February 1999.
180. A 1997 study by Nichols Research, Inc., found that TRW's EKV program was "no Nobel Prize winner," but was good enough to satisfy TRW's contractual obligations (quoted in Broad, "Ex-Employee," 1). In February 1999 the BMDO's Phase One Engineering Team (POET) reviewed Schwartz's charges and produced what *New York Times* writer William J. Broad called an "equivocal" report (Broad, "Ex-Employee," 1). Although the POET team concluded that TRW's actions did not warrant Department of Justice prosecution under the False Claims Act, it also offered some bleak assessments about the quality of TRW's scientific work, saying that the "performance of the discrimination architecture may be fragile," and that "it would be desirable to expand the current discrimination architecture to make it more robust" (quoted in Samuel Reed, letter to Keith Englander, 11 August 1999, copy on file with the author).
181. Original plans for the first four NMD flight tests called for nine decoys to be used in each flight test (see Nichols Research viewgraph, "TSRD Target Requirements Summary (IFT-1 ‹ IFT-4) (U)," quoted in Theodore A. Postol, "Technical Discussion of the Misinterpreted Results of the IFT-1A Experiment Due to Tampering With the Data and Analysis and Error in the Interpretation of the Data," Attachment B of letter to John Podesta, 11 May 2000, Internet, Federation of American Scientists website, online at http://www.fas.org/spp/starwars/program/news00/postol_051100.html; see also William J. Broad, "Antimissile System's Flaw Was Covered Up, Critic Says," *New York Times,* 18 May 2000, A21.
182. Quoted in Stephen Green, "Pentagon Lowers, Meets Criteria for Missile Defense," *San Diego Union-Tribune,* 22 March 2000, online, Lexis/Nexis.
183. Theodore A. Postol, letter to John Podesta, 11 May 2000, Internet, Federation of American Scientists website. Online at http://www.fas.org/spp/starwars/program/news00/postol_051100.html.
184. Postol, letter to Podesta, 1.
185. Postol, letter to Podesta, 2.
186. William J. Broad, "Pentagon Classifies a Letter Critical of Antimissile Plan," *New York Times,* 20 May 2000, A10.
187. Broad, "Pentagon Classifies," A10.
188. Dennis Kucinich, statement, *Congressional Record,* 18 May 2000, H3395.
189. Kenneth H. Bacon, statement, DOD News Briefing, Office of the Assistant Secretary of Defense, 18 May 2000, Internet, Federation of American Scientists website, online at http://www.fas.org/spp/starwars/program/news00/t05182000_t0518asd.html; see also William J. Broad, "U.S. Defends Antimissile Plan," *New York Times,* 26 May 2000, A10.
190. Lisbeth Gronlund, statement, "National Missile Defense: The First Intercept Tests," Union of Concerned Scientists Press Breakfast, 29 September 1999, Internet, Union of

Concerned Scientists website, online at http://www.ucsusa.org/publications/pubs-home.html. According to Philip Coyle, the Pentagon's Director, Operational Test and Evaluation, "Test targets of the current program do not represent the complete 'design-to' threat space and are not representative of the full sensor requirements spectrum (e.g., discrimination requirements)" (Philip Coyle, *National Missile Defense (NMD) FY99 Annual Report,* Internet, Director, Operational Test and Evaluation website, online at http://www.dote.osd.mil/reports/FY99/other/99nmd.html).

191. Robert Scheer, "'Missile Defense' is Offensive," Pittsburgh Post-Gazette, 25 May 2000, A31.
192. See C. F. Barnaby and A. Boserup, "Implications of Anti-Ballistic Missile Systems," in *Implications of Anti-Ballistic Missile Systems,* ed. C. F. Barnaby and A. Boserup (New York: Humanities Press, 1969), 207–30.
193. Ibid., 230.
194. A. Jardé, *The Formation of the Greek People* (New York: Alfred Knopf, 1926), 123.
195. Previous attempts have been made to establish this historical linkage. Rawson notes that "[i]n the debates on the admission of Missouri to the Union in 1820 Southerners praised Rome, Athens, and Sparta in turn for combining liberty, even democracy, with slavery. . . . Outsiders have occasionally been ready to accept the comparison of the American South with Sparta" (Elizabeth Rawson, *The Spartan Tradition in European Thought,* [Oxford: Clarendon Press, 1969], 370).
196. Anton Powell, "Plato and Sparta," in *The Shadow of Sparta,* ed. Anton Powell and Stephen Hodkinson (New York: Routledge, 1994), 284, emphasis added. Indeed, as Powell observes elsewhere, "the Spartans were masters of deception. . . . Xenophon records two cases in which a Spartan general, on learning of a defeat for Spartan forces elsewhere, announced it to his troops as a victory, to sustain morale. Thucidydies writes of the helots being deceived with attractive promises by the Spartan authorities, as a preliminary to massacre. The seditious Kinadon was removed from Sparta by means of a lie, according to Xenophon. In these cases we are not dealing with some untruth uttered briefly by a cornered politician. Rather, each deception was supported by careful arrangements and appears to have been successfully maintained for as long as necessary" (Anton Powell, *Athens and Sparta: Constructing Greek Political and Social History from 478 B.C.* [London: Routledge, 1988], 215, emphasis added).
197. Powell, "Plato and Sparta," 285.
198. See Werckmeister, *Citadel Culture.*
199. See Greider, *Fortress America.*
200. See John E. Pike, Bruce G. Blair, and Stephen I. Schwartz, "Defending Against the Bomb," in *Atomic Audit: The Costs and Consequences of U.S. Nuclear Weapons Since 1940,* ed. Stephen I. Schwartz (Washinton, D.C.: Brookings Institute Press, 1998), 296; Broad, "After Many Misses," A23.
201. Quoted in Tom Gervasi, *The Myth of Soviet Military Supremacy* (New York: Harper and Row, 1986), vii.
202. Paulo Freire, *Pedagogy of Hope* (New York: Continuum Press, 1994).
203. See James Glanz, "Patriots Missed," 413–16.

204. Garwin, "Rumsfeld Report."
205. Garwin, letter to Sen. Carl Levin.
206. As Collina explains, "This is all classified. You can only hear so much about it. And these people are saying, you know, 'Don't worry, we've thought about it' are the same people that are benefiting from the program. So what we really need here is an independent assessment of the effectiveness of the national missile defense against these kind of countermeasures. An independent group that we could all trust. And if they say, Yes, go for it. This sytem'll work,' then fine. But we're far from that" (Tom Collina, interview with Bob Edwards, *Morning Edition,* National Public Radio, Internet, online, Lexis/Nexis, 18 March 1999).
207. In pursuit of this goal, Garwin suggests establishment of a standing "Red Team" of independent countermeasure experts, working separately from BMDO to generate realistic countermeasure designs. See Garwin, statement, *Ballistic Missile Defense.*
208. Garwin, statement, *Ballistic Missile Defense.*
209. Gottlieb, *Defense Addiction,* 166.
210. The "born secret" procedure has its roots in the 1946 Atomic Energy Act. See Herbert N. Foerstel, *Secret Science* (Westport, Conn.: Praeger, 1993); Burr, Blanton, and Schwartz, "Costs and Consequences," 443.
211. Burr, Blanton, and Schwartz, "Costs and Consequences," 455.
212. See Mitchell, "Another Strategic Deception Initiative," 22.
213. National Research Council, *A Review of the Department of Energy Classification Policy and Practice* (Washington, D.C.: National Academy Press, 1995), 4. This recommendation was eventually carried out in the DOE's Openness Initiative. See United States Department of Energy, *Openness,* Press Conference Fact Sheets, 6 February 1996 (Washington, D.C.: U.S. DOE, 1996), i.
214. Along these lines, Sissella Bok suggests that in decisions about the appropriate scope of institutional secrecy, officials should engage in the "test of publicity." This test involves entertaining a counterfactual hypothesis in which disclosure of information is assumed. Such a standard, writes Bok, "helps correct biases, errors, and ignorance—and thus the misjudgments they bring about—by asking how justifications and excuses would hold up in open debate and what would happen if, as often occurs, those responsible for planning, authorizing, and carrying out clandestine actions were exposed to public criticism" (Bok, *A Strategy for Peace,* 95). For discussion of a similar counterfactual mechanism for making classification decisions in the context of covert military operations, see Michael W. Reisman and James E. Baker, *Regulating Covert Action* (New Haven, Conn.: Yale University Press, 1992).
215. Daniel Patrick Moynihan, "The Science of Secrecy," paper delivered at the MIT Conference on Secrecy in Science, 29 March 1999, Cambridge, Mass., Internet, American Association for the Advancement of Science Website, online at http://www.aaas.org/spp/secrecy/Presents/Moynihan.htm.
216. See Bok, *A Strategy for Peace,* 82.
217. See United States Department of Energy, *Openness.*
218. See Tim Weiner, "C.I.A.'s Openness Derided as 'Snow Job,'" *New York Times,* 20 May 1997, A16.

219. Quoted in ibid.
220. Garwin, statement, *Ballistic Missile Defense.*
221. See Jerry W. Markham, "The Federal Advisory Committee Act," *University of Pittsburgh Law Review* 35 (1974): 557–608.
222. See Carl Levin, statement, *Department of Defense/Strategic Defense Initiative Organization Compliance with Federal Advisory Committee Act,* 100th Cong., 2d sess., Senate Hearings, Committee on Governmental Affairs, 19 April 1988 (Washington, D.C.: GPO, 1988).
223. See Gordon R. Mitchell, "Sparta's TMD Footprint Advice to BMDO: Overstepping the Boundaries of FACA?" issue brief for Senator David Pryor and Senator Carl Levin, 24 July 1996, subsequently published as Space Policy Project E-print, Internet, Federation of American Scientists Website, 8 August 1996, online at http://www.fas.org/spp/eprint/gm2.htm.
224. Burr, Blanton, and Schwartz, "Costs and Consequences," 469.
225. See *Furman v Georgia* 408 U.S. 238 (1972).
226. These procedural safeguards included automatic appeals, bifurcated juries, and statutory definition of aggravating factors in capital cases. See *Gregg v Georgia* 428 U.S. 153 (1976).
227. Adalberto Aguirre Jr. and David V. Baker, *Race, Racism and the Death Penalty in the United States* (Berrien Spring, Mich.: Vande Vere Publishing, 1991), 102.
228. *McCleskey v Kemp* 481 U.S. 1782 (1987).
229. Karl Grossman, "Master of Space," *The Progressive* 64 (January 2000): 27; see also Mike Moore, "Unintended Consequences," *Bulletin of the Atomic Scientists* 56 (January/February 2000): 58-65.
230. Charles Robb, "Star Wars II," *Washington Quarterly* 22 (Winter 1999): 85.
231. Carol Rosin, telephone interview with the author, 26 September 1999.
232. Ibid.
233. According to the *Webster's Sports Dictionary,* aikido is a "Japanese art of self-defense developed in the 20th century which emphasizes dodging and leading an attacker in the direction his momentum takes him to subdue him without causing injury" (*Webster's Sports Dictionary* [Springfield, Mass.: Merriam-Webster, 1976], 3).
234. Rosin, telephone interview with the author.
235. Burr, Blanton, and Schwartz, "Costs and Consequences," 471.
236. Aftergood, statement, 175.
237. Bowman, telephone interview with the author.
238. Scott Armstrong, "The War Over Secrecy: Democracy's Most Important Low-Intensity Conflict," in *A Culture of Secrecy,* ed. Athan G. Theoharis (Lawrence: University Press of Kansas, 1998), 172.
239. For discussion of how inflated intelligence assessments led to official endorsement of the "missile gap" and subsequent Pentagon spending hikes, see chapter one. In the case of the "window of vulnerability," former Center for Defense Information analyst Sanford Gottlieb notes that "[t]he GAO found no fewer than three counts on which the Pentagon overstated its claim of a 'window of vulnerability' caused by improved

Soviet missile capability. The exaggerated threat projections drove U.S. spending on nuclear weapons, which has accounted for more than a quarter of all military expenditures since World War II" (Gottlieb, *Defense Addiction,* 8).

240. Jonathan Schell, *The Abolition* (New York: Alfred A Knopf, 1984), 116. Freeman Dyson made similar arguments, although he endorsed spaceborne weapons as tools to strengthen a world-wide nuclear disarmament regime (see Freeman Dyson, *Weapons and Hope* [New York: Harper and Row, 1984], 71).
241. Schell, *Abolition,* 117, emphasis added.
242. Heritage Foundation, "A Proposed Plan."
243. Edward Teller, "The Feasibility of Arms Control and the Principle of Openness," *Proceedings of the American Academy of Arts and Sciences* 89 (1960): 795-96.
244. Schell, *Abolition,* 116.
245. Quoted in Jonathan Alter, "Swords vs. Shields," *Newsweek,* 8 May 2000, 44.
246. The text of these "Talking Points" is archived online at the *Bulletin of Atomic Scientists* website at http://www.bullatomsci.org/issues/2000/mj00/treaty_doc.html.
247. See Mitchell, "National Missile Defense Fallacy," A17.
248. See "Excerpts from Bush's Remarks on National Security and Arms Policy," *New York Times,* 24 May 2000, A19.
249. See John M. Broder, "Breaking Cold War Mold: Bush's Proposal to Reduce Nuclear Aresenal Injects Arms Into Race for the White House," *New York Times,* 26 May 2000, A1.
250. "Excerpts from Bush's Remarks," A19.
251. See Steven Lee Myers, "Bush's Missile Defenses Could Limit Warhead Cuts, Experts Warn," *New York Times,* 24 May 2000, A19; Richard Sisk, "Bush Vows to Beef Up Star Wars," *New York Daily News,* 24 May 2000, Internet, Federation of American Scientists website, online at http://www.fas.org/spp/starwars/program/news00/000523-nmd-gwb.htm.
252. Initial European reactions to these plans have been less than welcoming. For example, on 4 March 2000, some 300 protesters staked out the Menwith Hill (U.K.) base, demonstrating against construction of the U.S.-backed SBIRS system (see London Press Association, "U.K. Anti-Nuclear Protesters Target U.S. Military Base," 4 March 2000, Foreign Broadcast Information Service [FBIS-WEU-2000-0304]). By May 2000, more than 270 British MPs had signed six separate Commons motions urging the government to have nothing to do with Washington's new anti-ballistic-missile project. Greenland's parliament declared in November 1999 that if U.S. NMD plans violate the ABM Treaty, then Greenland "can't support plans for an upgrade of the Thule radar." Danish Foreign Minister Niels Helveg Peterson backed up this declaration on 25 February 2000, clarifying that a "firm component" of Denmark's policy is that use of the Thule radar not be "in violation of international rules" (both quoted in Joergen Dragsdahl, "Use of Key Radar Facility Conditioned on Russian Approval of ABM-Treaty Changes," 7 March 2000, Internet, British American Security Council Website, online at http://www.basicint.org).
253. Heritage Foundation, "A Proposed Plan."

254. See Gordon R. Mitchell, "U.S. National Missile Defence: Technical Challenges, Political Pitfalls and Disarmament Opportunities," ISIS-Europe Briefing Paper No. 23 (May 2000), Internet, ISIS-Europe Website, online at http://www.fhit.org/isis. For discussion of peaceful space exploration and development projects that might be pursued as alternatives to space weaponization, see Charles Sheffield and Carol Rosin, *Space Careers* (New York: William and Morrow, 1985); Carol Rosin, statement, *Controlling Space Weapons*, 98th Cong., 1st sess., Senate Hearings, Committee on Foreign Affairs, 18 May 1983, 152-158.
255. Joseph Romm, "Pseudo-Science and SDI," *Arms Control Today* 19 (1989): 21.

References

Books

Addelson, K. P. "The Man of Professional Wisdom." In *Discovering Reality,* edited by S. Harding and M. Hintikka. Dordrecht, Netherlands: Reidel, 1983.

Aguirre Jr., Adalberto, and David V. Baker. *Race, Racism and the Death Penalty in the United States.* Berrien Springs, Mich.: Vande Vere Publishing, 1991.

Alford, Jonathan, ed. *Arms Control and European Security.* Hampshire, England: Gower/ International Institute for Strategic Studies, 1984.

Allin, Dana H. *Cold War Illusions.* New York: St. Martin's Press, 1994.

Anderson, Martin. *Revolution.* San Diego, Calif.: Harcourt Brace Jovanovich, 1988.

Anderson, Poul. "Star-Flights and Fantasies: Sagas Still to Come." In *The Craft of Science Fiction,* edited by Reginald Bretnor, 22–36. New York: Harper and Row, 1976.

Andrewes, A. "The Government of Classical Sparta." In *Ancient Society and Institutions: Studies Presented to Victor Ehrenberg on His 75th Birthday,* 1–20. New York: Barnes and Noble, 1967.

Annas, George J. "Questing for Grails: Duplicity, Betrayal, and Self-Deception in Postmodern Medical Research." In *Health and Human Rights: A Reader,* edited by Jonathan M. Mann, Sofia Gruskin, Michael A. Grodin, and George J. Annas, 312–35. New York: Routledge, 1999.

Anzovin, Steven, ed. *The Star Wars Debate.* New York: H.W. Wilson, 1986.

"Appeal by American Scientists to Ban Space Weapons." In *Perspectives on Strategic Defense,* edited by Steven W. Guerrier and Wayne C. Thompson, 327. Boulder, Colo.: Westview, 1987.

Arbatov, Alexei. "Deep Cuts and De-alerting: A Russian Perspective." In *The Nuclear Turning Point: A Blueprint for Deep Cuts and De-alerting of*

Nuclear Weapons, edited by Harold A. Feiveson, 305–24. Washington, D.C.: Brookings Institution, 1999.

Aristotle. *On Rhetoric.* Translated by George A. Kennedy. New York: Oxford University Press, 1991.

Armstrong, Scott. "The War over Secrecy: Democracy's Most Important Low-Intensity Conflict." In *A Culture of Secrecy,* edited by Athan G. Theoharis, 140–85. Lawrence: University Press of Kansas, 1998.

Atkinson, Rick. *Crusade: The Untold Story of the Persian Gulf War.* Boston: Houghton Mifflin, 1993.

Barnes, Trevor. "Democratic Deception: American Covert Operations in Post-War Europe." In *Deception Operations,* edited by David A. Charters and Maurice A. J. Tugwell, 297–324. London: Brassey's, 1990.

Baucom, Donald R. *The Origins of SDI, 1944–1983.* Lawrence: University of Kansas Press, 1992.

Baudrillard, Jean. *Simulations.* New York: Semiotext[e], 1983.

———. "Simulacrum and Simulations." In *Baudrillard: Selected Writings,* edited by Mark Poster. Stanford, Calif.: Stanford University Press, 1988.

———. *the gulf war did not take place.* Bloomington, Ind.: Indiana University Press, 1995.

Baxter, James Phinney. *Scientists Against Time.* Boston: Little, Brown and Company, 1946.

Barnaby, C. F., and A. Boserup. "Implications of Anti-Ballistic Missile Systems." In *Implications of Anti-Ballistic Missile Systems,* edited by C. F. Barnaby and A. Boserup, 207–30. New York: Humanities Press, 1969.

Baylis, John, and John Garnett. *Makers of Nuclear Strategy.* London: Pinter Publishers, 1991.

Bazerman, Charles. *Shaping Written Knowledge: The Genre and Activity of the Experimental Article in Science.* Madison: University of Wisconsin Press, 1988.

Beinhart, Larry. *American Hero.* New York: Ballantine, 1993.

Benveniste, Guy. *The Politics of Expertise.* Berkeley, Calif.: Glendesary Press, 1972.

Birrer, Frans. "Counteranalysis: Toward Social and Normative Constraints on the Production and Use of Scientific and Technological Knowledge." In *Controversial Science: From Content to Contention,* edited by Thomas Brante, Steve Fuller, and William Lynch, 41–55. New York: SUNY Press, 1993.

Bittman, Ladislav. *The Deception Game: Czechoslovak Intelligence in Soviet Political Warfare*. New York: Syracuse University Research Corporation, 1972.

———. *The KGB and Soviet Disinformation*. Washington, D.C.: Pergamon-Brassey's, 1985.

Bjork, Rebecca. *The Strategic Defense Initiative: Symbolic Containment of the Nuclear Threat*. New York: SUNY Press, 1992.

Blanton, W. B. *Medicine in Virginia in the Eighteenth Century*. Richmond, Va.: Garrett and Massie, 1931.

Blair, Bruce G., John E. Pike, and Stephen I. Schwartz. "Targeting and Controlling the Bomb." In *Atomic Audit: The Costs and Consequences of U.S. Nuclear Weapons Since 1940*, edited by Stephen I. Schwartz, 197–268. Washington, D.C.: Brookings Institution Press, 1998.

Blumberg, Stanley A., and Louis G. Panos. *Edward Teller: Giant of the Golden Age of Physics*. New York: Scribners, 1990.

Boffey, Philip M., William J. Broad, Leslie H. Gelb, Charles Mohr, and Holcomb B. Noble. *Claiming the Heavens*. New York: Times Books, 1988.

Bok, Sissela. *Secrets*. New York: Pantheon Press, 1982.

———. *A Strategy for Peace*. New York: Vintage Books, 1989.

Bordieu, Pierre. *On Television*. Translated by Priscilla Parkhurst Ferguson. New York: New Press, 1998.

Bostdorff, Denise M. *The Presidency and the Rhetoric of Foreign Crisis*. Columbia: University of South Carolina Press, 1994.

Bottome, Edgar. *The Balance of Terror*. Boston: Beacon Press, 1986.

Boulding, Kenneth, ed. *Peace and the War Industry*. New Brunswick, N.J.: Transaction Press, 1973.

———. "Comment." In *Peace and the War Industry*, edited by Kenneth Boulding, 74–76. New Brunswick, N.J.: Transaction Books, 1973.

Bowman, Robert M. *Star Wars: A Defense Insider's Case against SDI*. New York: St. Martin's Press, 1986.

Brante, Thomas. "Reasons for Studying Scientific and Science-Based Controversies." In *Controversial Science: From Content to Contention*, edited by Thomas Brante, Steve Fuller, and William Lynch, 177–92. New York: SUNY Press, 1993.

Brante, Thomas, Steve Fuller, and William Lynch. Introduction to *Controversial Science: From Content to Contention*, edited by Thomas Brante, Steve Fuller, and William Lynch, ix–xix. New York: SUNY Press, 1993.

Brennan, Donald G. "The Case for the ABM." In Center for the Study of Democratic Institutions, *Anti-Ballistic Missile: Yes or No? A Special Report from the Center for the Study of Democratic Institutions,* 27–35. New York: Hill and Wang, 1968.

Bretnor, Reginald. "SF: The Challenge to the Writer." In *The Craft of Science Fiction,* edited by Reginald Bretnor, 3–20. New York: Harper and Row, 1976.

Broad William. *Star Warriors.* New York: Simon and Schuster, 1985.

———. *Teller's War.* New York: Simon and Schuster, 1992.

Brody, Howard. *Placebos and the Philosophy of Medicine: Clinical, Conceptual, and Ethical Issues.* Chicago: University of Chicago Press, 1980.

Brown, Harold. "Is SDI Technically Feasible?" In *The Search for Security in Space,* edited by Kenneth Luongo and W. Thomas Wander, 205–22. Ithaca, N.Y.: Cornell University Press, 1989.

Bundy, McGeorge. *Danger and Survival: Choices about the Bomb in the First Fifty Years.* New York: Random House, 1988.

———, William J. Crowe, and Sidney D. Drell. *Reducing Nuclear Danger.* New York: Council on Foreign Relations, 1993.

Bunn, Matthew. *Foundation for the Future: The ABM Treaty and National Security.* Washington, D.C.: Arms Control Association, 1990.

Burr, William, Thomas S. Blanton, and Stephen I. Schwartz. "The Costs and Consequences of Nuclear Secrecy." In *Atomic Audit: The Costs and Consequences of U.S. Nuclear Weapons Since 1940,* edited by Stephen I. Schwartz, 433–84. Washington, D.C.: Brookings Institution Press, 1998.

Cabot, R. C. *Social Service and the Art of Healing.* New York: Moffat, Yard, and Co., 1909.

Cahn, Anne Hessing. "American Scientists and the ABM: A Case Study in Controversy." In *Scientists and Public Affairs,* edited by Albert H. Teich, 41–120. Cambridge: Massachusetts Institute of Technology Press, 1974.

Camilleri, Joseph A. "The Cold War . . . and After: A New Period of Upheaval in World Politics." In *Why the Cold War Ended,* edited by Ralph Summy and Michael E. Salla, 233–47. Westport, Conn.: Greenwood Press, 1995.

Campbell, David. *Politics without Principle: Sovereignty, Ethics and the Narratives of the Gulf War.* Boulder, Colo.: Lynne Rienner, 1993.

Campbell, John Angus. "Charles Darwin: Rhetorician of Science." In *Rhetoric of the Human Sciences: Language and Argument in Scholarship and Public Affairs,* edited by John S. Nelson, Allan Megill, and Donald N. McCloskey, 69–86. Madison: University of Wisconsin Press, 1987.

———. “Strategic Reading: Rhetoric, Invention and Interpretation.” In *Rhetorical Hermeneutics: Invention and Interpretation in the Age of Science,* edited by Alan G. Gross and William M. Keith, 113–37. New York: SUNY Press, 1997.

Chace, James, and Caleb Carr. *America Invulnerable.* New York: Summit, 1988.

Chaloupka, William. *Knowing Nukes: The Politics and Culture of the Atom.* Minneapolis: University of Minnesota Press, 1992.

Chapman, Gary, and Joel Yudkin. *Briefing Book on the Military-Industrial Complex.* Washington, D.C.: Council for a Livable World, 1992.

Charters, David A., and Maurice A.J. Tugwell, eds. *Deception Operations.* London: Brassey’s, 1990.

Chomsky, Noam. *Deterring Democracy.* London: Verso, 1991.

———. *World Orders Old and New.* New York: Columbia University Press, 1994.

Cirincione, Joseph, and Frank von Hippel, eds. *The Last 15 Minutes: Ballistic Missile Defense in Perspective.* Washington, D.C.: Coalition to Reduce Nuclear Dangers, 1996.

Clark, Ramsey. *The Fire This Time: U.S. War Crimes in the Gulf.* New York: Thunder’s Mouth Press, 1992.

Cloud, Dana. “Operation Desert Comfort.” In *Seeing through the Media: The Persian Gulf War,* edited by Lauren Rabinowitz and Susan Jeffords, 155–71. New Brunswick, N.J.: Rutgers University Press, 1994.

Code, Lorraine. *rhetorical spaces.* New York: Routledge, 1995.

Codevilla, Angelo. *While Others Build.* New York: Free Press, 1988.

Collins, Harry M., and Trevor J. Pinch. “The Construction of the Paranormal: Nothing Unscientific Is Happening.” In *On the Margins of Science: The Social Construction of Rejected Knowledge,* edited by Roy Wallis, 237–70. Keele, England: University of Keele Press, 1979.

———. *The Golem at Large: What You Should Know about Technology.* Cambridge: Cambridge University Press, 1998.

Currie-McDaniel, Ruth. *The U.S. Army Strategic Defense Command: Its History and Role in the Strategic Defense Initiative.* Huntsville, Ala.: U.S. Army Strategic Defense Command, 1987.

Cuthbertson, Ian M. *The Anti-Tactical Ballistic Missile Issue and European Security.* New York: Institute for East-West Security Studies, 1990.

D’Antonio, Michael. *Atomic Harvest.* New York: Crown Press, 1993.

D'Souza, Dinesh. *Ronald Reagan: How an Ordinary Man Became an Extraordinary Leader.* New York: Simon and Shuster, 1997.

Deane, Michael J. *The Role of Strategic Defense in Soviet Strategy.* Miami: Advanced International Studies Institute, 1980.

Denning, Dorothy E. *Information Warfare and Security.* Reading, Mass.: Addison-Wesley, 1999.

DeParle, Jason. "Keeping the News in Step: Are Pentagon Rules Here to Stay?" In *The Media and the Gulf War: The Press and Democracy in Wartime,* edited by Hedrick Smith, 381–90. Washington, D.C.: Seven Locks Press, 1992.

Dickson, David. *The New Politics of Science.* Chicago: University of Chicago Press, 1988.

Druzhinin, V. V. and D. S. Kontorov. *Concept, Algorithm, Decision (A Soviet View).* Translated by the U.S. Air Force. Washington, D.C.: GPO, 1972.

Duhem, Pierre. *The Aim and Structure of Physical Theory.* Princeton, N.J.: Princeton University Press, 1954.

Durch, William J. *The ABM and Western Security.* Cambridge, Mass.: Ballinger, 1988.

Dyson, Freeman. *Weapons and Hope.* New York: Harper and Row, 1984.

Eisenhower, Dwight. "Farewell Address to the Nation." 17 January 1961. In *Bureaucratic Politics and National Security,* edited by David C. Kozak and James M. Keagle, 278–82. Boulder, Colo.: Lynne Rienner, 1988.

Elzinga, Aant. "Science as Continuation of Politics by Other Means." In *Controversial Science: From Content to Contention,* edited by Thomas Brante, Steve Fuller, and William Lynch, 127–52. New York: SUNY Press, 1993.

Engelhardt, H. Tristram, Jr., and Arthur L. Caplan, eds. *Scientific Controversies: Case Studies in the Resolution and Closure of Disputes in Science and Technology.* Cambridge: Cambridge University Press, 1987.

Engelhardt, Tom. "The Gulf War as Total Television." In *Seeing through the Media: The Persian Gulf War,* edited by Lauren Rabinowitz and Susan Jeffords, 81–96. New Brunswick, N.J.: Rutgers University Press, 1994.

Enthoven, Alain C., and K. Wayne Smith. *How Much Is Enough? Shaping the Defense Program, 1961–1969.* New York: Harper Colophon Books, 1971.

Epstein, Steven. *Impure Science.* Berkeley: University of California Press, 1996.

Evangelista, Matthew. *Unarmed Forces: The Transnational Movement to End the Cold War.* Ithaca, N.Y.: Cornell University Press, 1999.

Farrell, Thomas B. *Norms of Rhetorical Culture.* New Haven, Conn.: Yale University Press, 1993.

———. "An Elliptical Postscript." In *Rhetorical Hermeneutics: Invention and Interpretation in the Age of Science,* edited by Alan G. Gross and William M. Keith, 317–29. New York: SUNY Press, 1997.

Fialka, John J. *Hotel Warriors: Covering the Gulf War.* Washington, D.C.: Woodrow Wilson Center Press, 1991.

Finney, John W. "A Historical Perspective." In *The ABM Treaty: To Defend or Not to Defend?,* edited by Walther Stützle, Bhupendra Jasani, and Regina Cowen, 29–44. Oxford: Stockholm International Peace Research Institute/Oxford University Press, 1987.

FitzGerald, Frances. *Way Out There in the Blue: Reagan, Star Wars and the End of the Cold War.* New York: Simon and Schuster, 2000.

Foerstel, Herbert N. *Secret Science.* Westport, Conn.: Praeger, 1993.

Foster, Gregory D. *In Search of a Post–Cold War Security Structure.* Washington, D.C.: National Defense University Institute for National Strategic Studies, 1994.

Franklin, Allan. *Can That Be Right? Essays on Experiment, Evidence, and Science.* Dordrecht, Netherlands: Kluwer, 1999.

Franklin, H. Bruce. *War Stars.* Oxford: Oxford University Press, 1988.

Fraser, Nancy. *Unruly Practices: Power, Discourse and Gender in Contemporary Social Theory.* Minneapolis: University of Minnesota Press, 1989.

Freedman, Lawrence, and Efraim Karsh. *The Gulf Conflict, 1990–1991: Diplomacy and War in the New World Order.* Princeton, N.J.: Princeton University Press, 1993.

Freire, Paulo. *Pedagogy of Hope.* New York: Continuum Press, 1994.

Fuller, Steve. *Philosophy, Rhetoric and the End of Knowledge.* Madison: University of Winconsin Press, 1993.

———. "Rhetoric of Science: Double the Trouble?" In *Rhetorical Hermeneutics: Invention and Interpretation in the Age of Science,* edited by Alan G. Gross and William M. Keith, 279–98. New York: SUNY Press, 1997.

Gaddis, John Lewis. *Strategies of Containment: A Critical Appraisal of Postwar American National Security Policy.* New York: Oxford University Press, 1982.

Gallagher, Carol. *American Ground Zero: The Secret Nuclear War.* New York: Random House, 1993.

Gaonkar, Dilip. "Close Readings of the Third Kind." In *Rhetorical Hermeneutics: Invention and Interpretation in the Age of Science,* edited by Alan G. Gross and William M. Keith, 330–56. New York: SUNY Press, 1997.

Garthoff, Raymond L. *The Great Transition: American-Soviet Relations and the End of the Cold War.* Washington, D.C.: Brookings, 1994.

Garwin, Richard L. "Theater Missile Defense, National ABM Systems, and the Future of Deterrence." In *Post–Cold War Conflict Deterrence,* edited by National Research Council, 182–200. Washington, D.C.: National Academy Press, 1997.

Gellman, Barton. "U.S. Bombs Missed 70 Percent of the Time." In *The Media and the Gulf War: The Press and Democracy in Wartime,* edited by Hedrick Smith, 197–99. Washington, D.C.: Seven Locks Press, 1992.

Gerbner, George. "Persian Gulf War, the Movie." In *Triumph of the Image: The Media's War in the Persian Gulf—A Global Perspective,* edited by H. Mowlana, G. Gerbner, and H. I. Schiller, 243–65. Boulder, Colo.: Westview Press, 1992.

Gervasi, Tom. *Arsenal of Democracy II.* New York: Grove Press, 1981.

———. *The Myth of Soviet Military Supremacy.* New York: Harper and Row, 1986.

Goodnight, G. Thomas. "Controversy." In *Argument in Controversy: Proceedings of the Seventh SCA/AFA Conference on Argumentation,* edited by Donn Parson, 1–13. Annandale, Va.: Speech Communication Association, 1992.

Gordon, Michael R., and Bernard E. Trainor. *The General's War.* Boston: Little, Brown and Company, 1995.

Gottlieb, Sanford. *Defense Addiction: Can America Kick the Habit?* Boulder, Colo.: Westview Press, 1997.

Graham, Daniel O. *The Non-Nuclear Defense of Cities: The High Frontier Space-Based Defense against ICBM Attack.* Cambridge, Mass.: Abt Books, 1983.

Graham, Daniel O., and Gregory A. Fossedal. *A Defense That Defends: Blocking Nuclear Attack.* Old Greenwich, Conn.: Devin-Adair Publishers, 1983.

Greider, William. *Fortress America.* New York: Public Affairs Press, 1998.

Gross, Alan G. *The Rhetoric of Science.* Cambridge, Mass.: Harvard University Press, 1990.

Gross, Alan G., and William M. Keith. Introduction to *Rhetorical Hermeneutics: Invention and Interpretation in the Age of Science,* edited by Alan G. Gross and William M. Keith, 1–24. New York: SUNY Press, 1997.

Gross, Paul R., and Norman Levitt. *Higher Superstitions: The Academic Left and Its Quarrels with Science.* Baltimore: Johns Hopkins University Press, 1994.

Guerrier, Steven W., and Wayne C. Thompson. *Perspectives on Strategic Defense.* Boulder, Colo.: Westview Press, 1987.

Guertner, Gary L. "What Is Proof?" In *The Search for Security in Space,* edited by Kenneth E. Luongo and W. Thomas Wander, 183–92. Ithaca, N.Y.: Cornell University Press, 1989.

Guertner, Gary L., and Donald M. Snow. *The Last Frontier: An Analysis of the Strategic Defense Initiative.* Lexington, Mass.: Lexington Books, 1986.

Habermas, Jürgen. *Toward a Rational Society: Student Protest, Science, and Politics.* Translated by Jeremy J. Shapiro. Boston: Beacon Press, 1970.

———. *Moral Consciousness and Communicative Action.* Translated by Christian Lenhardt and Shierry Weber Nicholsen. Cambridge: Massachusetts Institute of Technology Press, 1990.

———. *Between Facts and Norms.* Translated by William Rehg. Cambridge: Massachusetts Institute of Technology Press, 1996.

Hallin, Daniel C., and Todd Gitlin, "The Gulf War as Popular Culture and Television Drama." In *Taken by Storm: The Media, Public Opinion, and Foreign Policy in the Gulf War,* edited by W. Lance Bennett and David L. Paletz, 149–66. Chicago: University of Chicago Press, 1994.

Hammond, Paul Y. "NSC-68: Prologue to Rearmament." In *Strategy, Politics, and Defense Budgets,* edited by Warner R. Schilling, Paul Y. Hammond, and Glenn H. Snyder, 267–78. New York: Columbia University Press, 1962.

Handel, Michael I. *War, Strategy, and Intelligence.* London: Frank Cass, 1989.

———. "A 'Runaway Deception': Soviet Disinformation and the Six-Day War, 1967." In *Deception Operations,* edited by David A. Charters and Maurice A. J. Tugwell, 159–69. London: Brassey's, 1990.

Harding, Sandra. *Whose Science? Whose Knowledge?* Ithaca, N.Y.: Cornell University Press, 1991.

Hart, David M. *Forged Consensus: Science, Technology, and Economic Policy in the United States, 1921–1953.* Princeton, N.J.: Princeton University Press, 1998.

Held, David, and John B. Thompson, eds. *Habermas: Critical Debates.* Cambridge: Massachusetts Institute of Technology Press, 1982.

Hempel, Carl. *Philosophy of Natural Science.* Englewood Cliffs, N.J.: Prentice Hall, 1966.

Herken, Gregg. *Counsels of War.* New York: Oxford University Press, 1987.

Herman, Edward S., and Noam Chomsky. *Manufacturing Consent: The Political Economy of the Mass Media.* New York: Pantheon Books, 1988.

Heritage Foundation. "Strategic Defense: The Technology That Makes It Possible." In *Anti-Missile and Anti-Satellite Technologies and Programs,* edited by U.S. Department of Defense, U.S. Office of Technology Assessment, and the Heritage Foundation, 2–16. Park Ridge, N.J.: Noyes, 1986.

Heuer, Richards J., Jr. "Soviet Organization and Doctrine for Strategic Deception." In *Soviet Strategic Deception,* edited by Brian D. Dailey and Patrick J. Parker, 21–54. Lexington, Mass.: D.C. Heath, 1987.

Hogan, J. Michael. *The Nuclear Freeze Campaign: Rhetoric and Foreign Policy in the Telepolitical Age.* East Lansing, Mich.: Michigan State University Press, 1994.

Hollihan, Thomas A., Patricia Riley, and James F. Klumpp. "Greed versus Hope, Self-Interest versus Community: Reinventing Argumentative Practice in Post-Free Marketplace America." In *Argument and the Postmodern Challenge,* edited by Ray McKerrow, 332–39. Annandale, Va.: Speech Communication Association, 1993.

Holub, Robert C. *Jürgen Habermas: Critic in the Public Sphere.* London: Routledge, 1992.

Horowitz, Irving. "Comment." In *Peace and the War Industry,* edited by Kenneth Boulding, 55–60. New Brunswick, N.J.: Transaction Books, 1973.

Hughes, Robert C. *SDI: A View from Europe.* Washington, D.C.: National Defense University Press, 1990.

Hunter, Allen. "The Limits of Vindicationist Scholarship." In *Rethinking the Cold War,* edited by Allen Hunter, 1–34. Philadelphia: Temple University Press, 1998.

IPCC 95. *Climate Change 1995: The Science of Climate Change.* Cambridge: Cambridge University Press, 1995.

Jardé, A. *The Formation of the Greek People.* New York: Alfred Knopf, 1926.

Jones, A. H. M. *Sparta.* Oxford: Basil Blackwell, 1967.

Jones, R. V. *The Wizard War.* New York: Coward, McCann and Geogheghan, 1978.

Joseph, Paul. *Search for Sanity: The Politics of Nuclear Weapons and Disarmament.* Cambridge, Mass.: South End Press, 1984.

Jospe, Michael. *The Placebo Effect in Healing.* Lexington, Mass.: Heath Publishers, 1978.

Kagan, Donald, Steven Ozment, and Frank M. Turner. *The Western Heritage to 1715.* New York: Macmillan, 1979.

Kahn, Herman. *On Thermonuclear War.* Princeton, N.J.: Princeton University Press, 1960.

Kaplan, Fred. *Wizards of Armageddon.* New York: Simon and Shuster, 1983.

Katz, Amrom H. "The Fabric of Verification: The Warp and the Woof." In *Verification and SALT: The Challenge of Strategic Deception,* edited by William C. Potter, 193–220. Boulder, Colo.: Westview Press, 1980.

Katz, Jay. *The Silent World of Doctor and Patient.* New York: Free Press, 1984.

Keller, Evelyn Fox. *Secrets of Life: Essays on Language, Gender, and Science.* New York: Routledge, 1992.

Kellner, Douglas. *Jean Baudrillard: From Marxism to Postmodernism and Beyond.* Palo Alto, Calif.: Stanford University Press, 1989.

———. *The Persian Gulf TV War.* Boulder, Colo.: Westview Press, 1992.

King, Andrew. "The Rhetorical Critic and the Invisible Polis." In *Rhetorical Hermeneutics: Invention and Interpretation in the Age of Science,* edited by Alan G. Gross and William M. Keith, 299–316. New York: SUNY Press, 1997.

Klare, Michael. "The Pentagon's New Paradigm." In *The Gulf War Reader,* edited by Micah L. Sifry and Christopher Serf, 466–76. New York: Random House, 1991.

Kuhn, Rick. "Whose Cold War?" In *Why the Cold War Ended,* edited by Ralph Summy and Michael E. Salla, 153–69. Westport, Conn.: Greenwood Press, 1995.

Kuhn, Thomas. *The Structure of Scientific Revolutions.* Chicago: University of Chicago Press, 1970.

Lackey, Douglas P. *Moral Principles and Nuclear Weapons.* Totowa, N.J.: Rowman and Allanheld, 1984.

Lapham, Lewis H. "Trained Seals and Sitting Ducks." In *The Media and the Gulf War: The Press and Democracy in Wartime,* edited by Hedrick Smith, 256–63. Washington, D.C.: Seven Locks Press, 1992.

Latour, Bruno. *Science in Action: How to Follow Scientists and Engineers through Society.* Cambridge, Mass.: Harvard University Press, 1987.

Latour, Bruno, and Steve Woolgar. *Laboratory Life.* London: Sage, 1979.

Lebow, Richard Ned, and Janice Gross Stein. *We All Lost the Cold War.* Princeton, N.J.: Princeton University Press, 1994.

Leff, Michael. "The Idea of Rhetoric as Interpretive Practice: A Humanist's Response to Gaonkar." In *Rhetorical Hermeneutics: Invention and*

Interpretation in the Age of Science, edited by Alan G. Gross and William M. Keith, 89–100. New York: SUNY Press, 1997.

Lenczowski, John. "Themes of Soviet Strategic Deception and Disinformation." In *Soviet Strategic Deception*, edited by Brian D. Dailey and Patrick J. Parker, 55–75. Lexington, Mass.: D.C. Heath, 1987.

Leslie, Stuart W. *The Cold War and American Science.* New York: Columbia University Press, 1993.

Lewin, Leonard C. *Report From Iron Mountain: On the Possibility and Desirability of Peace.* New York: Free Press, 1996.

Lichter, Robert. "The Instant Replay War." In *The Media and the Gulf War: The Press and Democracy in Wartime*, edited by Hedrick Smith, 224–30. Washington, D.C.: Seven Locks Press, 1992.

Linenthal, Edward Tabor. *Symbolic Defense.* Chicago: University of Illinois Press, 1989.

Longino, Helen. *Science as Social Knowledge: Values and Objectivity in Scientific Inquiry.* Princeton, N.J.: Princeton University Press, 1990.

Lyne, John. "Bio-rhetorics: Moralizing the Life Sciences." In *The Rhetorical Turn: Invention and Persuasion in the Conduct of Inquiry*, edited by Herbert W. Simons, 35–57. Chicago: University of Chicago Press, 1990.

MacArthur, John R. *Second Front: Censorship and Propaganda in the Gulf War.* New York: Hill and Wang, 1992.

MacKenzie, Donald. "From Kwajalein to Armageddon? Testing and the Social Construction of Missile Accuracy." In *The Uses of Experiment*, edited by David Gooding, Trevor Pinch, and Simon Shaffer, 409–37. Cambridge: Cambridge University Press, 1989.

———. *Inventing Accuracy: A Historical Sociology of Nuclear Missile Guidance.* Cambridge, Mass.: Massachusetts Institute of Technology Press, 1990.

Makhijani, Arjun, Howard Hu, and Katherine Yih, eds. *Nuclear Wastelands.* Cambridge: Massachusetts Institute of Technology Press, 1995.

Makhijani, Arjun, and Stephen I. Schwartz. "Victims of the Bomb." In *Atomic Audit: The Costs and Consequences of U.S. Nuclear Weapons Since 1940*, edited by Stephen I. Schwartz, 395–432. Washington, D.C.: Brookings Institution Press, 1998.

Makhijani, Arjun, Stephen I. Schwartz, and William J. Weida. "Nuclear Waste Management and Environmental Remediation." In *Atomic Audit: The Costs and Consequences of U.S. Nuclear Weapons Since 1940*, edited by Stephen I. Schwartz, 353–94. Washington, D.C.: Brookings Institution Press, 1998.

Marcuse, Herbert. *One-Dimensional Man.* Boston: Beacon Press, 1964.

Martin, Brian. *Scientific Knowledge in Controversy: The Social Dynamics of the Fluoridation Debate.* New York: SUNY Press, 1991.

Mazarr, Michael J. *Missile Defenses and Asian-Pacific Security.* London: Macmillan Press, 1989.

Mazur, Alan. *The Dynamics of Technical Controversy.* Washington, D.C.: Communications Press, 1981.

McBride, James. *War, Battering and Other Sports.* Atlantic Highlands, N.J.: Humanities Press, 1995.

McGee, Michael Calvin, and John R. Lyne. "What Are Nice Folks Like You Doing in a Place Like This?" In *The Rhetoric of the Human Sciences,* edited by John S. Nelson, Allan Megill, and Donald N. McCloskey, 381–406. Madison: University of Wisconsin Press, 1987.

McMullin, Ernan. "Scientific Controversy and Its Termination." In *Scientific Controversies,* edited by H. T. Engelhardt and L. A. Caplan, 49–92. Cambridge: Cambridge University Press, 1987.

———. "The Social Dimensions of Science." In *The Social Dimensions of Science,* edited by Ernan McMullin, 1–27. Notre Dame, Ind.: Notre Dame University Press, 1992.

McNamara, Robert S. *The Essence of Security: Reflections in Office.* New York: Harper and Row, 1968.

Mikheev, Dmitri. *The Soviet Perspective on the Strategic Defense Initiative.* New York: Pergamon-Brassey's, 1987.

Miller, Richard Lawrence. *Heritage of Fear.* New York: Walker Press, 1988.

Mische, Patricia M.. *Star Wars and the State of Our Souls.* Minneapolis, Minn.: Winston Press, 1985.

———. "Star Wars and the State of Our Souls." In *Securing Our Planet,* edited by Don Carlson and Craig Comstock, 210–35. Los Angeles: Jeremy P. Tarcher, 1986.

Montagu, Ewen. *Beyond Top Secret ULTRA.* New York: Coward, McCann and Geoghegan, 1978.

Moynihan, Daniel Patrick. *Secrecy: The American Experience.* New Haven, Conn.: Yale University Press, 1998.

Mueller, John. *Policy and Opinion in the Gulf War.* Chicago: University of Chicago Press, 1994.

Mukerji, Chandra. *A Fragile Power: Scientists and the State.* Princeton, N.J.: Princeton University Press, 1990.

Nagel, Thomas. *The View from Nowhere.* New York: Oxford University Press, 1986.

"The National Pledge of Non-Participation: A Boycott of SDI Research." In *Perspectives on Strategic Defense,* edited by Steven W. Guerrier and Wayne C. Thompson, 323–25. Boulder, Colo.: Westview Press, 1987.

Naureckas, Jim. "Gulf War Coverage: The Worst Censorship Was at Home." In *The FAIR Reader: An EXTRA! Review of Press and Politics in the '90s,* edited by Jim Naureckas and Janine Jackson, 28–42. Boulder, Colo.: Westview, 1996.

———, and Janine Jackson, eds. *The FAIR Reader: An EXTRA! Review of Press and Politics in the '90s.* Boulder, Colo.: Westview Press, 1996.

Navasky, Victor. Introduction to *Report From Iron Mountain,* edited by Leonard C. Lewin, v–xvi. New York: Free Press, 1996.

Nelkin, Dorothy. *Controversy: Politics of Technical Decisions.* Longon: Sage, 1979.

———. *Science as Intellectual Property: Who Controls Scientific Research?* New York: Macmillan, 1984.

———. "Science, Technology, and Political Conflict: Analyzing the Issues." In *Politics of Technical Decisions,* 2d ed., edited by Dorothy Nelkin, 9–24. London: Sage, 1984.

———. *Selling Science.* New York: Freeman, 1995.

Nelson, Lynn Hankinson. *Who Knows: From Quine to a Feminist Empiricism.* Philadelphia: Temple University Press, 1990.

Nickles, Thomas. "Justification and Experiment." In *The Uses of Experiment,* edited by David Gooding, Trevor Pinch, and Simon Schaffer, 299–334. Cambridge: Cambridge University Press, 1989.

Norris, Christopher. *Uncritical Theory: Postmodernism, Intellectuals, and the Gulf War.* Amherst: University of Massachusetts Press, 1992.

O'Neill, Kevin. "Building the Bomb." In *Atomic Audit: The Costs and Consequences of U.S. Nuclear Weapons Since 1940,* edited by Stephen I. Schwartz, 33–104. Washington, D.C.: Brookings Institution Press, 1998.

Paletz, David L. "Just Deserts?" In *Taken by Storm: The Media, Public Opinion, and Foreign Policy in the Gulf War,* edited by W. Lance Bennett and David L. Paletz, 277–92. Chicago: University of Chicago Press, 1994.

Parrish, Thomas. The *ULTRA Americans: The U.S. Role in Breaking the Nazi Codes.* New York: Stein and Day, 1986.

Patton, Paul. Introduction to *the gulf war did not take place,* by Jean Baudrillard, 1–22. Bloomington: Indiana University Press, 1995.

Pawle, Gerald. *The Secret War: 1939–1945.* New York: William Sloane, 1957.

Pera, Marcello. *The Discourses of Science.* Translated by Clarissa Botsford. Chicago: University of Chicago Press, 1994.

Pike, John E., Bruce G. Blair, and Stephen I. Schwartz. "Defending Against the Bomb." In *Atomic Audit: The Costs and Consequences of U.S. Nuclear Weapons Since 1940,* edited by Stephen I. Schwartz, 269–325. Washington, D.C.: Brookings Institution Press, 1998.

Pilisuk, Marc. "Comment." In *Peace and the War Industry,* edited by Kenneth Boulding, 68–74. New Brunswick, N.J.: Transaction Books, 1973.

Popper, Karl. *The Myth of Framework.* London: Routledge, 1994.

Porter, Theodore. *Trust in Numbers: The Pursuit of Objectivity in Science and Public Life.* Princeton, N.J.: Princeton University Press, 1995.

Pournelle, Jerry. "The Construction of Believable Societies." In *The Craft of Science Fiction,* edited by Reginald Bretnor, 104–20. New York: Harper and Row, 1976.

Powaski, Ronald E.. *The March to Armageddon.* New York: Oxford University Press, 1987.

———. *The Cold War.* New York: Oxford University Press, 1998.

Powell, Anton. *Athens and Sparta: Constructing Greek Political and Social History from 478 B.C.* London: Routledge, 1988.

———. "Plato and Sparta." In *The Shadow of Sparta,* edited by Anton Powell and Stephen Hodkinson, 273–321. New York: Routledge, 1994.

Powell, Anton, and Stephen Hodkinson, eds. *The Shadow of Sparta.* London: Routledge, 1994.

Pratt, Erik K. *Selling Strategic Defense.* Boulder, Colo.: Lynne Rienner, 1990.

Prelli, Lawrence. *A Rhetoric of Science: Inventing Scientific Discourse.* Columbia: University of South Carolina Press, 1989.

Pressler, Larry. *Star Wars: The Strategic Defense Initiative Debates in Congress.* Westport, Conn.: Greenwood Press, 1986.

Quine, Willard. *Word and Object.* Cambridge: Massachusetts Institute of Technology Press, 1960.

Rawson, Elizabeth. *The Spartan Tradition in European Thought.* Oxford: Clarendon Press, 1969.

Reagan, Ronald. *An American Life.* New York: Simon and Schuster, 1990.

Reilly, Kevin. *The West and the World.* New York: Harper and Row, 1980.

Reisman, Michael W., and James E. Baker. *Regulating Covert Action.* New Haven, Conn.: Yale University Press, 1992.

Reiss, Edward. *The Strategic Defense Initiative.* Cambridge: Cambridge University Press, 1992.

Ris, Howard C., Jr. Preface to *The Fallacy of Star Wars: Why Space Weapons Can't Protect Us,* edited by John Tirman. New York: Vintage Books, 1984.

Rouse, Joseph. *Knowledge and Power: Toward a Political Philosophy of Science.* Ithaca, N.Y.: Cornell University Press, 1987.

———. *Engaging Science.* Ithaca, N.Y.: Cornell University Press, 1996.

Salla, Michael E. "The End of the Cold War: A Political, Historical, and Mythological Event." In *Why the Cold War Ended,* edited by Ralph Summy and Michael E. Salla, 249–59. Westport, Conn.: Greenwood Press, 1995.

Sarewitz, Daniel. *Frontiers of Illusion: Science, Technology, and the Politics of Progress.* Philadelphia: Temple University Press, 1996.

Schell, Jonathan. *The Abolition.* New York: Alfred A. Knopf, 1984.

Schelling, Thomas. *The Strategy of Conflict.* Cambridge, Mass.: Harvard University Press, 1960.

Schmitt, Eric. "Army Is Blaming Patriot's Computer for Failure to Stop the Dhahran Scud." In *The Media and the Gulf War: The Press and Democracy in Wartime,* edited by Hedrick Smith, 212–16. Washington, D.C.: Seven Locks Press, 1992.

Schonauer, Klaus. *Semiotic Foundations of Drug Therapy: The Placebo Problem in a New Perspective.* Berlin: Mouton de Gruyter, 1994.

Schwartz, David N. "Past and Present: The Historical Legacy." In *Ballistic Missile Defense,* edited by Ashton B. Carter and David N. Schwartz, 330–49. Washington, D.C.: Brookings Institution Press, 1984.

Schweizer, Peter. *Victory.* New York: Atlantic Monthly Press, 1994.

Scortia, Thomas N. "Science Fiction as the Imaginary Experiment." In *Science Fiction, Today and Tomorrow,* edited by Reginald Bretnor, 135–50. New York: Harper and Row, 1974.

Sealey, Raphael. *History of the Greek City States 700–338 B.C.* Berkeley: University of California Press, 1976.

Seth, Ronald. *The Truth Benders: Psychological Warfare in the Second World War.* London: Leslie Frewin, 1969.

Shapere, Dudley. "On Deciding What to Believe and How to Talk about Nature." In *Persuading Science: The Art of Scientific Rhetoric,* edited by M. Pera and W. Shea, 70–98. Canton, Mass.: Science History Publications, 1991.

Sheffield, Charles, and Carol Rosin. *Space Careers.* New York: William and Morrow, 1985.

Sherwin, Ronald G., and Barton Whaley, "Understanding Strategic Deception: An Analysis of 93 Cases." In *Strategic Military Deception,* edited by Donald C. Daniel and Katherine L. Herbig, 177–94. New York: Pergamon Press, 1982.

Shils, Edward. *The Torment of Secrecy.* Glencoe, Ill.: Free Press, 1976.

Shukman, David. *Tomorrow's War.* New York: Harcourt Brace Jovanovich, 1996.

Shultz, Richard H., and Roy Godson. *Dezinformatsia: Active Measures in Soviet Strategy.* Washington, D.C.: Pergamon-Brassey's, 1984.

Simon, Jeffrey, ed. *Overview to Security Implications of SDI.* Washington, D.C.: National Defense University Press, 1990.

Simpson, Christopher. *Science of Coercion.* Oxford: Oxford University Press, 1994.

———. "Commentary on NSDD 172: Publicizing the Strategic Defense Initiative." In *National Security Decision Directives of the Reagan and Bush Administrations,* edited by Christopher Simpson, 449. Boulder, Colo.: Westview, 1995.

———. Introduction to *National Security Decision Directives of the Reagan and Bush Administrations,* edited by Christopher Simpson, 1–7. Boulder, Colo.: Westview, 1995.

Smith, Gerard. Preface to *Foundation for the Future: The ABM Treaty and National Security,* edited by Matthew Bunn, vii–viii. Washington, D.C.: Arms Control Association, 1990.

Smith, Hedrick. *The Power Game.* New York: Random House, 1988.

Snitow, Ann. "A Gender Diary." In *Conflicts in Feminism,* edited by Marianne Hirsch and Evelyn Fox Keller, 9–43. New York: Routledge, 1990.

Sorrell, Tom. *Scientism.* London: Routledge, 1991.

Stares, Paul B. *Space and National Security.* Washington, D.C.: Brookings Institution Press, 1987.

Stein, Jonathan B. *From H-Bomb to Star Wars.* Lexington, Mass.: Lexington Books, 1984.

Stolfi, Russel H. S. "Barbarossa: German Grand Deception and the Achievement of Strategic and Tactical Surprise against the Soviet Union, 1940–1941." In *Strategic Military Deception,* edited by Donald C. Daniel and Katherine L. Herbig, 195–223. New York: Pergamon Press, 1982.

Talbott, Strobe. *Master of the Game: Paul Nitze and the Nuclear Peace.* New York: Alfred A. Knopf, 1988.

Taylor, Philip M. *War and the Media: Propaganda and Persuasion in the Gulf War.* Manchester: Manchester University Press, 1992.

Teller, Edward. *Better a Shield than a Sword.* New York: Free Press, 1987.

Thomas, Carol G., ed. *Paths from Ancient Greece.* Leiden, Netherlands: E.J. Brill, 1988.

Tirman, John. "The Politics of Star Wars." In *Empty Promise: The Growing Case Against Star Wars,* edited by John Tirman, 1–33. Boston: Beacon Press, 1986.

———, ed. *The Militarization of High Technology.* Cambridge, Mass.: Ballinger, 1984.

Traweek, Sharon. *Beamtimes and Lifetimes.* Cambridge, Mass.: Harvard University Press, 1988.

———. "Border Crossings: Narrative Strategies in Science Studies and among Physicists in Tsukuba Science City, Japan." In *Science as Practice and Culture,* edited by Andrew Pickering, 429–67. Chicago: University of Chicago Press, 1992.

Tucker, Jonathan B. "Scientists and Star Wars." In *Empty Promise: The Growing Case Against Star Wars,* edited by John Tirman, 34–61. Boston: Beacon Press, 1986.

Tzu, Sun. *The Art of War.* Translated by Samuel B. Griffith. New York: Oxford University Press, 1973.

Udall, Stewart L. *The Myths of August.* New York: Pantheon, 1994.

Union of Concerned Scientists. *The Fallacy of Star Wars.* New York: Vintage Books, 1984.

———. *Empty Promise: The Growing Case Against Star Wars,* edited by John Tirman. Boston: Beacon Press, 1986.

United States Congress. Office of Technology Assessment. *Strategic Defenses: Ballistic Missile Defense Technologies, Anti-Satellite Weapons, Countermeasures, and Arms Control.* Princeton, N.J.: Princeton University Press, 1986.

———. Strategic Defense Initiative Organization. "SDI: Goals and Technical Objectives." In *The Search for Security in Space,* edited by Kenneth N. Luongo and W. Thomas Wander, 144–59. Ithaca, N.Y.: Cornell University Press, 1989.

Waldman, Harry. *The Dictionary of SDI.* Wilmington, Del.: Scholarly Resources, Inc., 1988.

Waller, Douglas C. *Congress and the Nuclear Freeze.* Amherst, Mass.: University of Massachusetts Press, 1987.

Watts, Barry D. "Effects and Effectiveness." In *Gulf War Air Power Survey. Vol. 2, Operations and Effects and Effectiveness,* edited by Eliot A. Cohen, 363. Washington, D.C.: GPO, 1993.

Webster's Sports Dictionary. Springfield, Mass.: Merriam-Webster, 1976.

Weiner, Tim. *Blank Check: The Pentagon's Black Budget.* New York: Warner Books, 1990.

Werckmeister, O. K. *Citadel Culture.* Chicago: University of Chicago Press, 1991.

Whaley, Barton. *Codeword Barbarossa.* Cambridge: Massachusetts Institute of Technology Press, 1973.

Whitby, Michael. "Two Shadows: Images of Spartans and Helots." In *The Shadow of Sparta,* edited by Anton Powell and Stephen Hodkinson, 87–126. London: Routledge, 1994.

Wicker, Tom. "An Unknown Casualty." In *The Media and the Gulf War: The Press and Democracy in Wartime,* edited by Hedrick Smith, 194–96. Washington, D.C.: Seven Locks Press, 1992.

Winik, Jay. *On the Brink: The Dramatic, Behind-the-Scenes Saga of the Reagan Era and the Men and Women Who Won the Cold War.* New York: Simon and Schuster, 1996.

Wohlforth, William Curti. *The Elusive Balance: Power and Perceptions during the Cold War.* Ithaca, N.Y.: Cornell University Press, 1993.

Woodward, Bob. *The Commanders.* New York: Simon and Schuster, 1991.

Woolley, Benjamin. *Virtual Worlds.* Cambridge, Mass.: Oxford University Press, 1992.

Yanarella, Ernest J. *The Missile Defense Controversy: Strategy, Technology, and Politics 1955–1972.* Lexington: University of Kentucky Press, 1977.

Yant, Martin. *Desert Mirage: The True Story of the Gulf War.* Buffalo, N.Y.: Prometheus Books, 1991.

York, Herbert F. *Race to Oblivion: A Participant's View of the Arms Race.* New York: Oxford University Press, 1970.

Journals

Adams, Benson. "An Early SDI that Saved Britain." *Naval War College Review* 38 (1985): 50–58.

Anderson, Christopher. "Classification Catch-22." *Nature* 256 (1992): 274.

Arkin, William M. "Baghdad: The Urban Sanctuary in Desert Storm?" *Air Power Journal* 11 (1997): 4–20.

Bethe, Hans A., Jeffrey Boutwell, and Richard L. Garwin. "BMD Technologies and Concepts in the 1980s." *Daedalus* 114 (1985): 53–71.

Bethe, Hans A., and Richard L. Garwin. "New BMD Technologies." *Daedalus* 114 (1985): 331–68.

Bok, Sissela. "The Ethics of Giving Placebos." *Scientific American* 231 (1974): 17–23.

Boutwell, Jeffrey, and F. A. Long. "The SDI and U.S. Security." *Daedalus* 114 (1985): 315–29.

Brown, Richard Harvey. "Modern Science: Institutionalization of Knowledge and Rationalization of Power." *Sociological Quarterly* 34 (1993): 153–68.

Brzezinski, Zbigniew. "The Cold War and Its Aftermath." *Foreign Affairs* 71 (1992): 32–46.

Campbell, John Angus. "The Invisible Rhetorician: Charles Darwin's Third Party Strategy." *Rhetorica* 7 (1989): 55–85.

Campbell, P. N. "The Personae of Scientific Discourse." *Quarterly Journal of Speech* 61 (1975): 491–505.

Carey, William D. "The Secrecy Syndrome." *Bulletin of the Atomic Scientists* 38 (1982): 9–10.

Chalk, Rosemary. "Security and Scientific Communication." *Bulletin of the Atomic Scientists* 39 (1983): 19–23.

———. "Overview: AAAS Project on Openness and Secrecy in Science and Technology." *Science, Technology and Human Values* 10 (1985): 28–35.

Chayes, Abram, Antonia Handler Chayes, and Eliot Spitzer. "Space Weapons: The Legal Context." *Daedalus* 114 (1985): 193–218.

Chubin, Daryl E. "Open and Closed Science: Tradeoffs in a Democracy." *Science, Technology and Human Values* 10 (1985): 73–81.

Cirincione, Joseph. "Why the Right Lost the Missile Defense Debate." *Foreign Policy* 106 (1997): 39–55.

Condit, Celeste. "How Bad Science Stays That Way: Brain Sex, Demarcation, and the Status of Truth in the Rhetoric of Science." *Rhetoric Society Quarterly* 26 (1996): 83–109.

"Conferences on Therapy: The Use of Placebos in Therapy." *New York State Journal of Medicine* 46 (1976): 1718–27.

Conyers, John, Jr. "The Patriot Myth: Caveat Emptor." *Arms Control Today* 22 (1992): 3–10.

———. Response to letter of Frank Horton. *Arms Control Today* 23 (1993): 27–29.

Deudney, Daniel C., and G. John Ikenberry. "Who Won the Cold War?" *Foreign Policy* 87 (1992): 123–38.

Dick, James. "The Strategic Arms Race, 1957–1961." *Journal of Politics* 34 (1972): 1059–64.

Dickins, Guy. "The Growth of Spartan Policy." *Journal of Hellenic Studies* 32 (1912): 1–42.

Douglass, Joseph D., Jr. "Soviet Disinformation." *Strategic Review* 9 (1981): 16–26.

Doxtader, Erik. "Learning Public Deliberation through the Critique of Institutional Argument." *Argumentation and Advocacy* 31 (1995): 185–204.

Drell, Sidney D. Letter to the Editor. *Science* 232 (6 June 1986): 1183.

Farrell, Thomas B. "Knowledge, Consensus, and Rhetorical Theory." *Quarterly Journal of Speech* 62 (1976): 1–14.

———. "On the Disappearance of the Rhetorical Aura." *Western Journal of Communication* 57 (1993): 147–58.

Farrell, Thomas B., and G. Thomas Goodnight. "Accidental Rhetoric: The Root Metaphors of Three Mile Island." *Communication Monographs* 48 (1981): 270–300.

Faure, David, and Matthew McKinnon. "The New Prometheus: Will Scientific Inquiry Be Bound by the Chains of Governmental Regulation?" *Duquesne Law Review* 19 (1981): 651–730.

Feuerwerger, Marvin. "Defense Against Missiles: Patriot Lessons." *Orbis* 36 (1992): 581–88.

Flax, Alexander. "Ballistic Missile Defense: Concepts and History" *Daedalus* 114 (1985): 33–71.

Fuller, Steve. "Who Hid the Body? Rouse, Roth, and Woolgar on Social Epistemology." *Inquiry* 34 (1991): 391–400.

Gaonkar, Dilip. "The Idea of Rhetoric in the Rhetoric of Science." *Southern Communication Journal* 58 (1993): 258–95.

Garrety, Karin. "Social Worlds, Actor-Networks and Controversy: The Case of Cholesterol, Dietary Fat and Heart Disease." *Social Studies of Science* 27 (1997): 727–73.

Garwin, Richard L. "The Rumsfeld Report—What We Did." *Bulletin of the Atomic Scientists* 54 (1998): 40–45.

Glanz, James. "Missile Defense Rides Again." *Science* 284 (16 April 1999): 413–16.

Goodnight, G. Thomas. "Ronald Reagan's Reformulation of the Rhetoric of War: Analysis of the 'Zero Option,' 'Evil Empire,' and 'Star Wars' Addresses." *Quarterly Journal of Speech* 72 (1986): 390–414.

Graubard, Stephen R. Preface to special issue, "'Weapons in Space,' vol. 1." *Daedalus* 114 (1985): v–vii.

Greb, G. A. Review of *Teller's War. Science* 256 (15 May 1992): 1043.

Gronlund, Lisbeth, George N. Lewis, Theodore Postol, and David Wright. "Highly Capable Theater Missile Defenses and the ABM Treaty." *Arms Control Today* 24 (1994): 3–8.

Gronlund, Lisbeth, and David Wright. "Missile Defense: The Sequel." *Technology Review* 100 (1997): 29–36.

Gross, Alan G. "Rhetoric of Science without Constraints." *Rhetorica* 9 (1991): 283–99.

Grossman, Karl. "Master of Space." *The Progressive* 64 (January 2000): 27–30.

Hafner, Donald L. "Assessing the President's Vision: The Fletcher, Hoffman and Miller Panels." *Daedalus* 114 (1985): 91–107.

Handel, Michael I. "Intelligence and the Problem of Strategic Surprise." *Journal of Strategic Studies* 7 (1984): 229–81.

Hartung, William D. "Reagan Redux." *World Policy Journal* 15 (1998): 17–25.

Hicks, Donald A. Statement. News and Comment. *Science* 231 (25 April 1986): 444.

———. Letter to the editor. *Science* 232 (6 June 1986): 1183.

Hollihan, Thomas A. "Evidencing Moral Claims: The Activist Rhetorical Critic's First Task." *Western Journal of Speech Communication* 58 (1994): 229–34.

Holloway, David. "The Strategic Defense Initiative and the Soviet Union." *Daedalus* 114 (1985): 257–78.

Holtzman, Franklyn D. "Politics and Guesswork: CIA and DIA Estimates of Soviet Military Spending." *International Security* 14 (1989): 101–31.

Horgan, John. "Lying by the Book." *Scientific American* 267 (October 1992): 20.

Howe, Henry, and John Lyne. "Gene Talk in Sociobiology." *Social Epistemology* 6 (1992): 109–64.

Isaacs, John. "Rumbles from Rumsfeld." *Bulletin of the Atomic Scientists* 54 (1998): 16–18.

———. "Missile Defense: It's Back." *Bulletin of the Atomic Scientists* 55 (1999): 23.

Jasanoff, Sheila. "Peer Review in the Regulatory Process." *Science, Technology, and Human Values* 10 (1985): 20–32.

———. "Contested Boundaries in Policy-Relevant Science." *Social Studies of Science* 17 (1987): 195–230.

———. "Beyond Epistemology: Relativism and Engagement in the Politics of Science." *Social Studies of Science* 26 (1996): 393–418.

Jeffery, L. H. "The Pact of the First Settlers at Cyrene." *Historia* 10 (1961): 139–47.

Kirkpatrick, Jeane J. "Beyond the Cold War." *Foreign Affairs* 69 (1989–90): 1–16.

LaFollette, Michael. "Mass Media News Coverage of Scientific and Technological Controversy." *Science, Technology, and Human Values* 6 (1981): 25.

Lasagna, Louis, Frederick Mosteller, John von Felsinger, and Henry K. Beecher. "A Study of the Placebo Response." *American Journal of Medicine* 16 (1954): 770–79.

Leitenberg, Milton. "The Numbers Game or 'Who's on First?'" *Bulletin of the Atomic Scientists* 38 (1982): 28.

Lewis, George N., Theodore A. Postol, and John Pike. "Why National Missile Defense Won't Work." *Scientific American* 281 (1999): 36–41.

Lopez, George. "The Gulf War: Not So Clean." *Bulletin of the Atomic Scientists* 47 (1991): 30–36.

Lown, Bernard. "Clearing the Debris." *Technology Review* (August/September 1995). Internet. Technology Review Website. Online at http://www.techreview.com/articles/aug95/AtomicLown.html.

Lyne, John. "Quantum Mechanics, Consistency, and the Art of Rhetoric: Response to H. Krips." *Cultural Studies* 10 (1996): 115–32.

Lyne, John, and Henry F. Howe. "Punctuated Equilibria." *Quarterly Journal of Speech* 72 (1986): 132–47.

Mahan, Grant. "Hawk Publisher Bombs Dove Editor." *Washington Journalism Review* 13 (1991): 14–15.

Markham, Jerry W. "The Federal Advisory Committee Act." *University of Pittsburgh Law Review* 35 (1974): 557–608.

Marshall, Eliot. "Fatal Error: How Patriot Overlooked a Scud." *Science* 255 (13 March 1992): 1347.

Martin, Brian. "The Critique of Science Becomes Academic." *Science, Technology, and Human Values* 18 (1993): 247–59.

———. "Peer Review and the Origin of AIDS: A Case Study in Rejected Ideas." *BioScience* 43 (1993): 627–42.

———. "Sticking a Needle into Science: The Case of Polio Vaccines and the Origin of AIDS." *Social Studies of Science* 26 (1996): 245–76.

McEvoy, John G. "Positivism, Whiggism, and the Chemical Revolution: A Study in the Historiography of Chemistry." *History of Science* 35 (1997): 1–33.

McLauchlan, Gregory, and Gregory Hooks. "Last of the Dinosaurs? Big Weapons, Big Science, and the American State from Hiroshima to the End of the Cold War." *Sociological Quarterly* 36 (1995): 749–76.

Melia, Trevor. "Review of Peter Dear's *The Literary Structure of Scientific Argument: Historical Studies;* Alan G. Gross's *The Rhetoric of Science;* Greg Myers's *Writing Biology: Texts in the Social Construction of Scientific Knowledge,* and Lawrence J. Prelli's *A Rhetoric of Science: Inventing Scientific Discourse.*" *ISIS* 83 (1992): 100–6.

Mendelsohn, Jack. "A Tenth Inning for Star Wars." *Bulletin of the Atomic Scientists* 52 (1996): 24–31.

Mendelsohn, Jack, and John B. Rhinelander. "Shooting Down the ABM Treaty." *Arms Control Today* 24 (1994): 8–10.

Mendelsohn, John. Statement. "An ACA Panel: The Future of Arms Control." *Arms Control Today* 25 (1995): 16–17.

Meyerson, Adam. "The Battle for the History Books: Who Won the Cold War?" *Policy Review* 52 (1990): 2–3.

Mitchell, Gordon R.. "Another Strategic Deception Initiative." *Bulletin of the Atomic Scientists* 53 (1997): 22–23.

———. "Pedagogical Possibilities for Argumentative Agency in Academic Debate." *Argumentation and Advocacy* 35 (1998): 41–60.

———. "Placebo Defense: The Rhetoric of Patriot Missile Accuracy in the 1991 Persian Gulf War." *Quarterly Journal of Speech* 86 (2000): 121–45.

———. "Whose Shoe Fits Best? Dubious Physics and Power Politics in the TMD Footprint Controversy." *Science, Technology, and Human Values* 25 (2000): 52–86.

Moore, Mike. "Unintended Consequences." *Bulletin of the Atomic Scientists* 56 (2000): 58–65.

Nadis, Steve. "After the Boycott." *Science for the People* 20 (1988): 21–26.

Nazarkin, Yuri K., and Rodney W. Jones. "Moscow's START II Ratification: Problems and Prospects." *Arms Control Today* 25 (1995): 8–14.

Nelkin, Dorothy. "The Political Impact of Technical Expertise." *Social Studies of Science* 5 (1975): 35–54.

Nolan, Jeanne E. "Patriot: Savior of SDI?" *Brookings Review* 9 (1991): 50.

Olson, Kathyrn M., and G. Thomas Goodnight, "Entanglements of Consumption, Cruelty, Privacy, and Fashion: The Social Controversy over Fur." *Quarterly Journal of Speech* 80 (1994): 249–76.

Pike, John. "Strategic 'Deception' Initiative." *Arms Control Today* 23 (1993): 3–8.

———. "Theater Missile Defense Programs: Status and Prospects." *Arms Control Today* 24 (1994): 11–14.

Pinch, Trevor. "Cold Fusion and the Sociology of Scientific Knowledge." *Technical Communication Quarterly* 3 (1994): 85–102.

Points, Kari. "Reporting Conventions Mask Sexual Politics." *Media and Values* 56 (1991): 19.

Postol, Theodore A. "Lessons of the Gulf War Experience with Patriot." *International Security* 16 (1991/1992): 119–71.

———. "Reply to letter of Robert Stein." *International Security* 17 (1992): 225–40.

"The President's Strategic Defense Initiative." *Department of State Bulletin* 85 (1985): 65–72.

Rathjens, George, and Jack Ruina. "BMD and Strategic Instability." *Daedalus* 114 (1985): 239–55.

Relyea, Harold C. "Shrouding the Endless Frontier—Scientific Communication and National Security; Considerations for a Policy Balance Sheet." *Government Information Quarterly* 1 (1984): 1–14.

"The Return of Reagan." *Progressive* 63 (March 1999): 8–11.

Robb, Charles. "Star Wars II." *Washington Quarterly* 22 (1999): 81–87.

Robins, Kevin, and Les Levidow. "Video Games and Virtual War: High Tech Imaging Transforms Social Perceptions." *New Statesman and Society* 8 (1995): 28–29.

Romm, Joseph. "Pseudo–Science and SDI." *Arms Control Today* 19 (1989): 15–21.

Ross, Andrew. Introduction to special issue, *Social Text* 46 (1996): 1–14.

Rouse, Joseph. "Policing Knowledge: Disembodied Policy for Embodied Knowledge." *Inquiry* 34 (1991): 353–64.

———. "What Are Cultural Studies of Scientific Knowledge?" *Configurations* 1 (1992): 1–22.

Rushefsky, Mark E. "The Misuse of Science in Governmental Decisionmaking." *Science, Technology, and Human Values* 9 (1984): 47–59.

Rushing, Janice Hocker. "Ronald Reagan's 'Star Wars' Address: Mythic Containment of Technical Reasoning." *Quarterly Journal of Speech* 72 (1986): 415–33.

Schuchardt, Elliott. "Walking a Thin Line: Distinguishing between Research and Medical Practice during Operation Desert Storm." *Columbia Journal of Law and Social Problems* 26 (1992): 77–115.

Shackley, Simon, and Brian Wynne. "Representing Uncertainty in Global Climate Change Science and Policy: Boundary-Ordering Devices and Authority." *Science, Technology, and Human Values* 21 (1996): 275–302.

Shapere, Dudley. "Talking and Thinking about Nature: Roots, Evolution, and Future Prospects." *Dialectica* 46 (1992): 281–96.

Shapiro, Arthur K. "A Contribution to the History of the Placebo Effect" *Behavioral Science* 5 (1960): 109.

———. "A Historic and Heuristic Definition of the Placebo." *Psychiatry* 27 (1964): 52–58.

Spain, Lauren. "The Dream of Missile Defense: So Where's the Peace Dividend?" *Bulletin of the Atomic Scientists* 51 (1995): 49–53.

Stares, Paul. "U.S. and Soviet Military Space Programs: A Comparative Assessment." *Daedalus* 114 (1985): 127–45.

Stein, Robert. Letter to the editor. *International Security* 17 (1992): 199–225.

Sweetman, Bill, and Nick Cook. "Getting to Grips with Missile Defense." *Interavia* (Geneva) 54 (1999): 32–35.

Taylor, Charles A. "Defining the Scientific Community: A Rhetorical Perspective on Demarcation." *Communication Monographs* 58 (1991): 402–20.

Teller, Edward. "The Feasibility of Arms Control." *Daedalus* 89 (1960): 749–99.

———. "The Feasibility of Arms Control and the Principle of Openness." *Proceedings of the American Academy of Arts and Sciences* 89 (1960): 781–99.

Tiffen, Rodney. "Marching to Whose Drum? Media Battles in the Gulf War." *Australian Journal of International Affairs* 46 (1992): 44–60.

Wallop, Malcolm. "Opportunities and Imperatives of Ballistic Missile Defense." *Strategic Review* 7 (1979): 13–21.

Wander, Philip. "The Rhetoric of Science." *Western Speech Communication Journal* 40 (1976): 226–35.

Weimer, Walter. "Science as a Rhetorical Transaction: Toward a Nonjustificational Conception of Rhetoric." *Philosophy and Rhetoric* 10 (1977): 1–29.

———. "Why All Knowing Is Rhetorical." *Journal of the American Forensic Association* 20 (1984): 63–71.

Wiener, Jon. "Domestic Political Incentives for the Gulf War." *New Left Review* 187 (1991): 72–78.

Wiesner, Jerome B., and Herbert F. York. "National Security and the Nuclear-Test Ban." *Scientific American* 211 (1964): 27–35.

Wohlstetter, Albert. "The Delicate Balance of Terror." *Foreign Affairs* 37 (1959): 211–34.

"Wonder Weapon." *Progressive* 58 (January 1994): 10–14.

Woodruff, Roy D. "We Need to Do Something: Review of Cardinal Choices." *Bulletin of the Atomic Scientists* 48 (1992): 44–47.

Woolgar, Steve. "The Very Idea of Social Epistemology: What Prospects for a Truly Radical 'Radically Naturalized Epistemology'?" *Inquiry* 34 (1991): 377–89.

Yonas, George. "The Strategic Defense Initiative." *Daedalus* 114 (1985): 73–90.

York, Herbert F. "Controlling the Qualitative Arms Race." *Bulletin of the Atomic Scientists* 29 (1973): 4–8.

———. "Deterrence by Means of Mass Destruction." *Bulletin of the Atomic Scientists* 30 (1974): 4–9.

———. "Nuclear Deterrence and the Military Uses of Space." *Daedalus* 114 (1985): 17–32.

Zagacki, Kenneth, and Andrew King. "Reagan, Romance and Technology: A Critique of 'Star Wars.'" *Communication Studies* 40 (1989): 1–12.

Newspapers and Magazines

Alter, Jonathan. "Swords vs. Shields." *Newsweek*, 8 May 2000, 42–44.

Associated Press Online. "Army Releases Gulf War Records." 11 May 2000. Internet. Online. Lexis/Nexis.

Auster, Bruce B. "Sobering Lessons from the Gulf." *U.S. News & World Report*, 10 May 1993, 52.

Bar-Yosef, Avino'am. "U.S. 'Not Crazy about' Peres Despite 'Show' During Visit." *Ma'ariv*, 2 May 1996, 3. Translated by Foreign Broadcast Information Service (NES-96-087).

Baudrillard, Jean. "The Reality Gulf: Why the Gulf War Will Not Take Place." *Guardian* (London), 11 January 1991, 25.

Bennett, Charles. "The Rush to Deploy SDI." *Atlantic Monthly,* April 1988, 53–61.

Biddle, Wayne. "The Untold Story of the Patriot." *Discover,* June 1991, 74–79.

"The Big Lie in the Sky." Editorial. *St. Petersburg Times,* 20 August 1993, 16A.

Blum, Deborah. "'Star Wars' Based on Lie?" *Sacramento Bee,* 25 October 1987, A1.

Bond, David F. "Army Scales Back Assessments of Patriot's Success in Gulf War." *Aviation Week and Space Technology,* 13 April 1992, 64.

Broad, William J. "The Secret behind Star Wars." *New York Times Magazine,* 11 August 1985, 32–51.

———. "30 Lawmakers Urge Delay in Laser Weapon Test." *New York Times,* 7 December 1985, A7.

———. "Anti-Missile Laser Given First Tests for Use in Space." *New York Times,* 3 January 1988, 1.

———. "Beyond the Bomb: Turmoil in the Labs." *New York Times,* 9 October 1988, 6–23.

———. "Dispute Settled at Weapons Lab." *New York Times,* 23 May 1990, A18.

———. "Rocket Run by Nuclear Power Being Developed for 'Star Wars.'" *New York Times,* 3 April 1991, A1.

———. "After Many Misses, Pentagon Still Pursues Missile Defense." *New York Times,* 24 May 1999, A1.

———. "Ex-Employee Says Contractor Faked Results of Missile Tests." *New York Times,* 7 March 2000, A1.

———. "Antimissile System's Flaw Was Covered Up, Critic Says." *New York Times,* 18 May 2000, A21.

———. "Pentagon Classifies a Letter Critical of Antimissile Plan." *New York Times,* 20 May 2000, A10.

———. "U.S. Defends Antimissile Plan." *New York Times,* 26 May 2000, A10.

Broder, John M. "Breaking Cold War Mold: Bush's Proposal to Reduce Nuclear Arsenal Injects Arms Into Race for the White House." *New York Times,* 26 May 2000, A1.

Browne, Malcolm W. "The Military vs. the Press." *New York Times Magazine,* 3 March 1991, 29.

Buchanan, Patrick. "Evidence of SDI's Potential." *Washington Times,* 30 January 1991, G2.

Buckley, William F. "Intercepting the Truth about SDI Testing." *Buffalo News,* 14 October 1993, 3+.

Buenneke, Richard, Jr. "The SDI Blackout: How Can We Debate Star Wars If the Program Is Secret?" *Washington Post,* 19 April 1987, D5.

Burger, Timothy J. "Defense Firm Attorney Files Lawsuit for 41 Members." *Roll Call,* 22 July 1996. Internet. Online. Lexis/Nexis.

Carrington, Tim. "Star Wars Plan Spurs Defense Firms to Vie for Billions in Orders." *Wall Street Journal,* 21 May 1985, 1.

Chandler, David. "No Letup in War of Words." *Boston Globe,* 4 April 1994, 25–28.

———. "Sorting Out the Claims: A Q & A." *Boston Globe,* 4 April 1994, 25+.

Charles, Dan. "Critics Home in on Patriot Missile." *New Scientist,* 2 February 1991, 39.

Christianson, Gale E. "Naysayers, Thriving in the Heat." *New York Times,* 8 July 1999, A25.

"CIA Failed to Tell White House Spy Data Was Questionable." *Congressional Quarterly Weekly Report,* 11 November 1995, 3469.

Cockburn, Alexander. "Dumb Bombs." *New Statesman and Society,* 14 June 1991, 17.

Cooper, Henry. "Trying to Get SDI Off the Ground." *Washington Times,* 14 September 1990, F3.

Cronkite, Walter. Letter to the editor. *New York Times,* 16 May 2000, A30.

Crossette, Barbara. "Russians Heat Up Dispute on ABM." *New York Times,* 26 April 2000: 1.

D'Antonio, Michael. "Scars and Secrets: The Atomic Trail." *Los Angeles Times,* 20 March 1994, 14+.

Davidson, Kay. "Nuke Lab Scientist Regains Security Status." *San Francisco Examiner,* 6 May 1994, A2.

"Defending the Indefensible: U.S. Defense Spending." *New Scientist,* 28 March 1992, 11.

"Defensive Technologies Study Sets Funding Profile Options." *Aviation Week and Space Technology,* 24 October 1983, 50.

DeWitt, Hugh. "The Selling of a Wonder Weapon." *Stanford,* March 1991, 28–33.

Diamond, John. "'Rogue States' Want Weapons for Leverage." *Pittsburgh Post–Gazette,* 21 February 2000, A1.

"Disinformation." *Aviation Week and Space Technology,* 28 July 1986, 15.

Donnelly, John. "THAAD Intercepts Were Unrealistic, Top Tester Says." *Defense Week,* 23 August 1999. Internet. Online. Lexis/Nexis.

"Do We Have to Buy Patriots?" *Tong-a Ilbo,* 5 April 1997, 3. Translated by Foreign Broadcast Information Service (TAC–97–095).

Dragsdahl, Joergen. "Use of Key Radar Facility Conditioned on Russian Approval of ABM-Treaty Changes." 7 March 2000. Internet. British American Security Council Website. Online at http://www.basicint.org.

Drew, Elizabeth. "Letter from Washington." *New Yorker,* 4 February 1991, 83.

Drolet, Robert A. "PEO Air Defense Responds to Patriot Criticisms." *Inside the Army,* 9 December 1991, 5.

Ealy, Charles. "Lucas Made American Missile System Sound Realistic: Popular Culture's Phantom Menace." *Toronto Star,* 14 May 1999, 17.

"Energy Weapons." *Aviation Week and Space Technology,* 5 September 1983, 50.

Erlich, Jeff. "Third THAAD Failure Clouds Timetable for Deployment." *Defense News,* 22–28 July 1997, 4+.

"Excerpts From Bush's Remarks on National Security and Arms Policy." *New York Times,* 24 May 2000, A19.

Farrell, John Aloysius. "U.S. Ordered a Stop to Patriot Criticism, Massachusetts Institute of Technology Professor Says." *Boston Globe,* 18 March 1992, 3+.

———. "Massachusetts Institute of Technology Physicist Cleared in Patriot Analysis." *Boston Globe,* 28 March 1992, 3+.

Feeney, Susan. "Timing Poses Questions about President's Motivation: Parallels to Movie 'Wag the Dog' Spur Talk." *Dallas Morning News,* 21 August 1998, A1.

Fenster, Bob. "Life Imitates Art: Who's Playing Wag the Dog?" *Arizona Republic,* 21 August 1998, B7.

Fetter, Steve, George N. Lewis, and Lisbeth Gronlund. "Why Were Scud Casualties So Low?" *Nature,* 28 January 1993, 293–96.

Fisk, Robert. "The Bad Taste of Dying Men." *Independent* (London), 24 February 1991, 19.

Fleeman, Michael. "Is Life Imitating Movie, Again? 'Wag the Dog' Back in Spotlight." *AP Online,* 21 August 1998. Internet. Elibrary Website. Online at http://www.elibrary.com.

Fossedal, Gregory A. "A Star Wars Caucus in the Freeze Movement?" *Wall Street Journal,* 14 February 1985, 1.

Garwin, Richard L. "Keeping Enemy Missiles at Bay." *New York Times,* 28 July 1998, A15.

Gedmin, Jeffrey and Jon Kyl. "The Case for Missile Defense." *Washington Post,* 31 December 1999, A31.

Genrich, Mark. "Suing the President; Sen. Kyl Leads Effort for Congress." *Phoenix Gazette,* 24 July 1996, B9.

Gertz, Bill. "Son of Star Wars." *Air Force Magazine,* September 1994, 52.

———. "Russia Wants New Curbs on U.S. Defenses." *Washington Times,* 15 July 1996, A1.

"Get Smarter on Smart Weapons." Editorial. *New York Times,* 11 July 1996, A22.

Geyelin, Milo. "'Class' Trial Finds Tobacco Firms Liable." *Wall Street Journal,* 8 July 1999, A3.

———. "Patriots Missed, but Critics Hit Home." *Science* 284 (16 April 1999): 413–16.

Gildea, Kerry. "SMDC Rejects Lockheed Martin Proposal for THAAD EMD Phase." *Defense Daily,* 21 April 2000, Internet, Online, Lexis/Nexis.

Glanz, James. "Military Experts Debate the Success of Warhead Missile Interceptor Test." *New Orleans Times-Picayune,* 14 January 2000, 3A.

———. "Israeli-American Laser Passes a Missile Defense Test, U.S. Says." *New York Times,* 4 May 2000, 11.

Goldsborough, James O. "Victims of a Perverted Policy: Compensate People Seriously Harmed by Nuclear Tests." *San Diego Union-Tribune,* 13 January 1994, 1+.

Gordon, Michael R. "Report Revives Criticism of General's Attack on Iraqis in 91." *New York Times,* 15 May 2000, A11.

Graham, Bradley. "Anti-missile Weapon Fails." *Pittsburgh Post-Gazette,* 30 March 1999, A15.

———. "Panel Faults Antimissile Program on Many Issues." *Washington Post,* 14 November 1999, A1.

———. "Missile Shield Still Drawing Friends, Fire; Verdict on Deployment Due in Political Climate." *Washington Post,* 17 January 2000. Internet. Online. Lexis/Nexis.

Graybeal, Sidney, and Keith Payne. "ABM Debate Holds Cold War Echoes." *Defense News,* 9 January 1995, 19–20.

Green, Stephen. "Pentagon Lowers, Meets Criteria for Missile Defense." *San Diego Union-Tribune.* 22 March 2000. Internet. Online. Lexis/Nexis.

Grossman, Ron, and Charles Leroux. "Radiation Tests: Needed or Horrid?" *Chicago Tribune,* 9 January 1994, A18.

Gugiotta, Guy, and Caryle Murphy. "Warplanes Roar Off in Darkenss in 'Desert Storm.'" *Washington Post,* 17 January 1991, A27.

"Hackers Steal SDI Information in Internet System." *Washington Times,* 24 December 1990, 3.

Hartung, William, and Rosy Nimroody. "Cutting Up the Star Wars Pie." *Nation,* 14 September 1995, 200–2.

Healey, Jon. "Congress Ponders Compensation for Radiation Test Subjects." *Congressional Quarterly,* 8 January 1994, 21–22.

Hecht, Jeff. "Blinded by the Light: The Strategic Defense Initiative." *New Scientist,* 20 March 1993, 29–34.

Heilbrunn, Jacob. "Playing Defense." *New Republic,* 18 August 1998, 16–18.

Helvarg, David. "The Greenhouse Spin." *Nation,* 16 December 1996, 21–24.

Hendrick, Bill. "Brainstorming, 1990s-Style." *Atlanta Journal and Constitution,* 6 May 1995, E1.

Hiatt, Fred, and Rick Atkinson. "In Strategic Defense, the Seeds of a New Industry are Planted." *Washington Post,* 20 October 1985, A1.

"Hill SDI Report Stamped Secret by Pentagon." *Washington Post,* 5 March 1986, A23.

Hilts, Philip J. "Dispute on Gulf War Missile Continues to Fester at M.I.T." *New York Times,* 24 February 1996, 11.

Hoagland, Jim, and Dusko Doder. "Moscow Won't 'Ape' SDI, Top Soviet General Says." *Washington Post,* 9 June 1985, A1.

Hoffman, David. "Moscow Proposes Extensive Arms Cuts; U.S., Russia Confer over Stalled Pacts." *Washington Post,* 20 August 1999, A29.

Holum, John D. "Don't Put Allies at Risk." *New York Times,* 25 October 1994, A21.

"House OK's Missile Defense Plan." CNN Interactive. Internet. CNN Website. Online at http://www.cnn.com/ALLPOLITICS/stories/1999/03/18/missile.defense/.

"How the Pentagon Fixed the Star Wars Test." *CounterPunch.* 9 January 2000. Internet. CounterPunch Website. Online at http://www.counterpunch.org/starwars.html.

Hughes, David. "Success of Patriot System Shapes Debate on Future Antimissile Weapons." *Aviation Week and Space Technology,* 22 April 1991, 90–91.

———. "Tracking Software Error Likely Reason Patriot Battery Failed to Engage Scud." *Aviation Week and Space Technology,* 10 June 1991, 26–27.

———. "Saudi Order to Keep Patriot Line Open as Performance in Israel Is Questioned." *Aviation Week and Space Technology,* 25 November 1991, 38.

"Hughes Assumes Lead in Team Competing for THAAD Contract." *Defense Daily,* 4 November 1994, 182.

"Hughes Establishes Organization for Air Defense/Anti-Missile Efforts." *Defense Daily,* 5 July 1991, 29.

"Hughes to Focus Air Defense on THAAD." *Defense Daily,* 24 January 1992, 117.

Hyo-son, Kang, and Pak Chong-hun. "Kantor Asks ROK to Purchase Theater Missile Defense System." *Choson Ilbo,* 2 July 1996, 2. Translated by Foreign Broadcast Information Service (EAS-96-128).

Isikoff, Michael. "Carving Out an SDI Niche: Local Defense Companies Scramble to Hitch Their Wagons to Star Wars." *Washington Post,* 13 October 1986, F1.

Jordan, Mary. "Cohen Urges Seoul to Buy U.S. Missiles." *Washington Post,* 8 April 1997, A1.

Keeny, Spurgeon M. "The Arms Race is On." *New York Times,* 12 September 1995, A23.

———. "Tread Lightly on the ABM Treaty." *Space News,* 16–22 October 1995, 19.

Keeter, Hunter. "Laser Master Plan Okays HEL Systems for Offensive and Defensive Applications." *Defense Daily,* 3 April 2000. Internet. Online. Lexis/Nexis.

Kessler, Felix. "Who Wrote It? A Fad in Political Comment Is Using Pseudonyms." *Wall Street Journal,* 13 November 1967, 1.

Klare, Michael T. "The New 'Rogue State' Doctrine." *The Nation,* 8 May 1995, 625–28.

Kiernan, Vincent. "Excess Hype of X-Ray Laser Causes Funding Gap for Weapons Research." *Space News,* 9 October 1989, 8.

———. "Beyond the Patriot." *Discover,* June 1991, 80–82.

———. "DOD Cancels Plans for Nuclear Rocket." *Space News,* 17–23 May 1993, 6.

Krepon, Michael. "The War between Sales and Science." *New York Times,* 23 February 1992, 2.

Landay, Jonathan S. "Pentagon Hit for Hiding Spending on 'Star Wars.'" *Christian Science Monitor,* 5 September 1995, 3.

Lardner, George. "Scientist Says Army Seeks to Fire Him for Criticizing SDI." *Washington Post,* 10 January 1992, A17.

———. "SDI Scientist Wins Reprieve from Firing." *Washington Post,* 7 March 1992, A8.

"The Last Battle of the Gulf War." *New York Times,* 18 May 2000, A30.

Lebow, Richard Ned, and Janice Gross Stein. "Reagan and the Russians." *Atlantic Monthly,* February 1994, 37.

Leo, John. "'Report' on Peace Gets Mixed Views: Some See Book as Hoax, Others Take It Seriously." *New York Times,* 5 November 1967, 1.

Lewin, Leonard C. "The Guest Word." *New York Times,* 19 March 1972, 47.

Lewis, Flora. "A 'Star Wars' Cover-Up." *New York Times,* 3 December 1985, A16.

"The Lies that Propelled Star Wars." Editorial. *Cleveland Plain Dealer,* 19 August 1993, B6.

London Press Association. "U.K. Anti-Nuclear Protesters Target US Military Base." 4 March 2000. Foreign Broadcast Information Service (FBIS-WEU-2000-0304).

Lovece, Joseph. "Pentagon Audit Blasts SDI Nuclear Rocket Classification." *Defense Week,* 11 January 1993, 1.

Lynch, David J. "Pentagon Drawing the Shades on Star Wars, Senators Fret." *Defense Week,* 24 February 1986, 1.

———. "DOE May Discourage Journalists from Talking to Most Federal Scientists: Inverviews on SDI Could Be Sharply Curtailed." *Defense Week,* 22 September 1986, 1.

MacFarland, Margo. "Congressional Investigators Charge Army Data Doesn't Back Patriot Success." *Inside the Army,* 9 March 1992, 1.

Man-won, Chi. "Missile Sovereignty." *Choson Ilbo,* 8 February 1996, 5. Translated by Foreign Broadcast Information Service (EAS-96-028).

Marshall, Eliot. "Fatal Error: How Patriot Overlooked a Scud." *Science* 255 (13 March 1992): 1347.

"Mass Legislators Call on Israel to Defend Patriot Performance." *Boston Globe,* 23 March 1994, 12+.

Maugh, Thomas H. "'Star Wars' Chemical Laser Is Unveiled." *Los Angeles Times,* 23 June 1991, A3.

McFarlane, Robert. "Consider What Star Wars Accomplished." *New York Times,* 24 August 1993, A15.

McNamara, Robert S. "Text of Address to Press Club of San Francisco." *New York Times,* 19 September 1967, 18.

Meier, Barry. "Tobacco Industry Loses First Phase of Broad Lawsuit." *New York Times,* 8 July 1999, A1.

Mendelsohn, Jack. "U.S. Takes Nuclear Gamble: Broader Policy for Weapon Use Could Backfire." *Defense News,* 4–10 April 1994, 24.

Miller, John J. "The Rocket Boys: Missile Defense on Its Way." *National Review,* 25 October 1999, 52.

Mintz, John. "Reagan Is Urged to Increase Research on Exotic Defenses against Missiles." *New York Times,* 5 November 1983, A32.

———. "Sasser Cites 'Abuses' by Defense Firms." *Washington Post,* 4 March 1994, B2.

———. "Arms Race with an Attitude?" *Washington Post,* 18 October 1994, C1–C4.

Mitchell, Gordon R. "The National Missile Defense Fallacy." *Pittsburgh Post-Gazette,* 29 April 2000, A17.

———. "About Those 'Normal Accidents.'" *Pittsburgh Post-Gazette,* 16 July 2000, E4.

Mohr, Charles. "Scientist Quits Antimissile Panel, Saying Task Is Impossible." *New York Times,* 12 July 1985, A6.

Morain, Dan. "Energy Secretary Warns Scientists Not to Disagree in Public." *Los Angeles Times,* 23 July 1988, 1.

Morgan, Dan, and George Lardner Jr. "Scud Damage Suggests Patriot Needs Refinement: U.S. Missiles Sometimes Fail to Destroy Iraqi Warheads in Midair Interceptions." *Washington Post,* 21 February 1991, A27.

Morrison, David C. "Patriot Games." *National Journal,* 25 April 1992, 1028.

Myers, Steven Lee. "Bush's Missile Defenses Could Limit Warhead Cuts, Experts Warn." *New York Times,* 24 May 2000, A19.

Neikirk, William. "Senate Overwhelmingly OK's Bill for Anti-Missile Defense." *Chicago Tribune* (internet edition). 18 March 1999. Chicago Tribune Website. Online at http://chicagotribune.com/news/nationworld/article/0,1051,SAV-9903180135,00.html.

"News Leaks." *Aerospace Daily,* 9 June 1986, 386.

"New Weapons Take Aim at Anti-Missile Treaty." *San Francisco Chronicle,* 25 June 1994, A24.

North, David. "U.S. Using Disinformation Policy to Impede Technical Data Flow." *Aviation Week and Space Technology,* 17 March 1986, 16–17.

"Nuclear Swords and Shields." *New York Times,* 26 May 2000, A22.

Olmos, David. "Cold War's Thaw Puts a Chill on SDI Contractors." *Los Angeles Times,* 10 December 1989, D1.

O'Neill, Bill. "Fear and Laughter in the Kremlin." *New Scientist,* 20 March 1993, 34–37.

Orman, Stanley. "A Space Community Held Hostage." *Space News,* 30 October–5 November 1995, 14–18.

Osborne, David. "Reagan's Great Lie in the Sky." *The Independent* (London), 29 August 1993, 13+.

Pedatzur, Reuven. "Failed at Time of Test: The U.S. Administration and the Patriot's Manufacturer Presented a Deceptive Picture of the Performance of Patriot during the War." *Ha'aretz,* 24 October 1991. Internet. Federation of American Scientists Website. Online at http://www.fas.org/spp/starwars/docops/rp911024.htm.

Peligal, Rona. "Weinberger Stands Up for Hicks' Right to 'Freedom of Expression.'" *Defense News,* 9 June 1986, 4.

"Pentagon Admits Problems with Missile Defense Test." *St. Louis Post-Dispatch,* 15 January 2000, 21.

"Pentagon Blacks out SDI Budget." *Military Space,* 7 July 1986, 1.

"Pentagon Homes in on Patriot Critic." *New Scientist,* 28 March 1992, 13–15.

Peterson, Scott. "High Hopes for Foolproof Missile Defense Fizzle." *Christian Science Monitor* (online edition). 30 July 1997. Internet. Christian Science Monitor Website. Online at http://www.csm.org.

Postol, Theodore A. Letter to the editor. *Washington Post,* 17 November 1994, A22.

PR Newswire. "Lockheed Martin's THAAD Missile Records First Target Intercept over White Sands Missile Range." 10 June 1999. Internet. Council for a Livable World Website. Online at http://www.clw.org/ef/bmdnews/thaadhit.html.

"Professor Revives Massachusetts Institute of Technology Missile Flap." *Boston Globe,* 26 February 1996, 15.

Quinn-Judge, Paul. "Reagan-Era Officials Deny Faking SDI Test." *Boston Globe,* 19 August 1993, 3.

Raum, Tom. "GOP, Democrats Agree on Defense." Associated Press News Feed. 29 June 1999. Internet. Council for a Livable World Website. Online at http://www.clw.org/ef/bmdnews/.

"Raytheon Awarded Deal to Sell Taiwan Patriot Antimissiles." *Wall Street Journal,* 1 July 1994, B4.

Reeves, Phil. "Star Wars 'Fake' Fooled the World." *Independent* (London), 19 August 1993, 8.

"Representative Conyers Wants Cheney to Probe Army Claims Regarding Patriot's Success." *Inside the Pentagon,* 12 March 1992, 14.

Richter, Paul. "Defense Secretary Warns Seoul Not to Buy Russian Arms." *Los Angeles Times,* 6 April 1997, A1.

Ricks, Delthia. "Scientists Say Teller Misled Government on Star Wars." *UPI Newswire,* 22 October 1987. Internet. Online. Lexis/Nexis.

Robinson, Clarence A. "BMD Homing Interceptor Destroys Reentry Vehicle." *Aviation Week and Space Technology,* 18 June 1984, 19–20.

Rogers, Keith. "DOE Wants Rules Tightened on 'Star Wars' Talk." *Valley Times,* 11 February 1986, 1.

———. "Lab 'Gag Order' by Energy Dept. Alleged." *Valley Times,* 7 February 1987, 1.

———. "Livermore Lab Director Disputes Teller Allegation." *Valley Times,* 24 October 1987, 3A.

"ROK Reaffirms 'No Plans' to Join U.S.-led TMD Plan." *Korea Herald,* 4 May 1999. Translated by Foreign Broadcast Information Service (EAS-1999-0503).

"ROK Selection of New SAM System Political 'Pandora's Box.'" *Korea Times,* 5 June 1998. Translated by Foreign Broadcast Information Service (EAS-98-156).

Rosenberg, Tina. "The Authorized Version." *Atlantic,* February 1986, 26.

Ruddy, Christoper. "'Father of H-Bomb' Worried About Status of U.S. Defenses." *Pittsburgh Tribune-Review,* 12 January 1997, A1.

Ryan, Randoloph. "Patriot Propaganda." *Boston Globe,* 21 March 1992, 9.

Sanger, David E. "Pentagon and Critics Dispute Roles of Space Arms Designers." *New York Times,* 5 November 1985, C1.

———. "North Korea Warns U.S. on Patriot Missiles." *New York Times,* 30 January 1994, 4.

Scheer, Robert. "Scientists Dispute Test of X-Ray Laser Weapon." *Los Angeles Times,* 12 November 1985, 1.

———. "Missile Defense is Offensive." *Pittsburgh Post-Gazette,* 25 May 2000, A31.

Schmitt, Eric. "Aspin Disputes Report of 'Star Wars' Rigging." *New York Times,* 10 September 1993, B8.

Schneider, Keith. "1950 Memo Shows Worries Over Human Radiation Tests." *New York Times,* 12 December 1993, A1.

———. "Military Spread Nuclear Fallout in Secret Tests." *New York Times,* 16 December 1993, A1.

———. "Nuclear Scientists Irradiated People in Secret Research." *New York Times,* 17 December 1993, A1.

"SDI Budget Cutbacks to Delay Near Term Weapons Deployments." *Aviation Week and Space Technology,* 22 May 1989, 22–24.

Selvin, Paul, and Charles Schwartz. "Publish and Perish." *Science for the People,* January/February 1988, 6–10.

"Senate Backs Missile Defense System." CNN Interactive. 17 March 1999. Internet. CNN Website. Online at http://www.cnn.com/ALLPOLITICS/stories/1999/03/17/missile.defense/.

Shenon, Philip. "Allied Fliers Jubilantly Tell of Early Control of the Skies as Iraq Planes Fled." *New York Times,* 18 January 1991, A10.

Sirak, Michael C. "House Waives Initial Testing Requirements on NMD Production." *Inside Missile Defense,* 16 June 1999, 1.

———. "BMDO Begins Orderly Phaseout of Boeing Backup NMD Kill Vehicle." *Inside Defense,* 19 May 2000. Internet. InsideDefense.com news service. Online at http://www.InsideDefense.com.

Sisk, Richard. "Bush Vows to Beef Up Star Wars. " *New York Daily News.* 24 May 2000. Internet. Federation of American Scientists website. Online at http://www.fas.org/spp/starwars/program/news00/00523-nmd-gwb.htm.

Smith, Jeffrey R. "3 'Star Wars' Tests Rigged, Aspin Says." *Washington Post,* 10 September 1993, A19.

———. "Officials Say U.S. Wants to Change ABM Treaty to Buttress Missile Defense." *Washington Post,* 4 December 1993, 22.

Sommer, Mark. "Reagan's Strategic Deception Initiative." *Christian Science Monitor,* 3 September 1993, 19.

Song-ho, Kim. "The Arrogant U.S. Missile Sales." *Hangyore 21,* 24 April 1997, 42–44. Translated by Foreign Broadcast Information Service (EAS-97-083).

Stanford, Phil. "The Automated Battlefield." *New York Times Magazine,* 23 February 1975, 12.

"Star Wars." *Newsweek,* 4 April 1983, cover.

"'Star Wars' Goal Cut, Quayle Says." *Los Angeles Times,* 7 September 1989, 1.

"'Star Wars' Plan Facing a Delay, Its Director Says." *New York Times,* 22 May 1989, A1.

Subrahmanyam, K. "The 'Star Wars' Delusion." *World Press Review,* June 1983, 21–24.

Tanner, Adam. "General Raps U.S. Anti-missile Effort; Says Changes Sought to ABM Pact Threaten Arms Talks." *Washington Times,* 21 August 1999, A5.

Teller, Edward. "SDI: The Last, Best Hope." *Washington Times,* 28 October 1985, 75.

"Teller Said to Urge Development of X-Ray Laser." *Aerospace Daily,* 1 December 1982, 158–59.

Thompson, Mark. "Star Wars: The Sequel." CNN Interactive 15 February 1999. Internet. CNN Website. Online at http:www.cnn.com/ALLPOLITICS/time/1999/02/15/star.wars.html.

"U.S. Pressure on Missile Purchase." *Chungang Ilbo,* 6 April 1997, 6. Translated by Foreign Broadcast Information Service (TAC-97-096).

van Voorst, Bruce. "The Ploy That Fell to Earth: Star Wars Suffers Another Blow with Charges That an Antimissile Test Was Faked." *Time,* 30 August 1993, 26+.

Wagner, John. "Drafting Rules of Deception: How and When." *Washington Post,* 6 August 1992, A23.

Walton, Paul. "'Top Secret' SDI Technologies Ban?" *Jane's Defence Weekly,* 15 February 1986, 246.

Wark, McKenzie. "Gulfest." *Australian Left Review,* February 1991, 7.

"Washington Whispers." *U.S. News and World Report,* 16 June 1986, 12.

Weinberger, Caspar. "SDI, Ethics and Soviet Hypocrisy." *Insight,* 13 January 1986, 1.

Weiner, Tim. "Army Squelches Article—After It's Published." *Philadelphia Inquirer,* 19 March 1992, 1+.

———. "Army Cracks Down on Patriot's Critic." *Chicago Tribune,* 20 March 1992, 16.

———. "Lies and Rigged 'Star Wars' Test Fooled the Kremlin, and Congress." *New York Times,* 18 August 1993, A1+.

———. "General Details Altered 'Star Wars' Test." *New York Times,* 27 August 1993, A19.

———. "C.I.A.'s Openness Derided as 'Snow Job.'" *New York Times,* 20 May 1997, A16.

Weise, Elizabeth. "Stranger than Fiction." *USA Today,* 22 July 1998, 5D.

Weisskopf, Michael. "Conflict in 'Star Wars' Work Possible." *Washington Post,* 30 April 1985, A5.

Weyrich, Paul M. "A True Patriot." *Washington Times,* 8 January 1996, 34.

"What We Have Learned from the V-2 Firings." *Aviation Week,* 26 November 1951, 23.

Wolfe, Frank. "Army Running Short of THEL Funds." *Defense Daily,* 3 April 2000. Internet. Online. Lexis/Nexis.

Yarynich, Valery. "The Doomsday Machine's Safety Catch." *New York Times,* 1 February 1994, A15.

Yong-hwan O, and Choe Sang-yon. "Officials' 'Displeasure' with U.S. Pressure on Weapons Deal." *Chungang Ilbo,* 13 June 1997, 3. Translated by Foreign Broadcast Information Service (TAC-97-164).

Zierdt, John G. "Nike-Zeus: Our Developing Missile Killer." *Army Information Digest,* 15 December 1960, 5–6.

Zimmerman, Peter D. "A Review of the Postol and Lewis Evaluation of the White Sands Missile Range Evaluation of the Suitability of TV Video Tapes to Evaluate Patriot Performance during the Gulf War." *Inside the Army,* 16 November 1992, 7–9.

Zitner, Aaron. "Rival Haunts Patriot Missile." *Boston Globe,* 23 March 1994, 1+.

GOVERNMENT DOCUMENTS

Advisory Committee on Human Radiation Experiments. *Final Report.* Washington, D.C.: GPO, 1995.

Aftergood, Steven. Statement. *Government Secrecy After the Cold War.* 102d Cong., 2d sess. House Hearing. Committee on Government Operations. 18 March 1992. Washington, D.C.: GPO, 1992, 174–88.

Atkins, George. Statement. *Performance of the Patriot Missile in the Gulf War.* 102d Cong., 2d sess. House Hearing. Committee on Government Operations. 7 April 1992. Washington, D.C.: GPO, 1992, 214.

Brown, George. Statement. ASAT Testing Certification. *Congressional Record,* 4 September 1985, H1610.

Carnesdale, Albert. Statement. *The Impact of the Persian Gulf War and the Decline of the Soviet Union on How the United States Does Its Defense Business.* 102d Cong., 1st sess. House Hearing. Committee on Armed Services. 16 April 1991. Washington, D.C.: GPO 1991, 476–87.

Carter, James, Stephen R. Stanvick, and Robert Stein. Statement. *Performance of the Patriot Missile in the Gulf War.* 102d Cong., 2d sess. House Hearing. Committee on Government Operations. 7 April 1992. Washington, D.C.: GPO, 1992, 243–46.

Cheney, Richard. Statement. *Department of Defense Authorization for Appropriations for Fiscal Years 1992 and 1993.* 102d Cong., 1st sess. Senate

Hearing. Committee on Armed Services. 21 February 1991. Washington, D.C.: GPO, 1991, 15.

Cohen, Eliot A., ed. *Gulf War Air Power Survey.* Vol. 2, *Operations and Effects and Effectiveness.* Washington, D.C.: GPO, 1993.

———. *Gulf War Air Power Survey.* Vol. 3, *Logistics and Support.* Washington, D.C.: GPO, 1993.

Conyers, John, Jr. Statement. *Performance of the Patriot Missile in the Gulf War.* 102d Cong., 2d sess. House Hearing. Committee on Government Operations. 7 April 1992. Washington, D.C.: GPO, 1992, 1–3.

Cooper, Henry. Statement. *Hearing Before the Subcommittee on the Department of Defense.* 102d Cong., 1st sess. House Hearing. 17 April 1991. Washington, D.C.: GPO, 1991, 613.

Coyle, Philip. *National Missile Defense (NMD) FY99 Annual Report,* Internet, Director, Operational Test and Evaluation Website, online at http://www.dote.osd.mil/reports/FY99/other/99nmd.html.

Davis, Richard. Statement. *Performance of the Patriot Missile in the Gulf War.* 102d Cong., 2d sess. House Hearing. Committee on Government Operations. 7 April 1992. Washington, D.C.: GPO, 1992, 77–98.

Dickinson, William. Statement. *The Impact of the Persian Gulf War and the Decline of the Soviet Union on How the United States Does Its Defense Business.* 102d Cong., 1st sess. House Hearing. Committee on Armed Services. 16 April 1991. Washington, D.C.: GPO, 1991, 418.

Doggett, Lloyd. Statement. *Congressional Record,* 20 May 1999, H3428–29.

Donnelly, Charles. Report. *United States Guided Missile Programs.* 86th Cong., 1st sess. Senate Hearing. Subcommittee on Preparedness Investigation of the Committee on Armed Services. Washington, D.C.: GPO, 1959, 1–9.

Garwin, Richard L. Statement. *Ballistic Missile Defense.* 108th Cong. 2d sess. Senate Hearing. Foreign Relations Committee. 4 April 1999. Washington, D.C.: GPO, 1999.

Gingrich, Newt. Statement. *Congressional Record,* 17 March 1997, H1023–H1030.

Graybeal, Sidney, and Keith Payne. Statement. *Department of Defense Authorization for Appropriations for Fiscal Year 1996 and the Future Years Defense Program.* 104th Cong., 1st sess. Senate Hearing. Committee on Armed Services. May 1995. Washington, D.C.: GPO, 1995, 797–800.

Helms, Jesse. "The President Turns a New Page in Defense." *Congressional Record,* 24 March 1983, 7155–57.

———. Statement. *Congressional Record,* 6 February 1996, S917–19.

———. "The Strategic Anti-Missile Revitalization and Security Act of 1996." *Congressional Record,* 6 February 1996, S917–19.

———. "Defend America Act of 1996." *Congressional Record,* 4 June 1996, S5741.

Hildreth, Steven A.. "Evaluation of U.S. Army Assessment of Patriot Antitactical Missile Effectiveness in the War against Iraq." Report. *Performance of the Patriot Missile in the Gulf War.* 102d Cong., 2d sess. House Hearing. Committee on Government Operations. 7 April 1992. Washington, D.C.: GPO, 1992, 22–76.

———. "Theater Ballistic Missile Defense Policy, Missions, and Current Status." CRS Issue Brief no. 95–585. 10 June 1993. Washington, D.C.: GPO, 1993.

———. "The ABM Treaty and Theater Missile Defense: Proposed Changes and Potential Implications." CRS Issue Brief no. 94–374 F. 2 May 1994. Washington, D.C.: GPO, 1994.

———. "Theater Missile Defense: Issues for the 104th Congress." CRS Issue Brief no. 95–012. 22 January 1996. Washington, D.C.: GPO, 1996.

———. "National Missile Defense: The Current Debate." CRS Issue Brief no. 96–441 F. 7 June 1996. Washington, D.C.: GPO, 1996.

Holum, John. Statement. *Administration's Proposal to Seek Modification of the 1972 Anti-Ballistic Missile Treaty.* 103d Cong., 2d sess. Senate Hearings. Foreign Relations Committee. 3 May 1994. Washington, D.C.: GPO, 1994, 8–11.

Horton, Frank. Statement. *Performance of the Patriot Missile in the Gulf War.* 102d Cong., 2d sess. House Hearing. Committee on Government Operations. 7 April 1992. Washington, D.C.: GPO, 1992, 5–8.

Israel. Government Press Office. *Text of Joint Peres-Clinton Settlement,* 30 April–1 May 1996. Translated by Foreign Broadcast Information Service (NES-96-086).

Kadish, Lt. Gen. Robert. Statement. *Military Procurement—Missile Defense Programs.* 106th Cong., 2d Sess. Committee on Armed Services. House Hearings. 16 February 2000 (Washington, D.C.: GPO, 2000). Internet. Online. Lexis/Nexis.

Kerrey, Bob. "Comments on the National Missile Defense Act of 1999." *Congressional Record,* 16 March 1999, S2707.

Kingsbury, Nancy B. Statement. *Development of Nuclear Thermal Propulsion Technology for Use in Space.* 102d Cong., 2d sess. House Hearings.

Committee on Science, Space, and Technology. 1 October 1991. Washington, D.C.: GPO, 1991, 4–11.

Kucinich, Dennis. Statement. *Congressional Record,* 18 May 2000, H3395.

Levin, Carl. Statement. *Department of Defense/Strategic Defense Initiative Organization Compliance with Federal Advisory Committee Act.* 100th Cong., 2d sess. Senate Hearings. Committee on Governmental Affairs. 19 April 1988. Washington, D.C.: GPO, 1988, 6.

Lewis, George. Statement. *Administration's Proposal to Seek Modification of the 1972 Anti-Ballistic Missile Treaty.* 103d Cong., 2d sess. Senate Hearings. Committee on Foreign Relations. 3 May 1994. Washington, D.C.: GPO, 1994, 80–91.

Markey, Edward. "Goldstone X-Ray Laser Test Should Be Delayed." *Congressional Record,* 5 December 1985, E5406.

McElroy, Neil H. *Program for Defense against the Intercontinental Ballistic Missile.* Memorandum to the secretary of the Air Force. In *Investigation of National Defense Missiles: Hearings Before the Committee Pursuant to H. Res.* 67. 85th Cong., 2d sess. House Hearings. Committee on Armed Services. Washington, D.C.: GPO, 1958, 4196–97.

Neal, Stephen. Statement. *Performance of the Patriot Missile in the Gulf War.* 102d Cong., 2d sess. House Hearing. Committee on Government Operations. 7 April 1992. Washington, D.C.: GPO, 1992, 96.

O'Neill, Lt. Gen. Malcolm. Statement. *Department of Defense Authorization for Appropriations for Fiscal Year 1994 and the Future Years Defense Program.* 103d Cong., 1st sess. Committee on Armed Services. Senate Hearings 10 June 1993. Washington, D.C.: GPO, 333–84.

———. Statement. *Hearings on National Defense Authorization Act for FY94: H.R. 2401 and Oversight of Previously Authorized Programs, Procurement of Aircraft, Missiles, Weapons, and Tracked Combat Vehicles, Ammunition, and Other Procurement.* 102d Cong., 2d sess. House Hearings. Committee on Armed Services. 10 June 1993. Washington, D.C.: GPO, 1993.

———. Statement. *Administration's Proposal to Seek Modification of the 1972 Anti-Ballistic Missile Treaty.* 103d Cong., 2d sess. Senate Hearings. Committee on Foreign Relations. 3 May 1994. Washington, D.C.: GPO, 1994, 11–15.

Pedatzur, Reuven. Statement. *Performance of the Patriot Missile in the Gulf War.* 102d Cong., 2d sess. House Hearing. Committee on Government Operations. 7 April 1992. Washington, D.C.: GPO, 1992, 118–21.

———. Statement. "Ballistic Missile Defense and HR 7-National Security Revitalization Act." Federal News Service. Online. Lexis/Nexis. 25 January 1995.

Postol, Theodore A. Statement. *The Impact of the Persian Gulf War and the Decline of the Soviet Union on How the United States Does its Defense Business.* 102d Cong., 1st sess. House Hearing. Committee on Armed Services. 16 April 1991. Washington, D.C.: GPO, 1991, 429–75.

———. "Improper Use of the Classification System to Suppress Public Debate on the Gulf War Performance of the Patriot Air-Defense System." *Government Secrecy After the Cold War.* 102d Cong., 2d sess. House Hearing. Committee on Government Operations. 18 March 1992. Washington, D.C.: GPO, 1992, 38–62.

———. "Optical Evidence Indicating Patriot High Miss Rates During the Gulf War." *Performance of the Patriot Missile in the Gulf War.* 102d Cong., 2d sess. House Hearing. Committee on Government Operations. 7 April 1992. Washington, D.C.: GPO, 1992, 131–49.

Povdig, Pavel. Statement. *Congressional Record,* 22 December 1995, S19210.

Raytheon, Inc. "Answers to Questions." Appendix. *Performance of the Patriot Missile in the Gulf War.* 102d Cong., 2d sess. House Hearing. Committee on Government Operations. 7 April 1992. Washington, D.C.: GPO, 1992, 316–23.

Reagan, Ronald. "Presenting the Strategic Defense Initiative." National Security Decision Directive 172. 30 May 1985. Reprinted in Christopher Simpson, ed., *National Security Decision Directives of the Reagan and Bush Administrations.* Boulder, Colo.: Westview, 1995, 535–48.

Rhinelander, John. Statement. *Administration's Proposal to Seek Modification of the 1972 Anti-Ballistic Missile Treaty.* 103d Cong., 2d sess. Senate Hearings. Committee on Foreign Relations. 3 May 1994. Washington, D.C.: GPO, 1994, 98.

Rosin, Carol. Statement. *Controlling Space Weapons.* 98th Cong., 1st sess. Senate Hearings. Committee on Foreign Relations. 14 April and 18 May, 1983. Washington, D.C.: GPO, 1983, 149–60.

Shuey, Robert D., and Steven A. Hildreth. "Missile Defense: Proposals, Programs, and Treaty Constraints." CRS Issue Brief no. 95–24 F. 21 December 1994. Washington, D.C.: GPO, 1994.

Smith, David J. Statement. *Senate Armed Services Strategic Forces FY99 Defense Budget: Ballistic Missile Defense.* 107th Cong., 1st sess. Senate

Hearings. Committee on Armed Services. 24 March 1998. Washington, D.C.: GPO, 1998.

Staff of the House Subcommittee on Energy Conservation and Power. *American Nuclear Guinea Pigs: Three Decades of Radiation Experiments on U.S. Citizens.* Committee Print. 99th Cong., 2d Sess. Washington, D.C.: GPO, 1986.

Stewart, Nina J. Letter to Hon. John Conyers. Reprinted in *Government Secrecy after the Cold War.* 102d Cong., 2d sess. House Hearing. Committee on Government Operations. 18 March 1992. Washington, D.C.: GPO, 1992, 72–73.

Sturm, Thomas A. *The USAF Scientific Advisory Board: Its First Twenty Years, 1944–1964.* Office of Air Force History Special Studies. Reprint of 1967 edition. Washington, D.C.: GPO, 1986.

Trudeau, Arthur. Statement. *Department of Defense Appropriations for Fiscal Year 1962, Hearings Before a Subcommittee of the Committee on Appropriations.* 83d Cong., 1st sess. House Hearings, Committee on Appropriations. Washington, D.C.: U.S. GPO, 1961, 205.

United States. Central Intelligence Agency. "Possible Soviet Responses to the U.S. Strategic Defense Initiative." NIC M 83–10017, copy 458 (12 September 1983). Formerly secret. Internet. Federation of American Scientists Website. Online at http://www.fas.org/spp/starwars/offdocs/m8310017.htm.

United States. Congress. *Department of Defense Appropriations for Fiscal 1961.* 86th Cong., 2d sess. House Hearings. Committee on Appropriations. Washington, D.C.: GPO, 1960.

———. Congress. *Strategic and Foreign Policy Implications of ABM Systems.* 91st Cong., 1st sess. Senate Hearings. Committee on Foreign Relations. 6 March 1969. Washington, D.C.: GPO, 1969.

———. Congress. *The Acquistion of Weapons Systems.* 91st Cong., 2d sess. Joint Hearings. Subcommittee on the Economy in Government of the Joint Economic Committee. Washington, D.C.: GPO, 1970.

———. Congress. *Changing National Priorities.* 91st Cong., 2d sess. Hearings. Joint Economic Committee. Washington, D.C.: GPO, 1970.

———. Congress. *Arms Control and the Militarization of Space.* 97th Cong., 2d sess. Senate Hearings. Committee on Foreign Relations. 20 September 1982. Washington, D.C.: GPO, 1982.

———. Congress. *Controlling Space Weapons.* 98th Cong., 1st sess. Senate Hearings. Committee on Foreign Relations. 14 April and 18 May 1983. Washington, D.C.: GPO, 1983.

———. Congress. *Department of Defense Authorization for Appropriations for Fiscal Year 1984.* 98th Cong., 1st sess. Senate Hearings. Committee on Armed Services. Washington, D.C.: GPO, 1983.

———. Congress. *Hearings Before the Subcommittee on Defense of the House Committee on Appropriations on Department of Defense Appropriations for 1984.* 98th Cong., 1st sess. House Hearings. Committee on Appropriations. Washington, D.C.: GPO, 1983.

———. Congress. *Department of Defense Appropriations, Fiscal Year 1985.* 98th Cong., 2d sess. Senate Hearings. Commitee on Appropriations. Washington, D.C.: GPO, 1984.

———. Congress. *Department of Defense Authorization for Appropriations for Fiscal Year 1985.* 98th Cong., 2d sess. Senate Hearings. Committee on Armed Services. Washington, D.C.: GPO, 1984.

———. Congress. *Strategic Defense and Anti-Satellite Weapons.* 98th Cong., 2d sess. Senate Hearing. Committee on Foreign Relations. Washington, D.C.: GPO, 1984.

———. Congress. *Performance of the Patriot Missile in the Gulf War.* 102d Cong., 2d sess. House Hearing. Committee on Government Operations. 7 April 1992. Washington, D.C.: GPO, 1992, 265–69.

———. Congress. *Department of Defense/Strategic Defense Initiative Organization Compliance with Federal Advisory Committee Act.* 100th Cong., 2d sess. Senate Hearings. Committee on Governmental Affairs. 19 April 1988. Washington, D.C.: GPO, 1998.

———. Congress. *Department of Defense Appropriations Bill, Fiscal Year 2000.* House Report 106–244. 109th Cong., 2d sess. Committee on Appropriations. 22 July 1999. Washington, D.C.: GPO, 1999.

———. Congressional Budget Office. *Budgetary and Technical Implications of the Administration's Plan for National Missile Defense. Congressional Budget Office Paper* Washington, D.C.: GPO, 2000.

———. Department of Defense. *Conduct of the Persian Gulf War, Final Report to Congress.* Washington, D.C.: GPO, 1992.

———. Department of Defense. Biography of Robert McNamara, past Secretary of Defense. Internet. DefenseLink Website. Posted 24 November 1998. Online at http://www.defenselink.mil/specials/secdef_histories/bios/mcnamara.htm.

———. Department of Defense. "THAAD Test Flight Does Not Achieve Intercept Target." Department of Defense Press Release. 29 March 1999. Internet. Coalition to Reduce Nuclear Danger Website. Online at http://www.clw.org/coalition/thaad99a.htm.

———. Department of Energy. *Openness.* Press Conference Fact Sheets. 6 February 1996. Washington, D.C.: U.S. DOE, 1996.

———. General Accounting Office. *SDI Program: Evaluation of DOE's Answers to Questions on X-Ray Laser Experiment.* Report NSIAD-88-181BR. Washington, D.C.: GPO, 1988.

———. General Accounting Office. *Operation Desert Storm: Data Does Not Exist to Conclusively Say How Well Patriot Performed.* Report to Congressional requesters. NSIAD-92-340. September 1992. Washington, D.C.: GPO, 1992.

———. General Accounting Office. *Ballistic Missile Defense: Evolution and Current Issues.* Report NSIAD-93-229. July 1993. Washington, D.C.: GPO, 1993.

———. General Accounting Office. *Information on Theater High Altitude Area Defense (THAAD) System.* Report NSIAD-94-107BR. January 1994. Washington, D.C.: GPO, 1994.

———. General Accounting Office. *Issues Concerning Acquisition of THAAD Prototype System.* Letter Report GAO/NSIAD-96-136. 9 July 1996. Washington, D.C.: GPO, 1996.

———. General Accounting Office. *Operation Desert Storm: Evaluation of the Air War.* PEMD-96-10. July 1996. Washington, D.C.: GPO, 1996.

van Tilbourg, Andre M. Memorandum. In *Department of Defense/Strategic Defense Initiative Organization Compliance with Federal Advisory Committee Act.* 100th Cong., 2d sess. Senate Hearings. Committee on Governmental Affairs. 19 April 1988. Washington, D.C.: GPO, 1988, 188.

Wade, Troy E., II. Statement. *Energy and Water Appropriations for 1989.* 100th Cong., 1st sess. House Hearings. Committee on Appropriations. Energy and Water Development Subcommittee. Washington, D.C.: GPO, 1988, 787.

Wakelin, James. Statement. *Department of Defense Fiscal Year 1962, Hearings Before a Subcommittee of the Committee on Appropriations.* 83d Cong., 1st sess. House Hearings. Committee on Appropriations. Washington, D.C.: U.S. GPO, 1961.

Westmoreland, William. Statement. *Congressional Record,* 13 July 1970, 23823–25.

Zimmerman, Peter D. Statement. *Performance of the Patriot Missile in the Gulf War.* 102d Cong., 2d sess. House Hearing. Committee on Government Operations. 7 April 1992. Washington, D.C.: GPO, 1992, 149–76.

Reports and papers

Bernstein, Alvin H. "Soviet Defense Spending: The Spartan Analogy." RAND Note N-2817-NA. Prepared for the director of net assessment, Office of the Secretary of Defense. Santa Monica, Calif.: RAND Corporation, 1989.

Blank, Stephen. "SDI and Defensive Doctrine: The Evolving Soviet Debate." Kennan Institute for Advanced Russian Studies Occasional Paper no. 240. Washington, D.C.: Woodrow Wilson Center, 1990.

Bowman, Robert M. "BMD & Arms Control." ISSS Issue Paper. January 1985.

Burt, Richard. "New Weapons Technologies: Debate and Directions." Adelphi Paper 126. London: Brassey's/International Institute for Strategic Studies, 1976.

Center for Security Policy. "Critical Mass #2: Senator Lott, Rumsfeld Commission Add Fresh Impetus to Case for Beginning Deployment of Missile Defenses." Center for Security Policy Decision Brief no. 98-D 133. 15 July 1998. Internet. Center for Security Policy Website. Online at http://www.security-policy.org/papers/1998/98-D133.html.

———. "*Wall Street Journal* Lauds Rumsfeld Commission Warning on Missile Threat; Reiterates Call for Aegis Option in Response." Decision Brief no. 98-P 134. 16 July 1998. Internet. Center for Security Policy Website. Online at http://www.security-policy.org/papers/1998/98-D134.html.

Center for Strategic and Budgetary Assessments. "The Rumsfeld Commission Report: Where Do We Go From Here?" 4 August 1998. Internet. Center for Strategic and Budgetary Assessments Website. Online at http://www.csbahome.com/Publications/Rumsfeld.html.

Cirincione, Joseph. "The Persistence of the Missile Defense Illusion." Paper presented at the Conference on Nuclear Disarmament, Safe Disposal of Nuclear Materials or New Weapons Development? Como, Italy. 2–4 July 1998. Internet. Carnegie Endowment for International Peace Website. Online at http://www.ceip.org/people/cirincio.htm.

"Charged Particle Beam: Anatomy of a Defense Scare." *F.A.S. Public Interest Report* 30 (1977): 1–2.

Coyle, Philip. Missile Defense and Related Programs Director, Operational Test & Evaluation (DOT&E) FY'99 Annual Report. February 2000. Internet. Director of Operational Test & Evaluation (DOT&E) Website. Online at http://www.dote.osd.mil/reports/FY99/index.html.

Danchick, Roy. Declaration filed under Civil Action CV96-3065. United States District Court. Central District of California. 18 May 1999. Internet. Federation of American Scientists Website. Online at http://www.fas.org/spp.starwars/program/news00/000203-trw.htm.

Dannreuther, Roland. "The Gulf Conflict: A Political and Strategic Analysis." Adelphi Paper 264. London: Brassey's/International Institute for Strategic Studies, 1992.

Fitzgerald, Mary C. "Soviet Views on SDI." Carl Beck Paper in Russian and East European Studies no. 601. Pittsburgh: University of Pittsburgh Center for Russian and East European Studies, 1987.

Freedman, Lawrence. "Strategic Defence in the Nuclear Age." Adelphi Paper 224. Oxford: Oxford University Press/International Institute for Strategic Studies, 1987.

———. "The Revolution in Strategic Affairs." Adelphi Paper 318. Oxford: Oxford University Press/International Institute for Strategic Studies, 1998.

Gold, Ted, and Rich Wagner. "Long Shadows and Virtual Swords: Managing Defense Resources in the Changing Security Environment." Unpublished paper. June 1990. Copy on file with the author.

Gronlund, Lisbeth. Statement. "National Missile Defense: The First Intercept Tests." Union of Concerned Scientists Press Breakfast. 29 September 1999. Internet. Union of Concerned Scientists website. Online at http://www.ucsusa.org/publications/pubs-home.html.

Heritage Foundation. "A Proposed Plan for Project on BMD and Arms Control." NSR #46. *High Frontier: A New Option in Space.* Labeled "Not for release." Copy on file with Carol Rosin.

Hildebrandt, Gregory G., ed. "Rand Conference on Models of the Soviet Economy, October 11–12, 1984." Santa Monica, Calif.: RAND Corporation, 1985.

Holmes, Kim. "Technology Speeds the Strategic Defense Initiative Timetable." Heritage Foundation Backgrounder no. 557, 13 January 1987.

Lee, Laura T., Martin T. Durbin, Rick B. Adelson, Sean K. Collins, and Wai H. Lee. *The Abuse of 'Footprints' for Theater Missile Defense and the ABM Treaty (U)*. Study by Sparta, Inc., under contract to the Ballistic Missile Defense Organization. September 1994. Cleared for public release to Theodore Postol, February 1995. Copy on file with the author.

Lewis, Kevin N. "Possible Soviet Responses to the Strategic Defense Initiative: A Functionally Organized Taxonomy." RAND Note N-2478-AF. Santa Monica, Calif.: RAND Corporation, 1986.

Maenchen, George. *Livermore Report* COPD 84–1993. 14 August 1984. Copy on file with the author.

Mitchell, Gordon R. "Sparta's TMD Footprint Advice to BMDO: Overstepping the Boundaries of FACA?" Issue brief for Senator David Pryor and Senator Carl Levin. 24 July 1996. Subsequently published as Space Policy Project E-print. 8 August 1996. Internet. Federation of American Scientists Website. Online at http://www.fas.org/spp/eprint/gm2.htm).

———. "The TMD Footprint Controversy." Space Policy Project E-print. 12 September 1996. Internet. Federation of American Scientists Website. Online at http://www.fas.org/spp/eprint/gm1.htm.

———. "Placebo Defense: The Legacy of Strategic Deception in Missile Defense Advocacy." Ph.D. diss., Northwestern University, December 1997.

———. "Two-Way Deception Traffic on the Road to Missile Defense." Space Policy Project E-print. December 1997. Internet. Federation of American Scientists Website. Online at http://www.fas.org/spp/eprint/gm3.htm.

———. "Missile Defence Policy: Strident Voices and Perilous Choices." ISIS-U.K. Briefing Series on Ballistic Missile Defence no. 1. April 2000.

———. "U.S. National Missile Defence: Technical Challenges, Political Pitfalls and Disarmament Opportunities." ISIS-Europe Briefing Paper no. 2. 3 May 2000. Internet. ISIS-Europe Website. Online at http://www.fhit.org/isis.

Mitchell, Gordon R., and Anthony Todero. "On Viewing Arguments for Japanese Missile Defense as Delicious but Deceptive American Exports." Paper presented at the First Tokyo Argumentation Conference, Tokyo Olympic Center. Tokyo, Japan. August 2000. Copy on file with the author.

Mitchell, Gordon R. (with the Cherub Study Group). "Return of the Death Star?" Space Policy Project E-print. 17 July 1998. Internet. Federation of

American Scientists Website. Online at: http://www.fas.org/spp/eprint/980731-ds.htm.

National Research Council. *A Review of the Department of Energy Classification Policy and Practice.* Washington, D.C.: National Academy Press, 1995.

O'Donnell, Timothy. "The Idea of Rhetoric in the Rhetoric of Science." M.A. thesis, Wake Forest University, 1997.

———. "The Rhetoric of American Science Policy." Ph.D. diss., University of Pittsburgh, October 2000.

Parnas, David Lorge and Danny Cohen. "SDI: Two Views of Professional Responsibility." Institute on Global Conflict and Cooperation Policy Paper no. 5. Berkeley, Calif.: Regents of the University of California, 1987.

Pike, John. "Star Wars: Clever Politics in the Service of Bad Policy." F.A.S. Public Interest Report 49 (1996). Internet. Federation of American Scientists Website. Online at http://www.fas.org/faspir/pir0996.html#starwars.

Quist, Arvin S. *Security Classification of Information.* Vol. 1, *Introduction, History, and Adverse Impacts.* Oak Ridge Gaseous Diffusion Plant, Oak Ridge, Tenn.: U.S. Department of Energy, 1989.

Reed, Samuel W. Defense Investigative Service Report (DCIS/DOD-IG). 1999. Internet. Federation of American Scientists Website. Online at http://www.fas.org/spp.starwars/program/news00/000203-trw.htm.

Schlafly, Phyllis. "ABM Should Be Republicans' Unifying Issue." Eagle Forum Column. November 1998. Internet. Eagle Forum Website. Online at http://www.eagleforum.org/column/1998/nov98/98–11–11.html.

Schroeer, Dietrich. "Directed-Energy Weapons and Strategic Defence: A Primer." Adelphi Paper 221. Oxford: Oxford University Press/International Institute for Strategic Studies, 1987.

Schubert, Frank N. and Theresa L. Kraus. "The Whirlwind War." Center of Military History Report. 1995. Internet. Federation of American Scientists Website. Online at http://www.fas.org/spp/starwars/docops/wwwapena.htm.

Schwartz, Nira. Third Amended Complaint for Violation of False Claims Act 31, USC 3729. Civil Action CV96-3065. United States District Court. Central District of California. 21 September 1999.

Spring, Baker. *Congress Bickered over Weapons Now Proving Themselves in the Gulf.* Heritage Foundation Backgrounder no. 808. Washington, D.C.: Heritage Foundation, 1991.

United States Congress. Strategic Defense Initiative Organization. *1989 Report to Congress on the Strategic Defense Initiative.* 13 March 1989. Washington, D.C.: GPO, 1989.

Welch, Larry, Task Leader. *Report of the Panel on Reducing Risk in Ballistic Missile Defense Flight Test Programs.* 27 February 1998. Internet. Federation of American Scientists Website. Online at http://www.fas.org/spp/starwars/program/welch/welch-1.htm.

Wood, Lowell. "Soviet and American X-Ray Laser Efforts: A Technological Race for the Prize of a Planet." Briefing presented to William Casey. 23 April 1985. Copy on file at the Federation of American Scientists Space Policy Project Archive, Washington, D.C.

PRESS RELEASES, SPEECHES, AND BROADCASTS

Abrahamson, James. Statement. *MacNeil/Lehrer NewsHour.* 12 March 1985. Internet. Online. Lexis/Nexis.

Bacon, Kenneth H. Statement. DOD News Briefing. Office of the Assistant Secretary of Defense. 18 May 2000. Internet. Federation of American Scientists website. Online at http://www.fas.org/spp/starwars/program/news00/t05182000_t0518asd.html.

Batzel, Roger. Written statement. Lawrence Livermore National Laboratory. 23 October 1987. Copy on file at the Federation of American Scientists Space Policy Project Archive, Washington, D.C.

Brown, George. Statement. "Representatives Release Teller X-Ray Laser Letters, Demand Release of Documents on DOE Investigation." In press release by Rep. Ed Markey. 1 August 1988. Copy on file with the author.

Bush, George. "Address to the People of Iraq on the Persian Gulf Crisis." *Weekly Compilation of Presidential Documents.* 16 September 1990. Washington, D.C.: GPO, 1990.

———. "Remarks Following Discussions With Amir Jabir al-Ahmad al-Jabir Al-Sabah of Kuwait." *Weekly Compilation of Presidential Documents.* 29 September 1990. Washington, D.C.: GPO, 1990.

———. "Exchange With Reporters in Andover, Massachusetts, on the Iraqi Offer To Withdraw From Kuwait, February 15, 1991." *Public Papers of the Presidents of the United States: George Bush, 1991.* Vol. 1, 147–48. Washington, D.C.: GPO, 1992.

———. "Remarks to Raytheon Missile System Plant Employees in Andover, Massachusetts, February 15, 1991." *Public Papers of the Presidents of the*

United States: George Bush, 1991. Vol. 1, 148–50. Washington, D.C.: GPO, 1992.

Churchill, Winston. Speech from the House of Commons Debate. "The Need for Air Defense Research." 28 November 1934. Reprinted in *Promise or Peril: The Strategic Defense Initiative*, edited by Zbigniew Brzezinski, 5–7. Washington, D.C.: Ethics and Public Policy Center, 1986.

Cosumano, Brig. Gen. Joseph. Statement. Department of Defense News Briefing (Subject: National Missile Defense Flight Test). Office of the Assistant Secretary of Defense. 2 July 1997. Internet. Federation of American Scientists Website. Online at http://www.fas.org/spp/starwars/program/news97/t07181997_t702bmdo.html.

DeFazio, Peter. Statement. "Ballistic Missile Defense: The Emperor's Newest Clothes." *American Defense Monitor* television show. Internet. Center for Defense Information Website. Online at http://www.cdi.org/adm/939/index.html.

Demurin, Mikhail. Quoted in Moscow ITAR-TASS World Service. "Spokesman Views U.S. Report on ABM Accords." Translated by Foreign Broadcast Information Service. 4 December 1995, 40.

Fox, Eugene. Statement. *Crossfire.* CNN, Atlanta. 18 August 1993. Internet. Online. Lexis/Nexis.

Garwin, Richard L.. "Scientist, Citizen, and Government: Ethics in Action (or Ethics Inaction)." Richard L. Horwitz Lecture. Harvard University. 4 May 1993. Garwin Archive. Internet. Federation of American Scientists Website. Online at http://www.fas.org/rlg/930504-imsa.htm.

Herrington, John. Press conference. Lawrence Livermore National Laboratory. 22 July 1988. Copy on file at the Federation of American Scientists Space Policy Project Archive, Washington, D.C.

Keeny, Spurgeon. Statement. "Arms Control Issues in the Fourth Year of the Clinton Administration." Annual Arms Control Association Membership Meeting and Luncheon. 8 March 1996. Washington: Carnegie Endowment for International Peace.

Markey, Ed. Press release, "Representatives Release Teller X-Ray Laser Letters, Demand Release of Documents on DOE Investigation." 1 August 1988. Copy on file with the author.

Martin, John, and David Ruppe. "Finding a Way to Pay: House Report Says Pentagon Ignored Spending Rules." Internet. ABC News Website. 23 July 1999. Online at http://abcnews.go.com:80/sections/us/DailyNews/pentagon990723.html.

McFarlane, Robert. Statement. "The Cold War." *CNN International News.* 21 March 1999. Transcript #99032100V16. Internet. Online. Lexis/Nexis.

Mendelsohn, John. Statement. "Extension of the Nuclear Non-Proliferation Treaty and the Clinton-Yeltsin Summit in Moscow." 17 May 1995. Washington: Arms Control Association Breakfast.

———. Statement. "Arms Control at the Clinton-Yeltsin and Clinton-Jiang Summit Meetings." Program on Nuclear Policy. 20 October 1995. Washington: Atlantic Council, 1995. Copy on file with the author.

———. "Arms Control Issues in the Fourth Year of the Clinton Administration." 8 March 1996. Annual Arms Control Association Membership Meeting and Luncheon. Washington: Carnegie Endowment for International Peace.

Moynihan, Daniel Patrick. "The Science of Secrecy." Paper delivered at the Massachusetts Institute of Technology Conference on Secrecy in Science. 29 March 1999. Cambridge, Mass.. Internet. American Association for the Advancement of Science Website. Online at http://www.aaas.org/spp/secrecy/Presents/Moynihan.htm.

Pike, John. Statement. *Arms Control Issues in the Fourth Year of the Clinton Administration.* Arms Control Association Panel Discussion. 8 March 1996. Washington: Carnegie Endowment for International Peace. Copy on file with the author.

Postol, Theodore A. Statement. "Patriot Wasn't Precise, Scientist Shows." Narr. Dan Charles. *All Things Considered.* National Public Radio. 19 March 1992. Online. Lexis/Nexis.

Raytheon, Inc. "Raytheon's Response to WGBH Frontline 'Gulf War.'" News Release. 1996. Internet. Frontline Website. Online at http://www.pbs.org/wgbh/pages/frontline/gulf/weapons/raytheontext.

Reagan, Ronald. "Remarks and a Question-and-Answer Session With Reporters on Domestic and Foreign Policy Issues, March 25, 1983." *Public Papers of the Presidents of the United States: Ronald Reagan, 1983.* Vol. 1, 447–50. Washington, D.C.: GPO, 1984.

———. "Address to the Nation on Defense and National Security." *Public Papers of the Presidents of the United States: Ronald Reagan, 1983.* Vol. 1, 437–43. Washington, D.C.: GPO, 1992.

Roston, Aram. Report. *Moneyline News Hour with Lou Dobbs.* 10 June 1999. Internet. Federation of American Scientists Website. Online at http://www.fas.org/spp/starwars/program/news99/990610-thaad-cnn.htm.

Stehr, John. "Nuclear-Powered X-Ray Laser Meets a Silent Death." *CBS Morning News.* 21 July 1992. Internet. Online: Lexis/Nexis.

Sununu, John. Statement. *Crossfire.* CNN, Atlanta. 18 August 1993. Online. Lexis/Nexis.

Talbott, Strobe. "America and Russia in a Changing World." Speech at 50th anniversary of Harriman Institute. 29 October 1996. Internet. Federation of American Scientists website. Online at http://www.fas.org.

"Technology of War." Video produced by *Popular Mechanics* magazine. New York: Goodtimes Home Video, 1994.

Teller, Edward. Statement. *Firing Line.* 19 October 1987. Internet. Online. Lexis/Nexis

Won-ung, Pak. "U.S. Applies 'Virtual' Pressure on ROK in Selling Weapons." Seoul KBS-1 Radio Network. 12 June 1997 broadcast. Translated by Foreign Broadcast Information Service (TAC-97-163).

INTERVIEWS AND CORRESPONDENCE

Arbatov, Alexei. Letter to Lieutenant General Malcolm O'Neill. Received 20 July 1995. Copy on file with the author.

Bendetsen, Karl. Letter to Edwin Meese. 20 October 1981. Copy on file with the author.

Bowman, Robert M. Telephone interview with the author. 21 July 1999. Transcript on file with the author.

Broad, William. Interview with Noah Adams. *All Things Considered.* National Public Radio. 6 March 1992. Internet. Online. Lexis/Nexis.

Collina, Tom. Interview with Bob Edwards. *Morning Edition.* National Public Radio. 18 March 1999. Internet. Online. Lexis/Nexis.

Conyers, John, Jr. Letter to Defense Secretary Cheney. 28 February 1991, copy on file with the author.

DeWitt, Hugh E.. Letter to David Saxon. 3 October 1982. Copy on file with the author.

———. Letter to Secretary Hazel O'Leary. 4 April 1994. Quoted in *Atomic Audit: The Costs and Consequences of U.S. Nuclear Weapons Since 1940,* ed. Stephen I. Schwartz (Washington, D.C.: Brookings Institution Press, 1998), 467.

Fallows, James. Interview with Neal Conan. *Morning Edition.* National Public Radio. 10 March 1993. Internet. Online. Lexis/Nexis.

Garwin, Richard L. FILE Memorandum. 10 October 1995. Copy on file with the author.

———. Letter to Senator Carl Levin. 2 September 1998. Copy on file with the author.

———. "Effectiveness of Proposed National Missile Defense Against ICBMs from North Korea." Letter prepared for distribution to Congress and the press. 17 March 1999. Garwin Archive. Internet. Federation of American Scientists Website. Online at http://www.fas.org/rlg/930504-imsa.htm.

Gold, Ted. Interview with the author. McLean, Va. 18 July 1996.

Goodman, Rick. Telephone conversation with the author. 21 January 1997.

Gronlund, Lisbeth, George Lewis, Theodore Postol, and David Wright. "Documentation of Technical Errors in the BMDO Study 'The Abuse of "Footprints" for Theater Missile Defenses and the ABM Treaty.'" Main Appendix of letter by Theodore Postol to Malcolm O'Neill, 31 March 1995, 1–7. Copy on file with the author.

Hildreth, Steven A. Memorandum to Rep. David Pryor. 1 August 1995, copy on file with the author.

Kane, Dennis. Telephone conversation with the author. 22 July 1996.

Lee, Laura. Interview with the author. Rosslyn, Va. 12 July 1996.

———. Telephone conversation with the author. 17 July 1997.

———. E-mail to the author. 22 August 1999.

Lesser, Marc. Telephone conversation with the author. 27 February 1997.

Lewis, George N. Telephone conversation with the author. 14 July 1999.

Lewis, George N., and Theodore A. Postol. Letter to the editor. *Christian Science Monitor,* 8 September 1997, 19.

———. "Search Solid Angle Requirements and Detection Ranges for Theater and Strategic Ballistic Missiles." Appendix A of letter by Theodore Postol to Malcolm O'Neill, 31 March 1995, 1–11. Copy on file with the author.

Lowder, J. E. "BMDO Request for Information Regarding SPARTA's 'Footprint Analysis.'" Memo to Peter Van Name. Sparta, Inc., Ref: JEL95–085 (Rev. 3). 5 October 1995. Released 7 November 1996 under FOIA request 96-F-1714, filed by the author.

Mische, Patricia. Letter to the author. 17 July 1999. Copy on file with the author.

Morgan, W. Lowell. Letter to Rep. George E. Brown Jr. 3 November 1987. Copy on file with the author.

Nacht, Michael. Interview with the author. Washington, D.C. 17 July 1996.

O'Neill, Malcolm. "One on One." Interview. *Defense News* (3–9 April 1994): 30.

———. Letter to Theodore Postol. 3 May 1995. Copy on file with the author.

Postol, Theodore A. Letter to Charles Bowsher, United States Comptroller General. 2 October 1992. Copy on file with the author.

———. Letter to Malcolm O'Neill. 31 March 1995. Copy on file with the author.

———. Letter to John R. Harvey. 2 May 1995. Copy on file with the author.

———. Interview with the author. Cambridge, Mass. 10 May 1995. Full transcript on file with the author.

———. Letter to Malcolm R. O'Neill. 8 September 1995. Copy on file with the author.

———. Telephone conversation with the author. 10 July 1996.

———. Letter to the author. 21 January 1997. Copy on file with the author.

———. "Technical Discussion of the Misinterpreted Results of the IFT-1A Experiment Due to Tampering With the Data and Analysis and Error in the Interpretation of the Data." Attachment B of letter to John Podesta. 11 May 2000. Internet. Federation of American Scientists website. Online at http://www.fas.org/spp/starwars/program/news00/postol_0511100.html.

Reagan, Ronald. Letter to Karl Bendetsen. 20 January 1982. Bendetsen Collection.

Reed, Samuel W, Jr. Letter to Keith Englander. 1 February 1999. Copy on file with the author.

———. Letter to Keith Englander. 11 August 1999. Copy on file with the author.

Rosin, Carol. Telephone interview with the author. 26 September 1999.

Rumsfeld, Donald. Interview with John McLaughlin. "One on One." Federal News Service. 14 February 1999. Internet. Online. Lexis/Nexis.

Sagdeev, Roald. Interview with Matthew Evangelista. Quoted in Matthew Evangelista, *Unarmed Forces: The Transnational Movement to End the Cold War.* Ithaca, N.Y.: Cornell University Press, 1999, 242.

Schwartz, Nira. Letter to Dennis C. Egan. 29 July 1998. Copy on file with the author.

Schwartz, Stephen I. Telephone conversation with the author. 3 May 2000.

Shelton, Gen. Harry H. Interview with *Sea Power Magazine.* Quoted in *Briefing Book on Ballistic Missile Defense.* Internet. Council for a Livable World Website. Online at htttp://www.clw.org/ef/bmdbook/legis.html.

Stein, Robert. "Patriot ATBM Experience in the Gulf War." Letter to *International Security* subscribers. 8 January 1992. Copy on file with the author.

Teller, Edward. Letter to George Keyworth. 22 December 1983. Copy on file with the author.

———. Letter to Paul Nitze. 28 December 1984. Copy on file with the author.

———. Letter to Robert McFarlane. 28 December 1984. Copy on file with the author.

Truman, Preston J. Letter to Rep. Howard Wolpe. Printed in House Hearings, *The Development of Nuclear Thermal Propulsion Technology for Use in Space.* 102d Cong, 2d sess. 1 October 1992. Washington, D.C.: GPO, 333–36.

Weiner, Tim. Interview with Noah Adams. *All Things Considered.* National Public Radio. 18 August 1993. Internet. Online. Lexis/Nexis.

Woodruff, Roy D. Letter to George Keyworth. 28 December 1983. Copy on file with the author.

———. Letter to Paul Nitze. 31 January 1985. Copy on file with the author.

———. Letter to Kenneth Withers. 6 February 1985. Copy on file with the author.

———. Letter to Roger Batzel. 19 October 1985. Copy on file with the author.

Yarmolinsky, Adam. Memorandum to McGeorge Bundy. 15 March 1963. Copy on file with the author.

Young, Robert A. Letter to Keith Englander. 25 March 1998. Copy on file with the author.

———. Letter to Keith Englander. 10 April 1998. Copy on file with the author.

Index

Figure 1. Launch of Homing Overlay Experiment vehicle. U.S. Army photograph.

Figure 2. Official Homing Overlay Experiment project team logo. U.S. Army photograph.

Figure 3. An actual HOE vehicle with its radial net unfurled hangs in the Alabama Space and Rocket Center in Huntsville, Alabama. Note the diagram showing "HOE 1984 INTERCEPT" in the background. U.S. Army photograph.

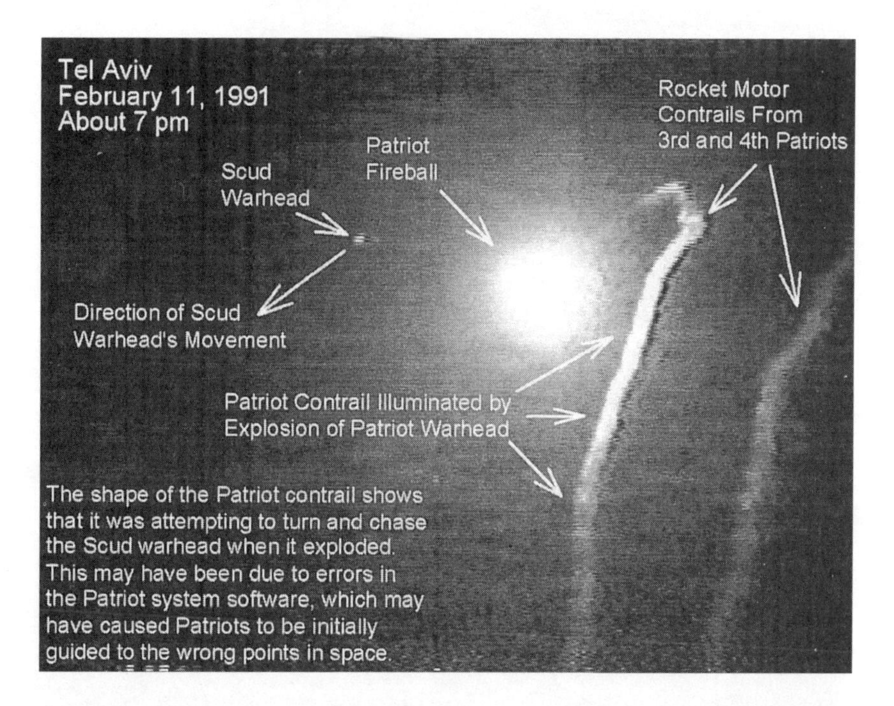

Figure 4. The above video frame shows the explosion of the last of four Patriots that engaged a Scud at Tel Aviv on 11 February 1991. All four Patriots missed the Scud by hundreds of meters. The fourth Patriot was observed by a video camera at sufficiently close range that the light-flash from its exploding warhead illuminated its contrail, showing that it was turning in an attempt to chase the Scud warhead. The turning of the Patriot suggests that it was sent to the wrong point by its Fire Unit Computer. When the Patriot was finally close enough to the Scud warhead to begin homing in on it, the Patriot was too distant to be able to make an intercept attempt. The Patriot's flight was then likely terminated by a self-destruct command sent from the Fire Unit on the ground. It is possible that the Patriot miss, observed in the above video frame, is due to a timing error which is believed to have also caused the failure of a Patriot Fire Unit to acquire, track, and engage a Scud at Dharan on 25 February 1991. The unengaged arriving Scud later hit a U.S. barrack, killing 28 and wounding 98, resulting in the largest single incident of U.S. deaths during the Gulf War. Photograph and caption courtesy of George Lewis.

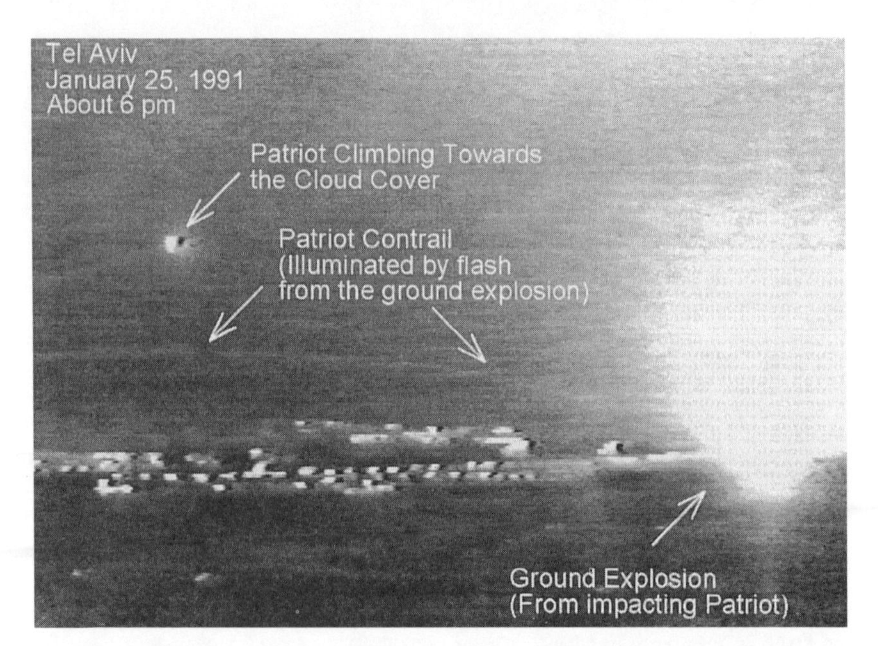

Figure 5. Ground explosion from errant Patriot missile diving into the ground at Tel Aviv, Israel, 25 January 1991. Photograph courtesy of George Lewis.

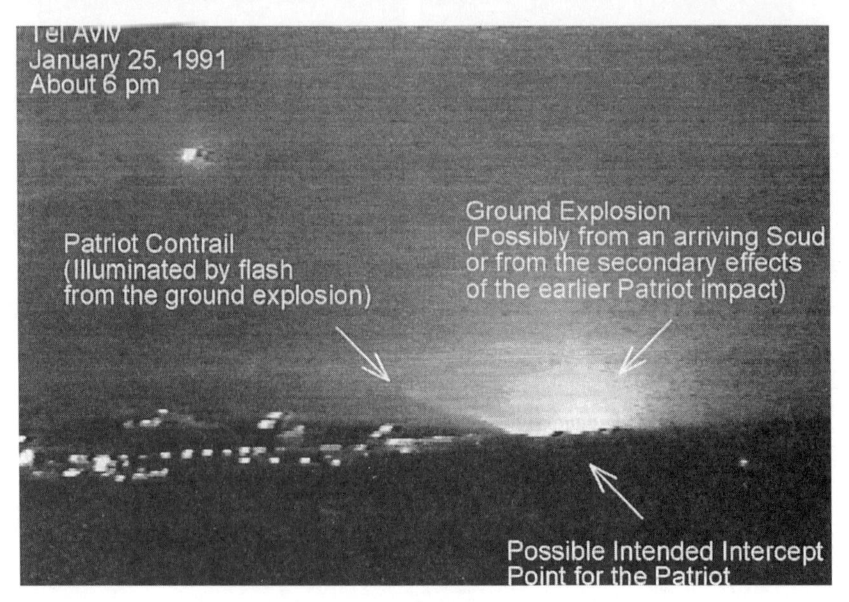

Figure 6. Secondary ground explosion following impact of an errant Patriot missile at Tel Aviv, Israel. The secondary explosion may have been caused by the Patriot and/or Scuds exploding on the ground. Photograph courtesy of George Lewis.

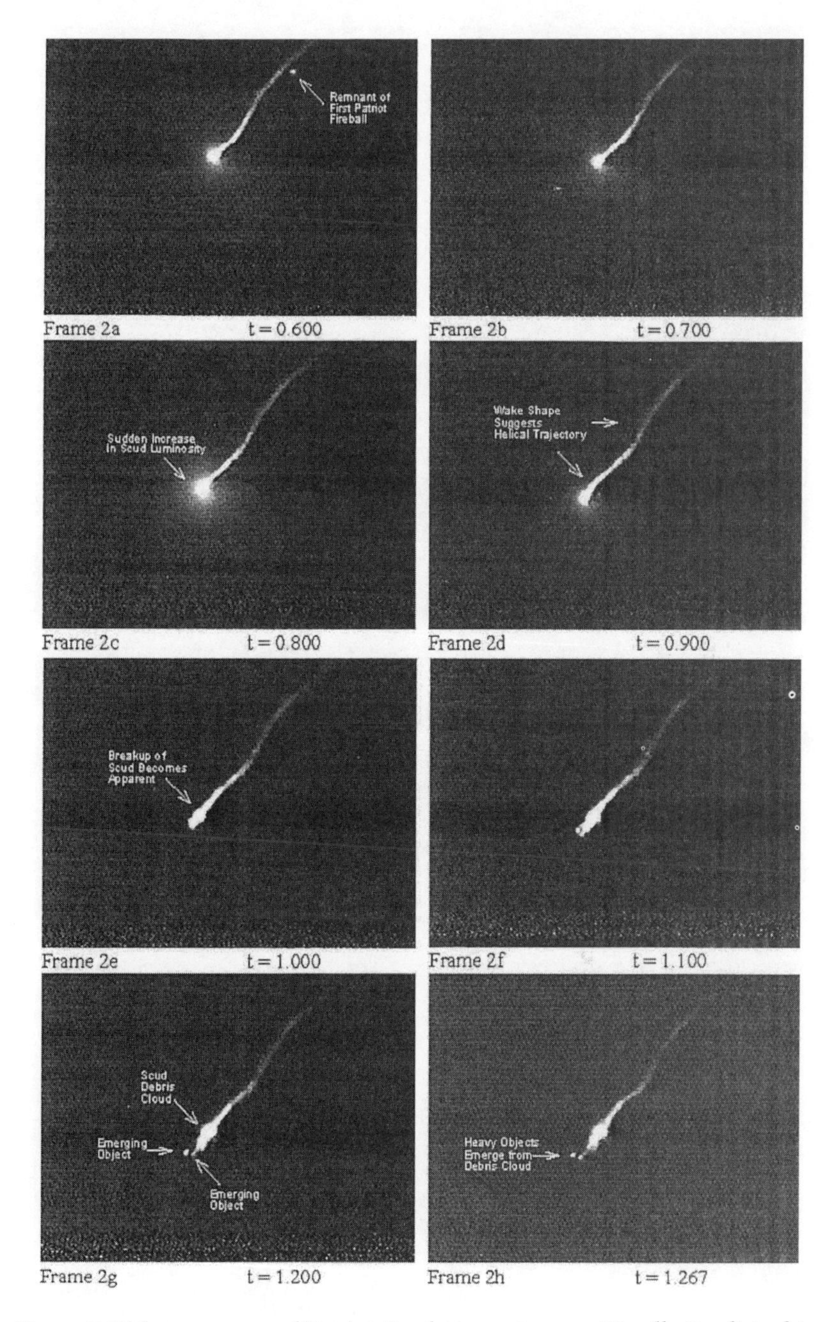

Figure 7. Video sequence of Patriot-Scud encounter over Riyadh, Saudi Arabia, 26 January 1991. The Scud breaks up due to aerodynamic forces and its warhead emerges from the breakup debris cloud. Photograph courtesy of George Lewis.

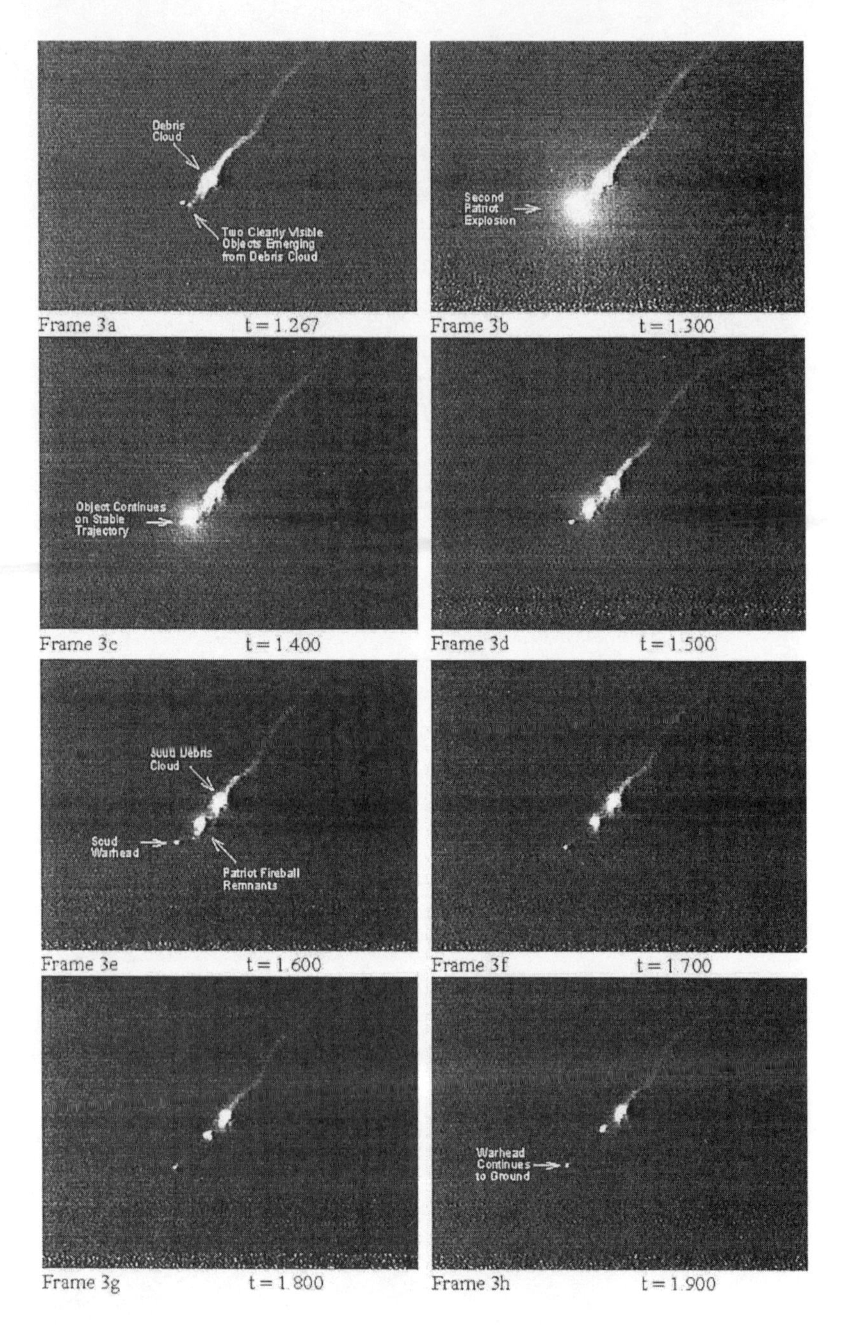

Figure 8. Video sequence showing the Patriot-Scud engagement over Riyadh, Saudi Arabia, 26 January 1991. The warhead on the second Patriot detonates, and the fireball overlaps the apparent position of the Scud warhead. However, the Scud warhead continues onward. Photograph courtesy of George Lewis.

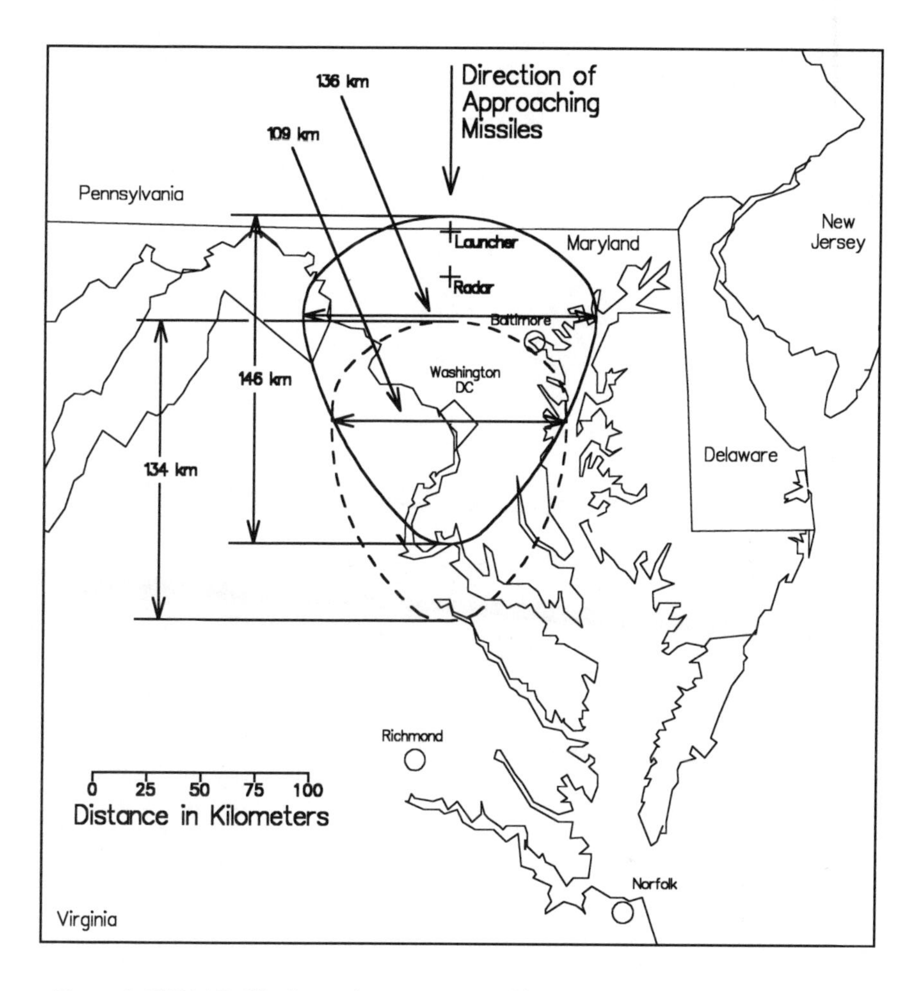

Figure 9. THAAD-like footprints constructed by Gronlund et al. The above figure shows the defended footprints calculated for an antitactical ballistic missile (ATBM) against a 3,000 kilometer range theater ballistic missile (solid line) and a 10,000 kilometer range strategic ballistic missile or SBM (dashed line). The calculations used to generate the footprints assume an ATBM site radar with a power aperture product of 500,000 watts-meter squared and an attacking missile (coming from the top of the figure) arriving on a minimum energy trajectory and having a radar cross section of .05 square meters. These assumptions result in a TBM detection range of about 295 kilometers and an SBM detection range of about 355 kilometers. The minimum intercept altitude is assumed to be 40 kilometers. Diagram and caption courtesy of Theodore Postol.

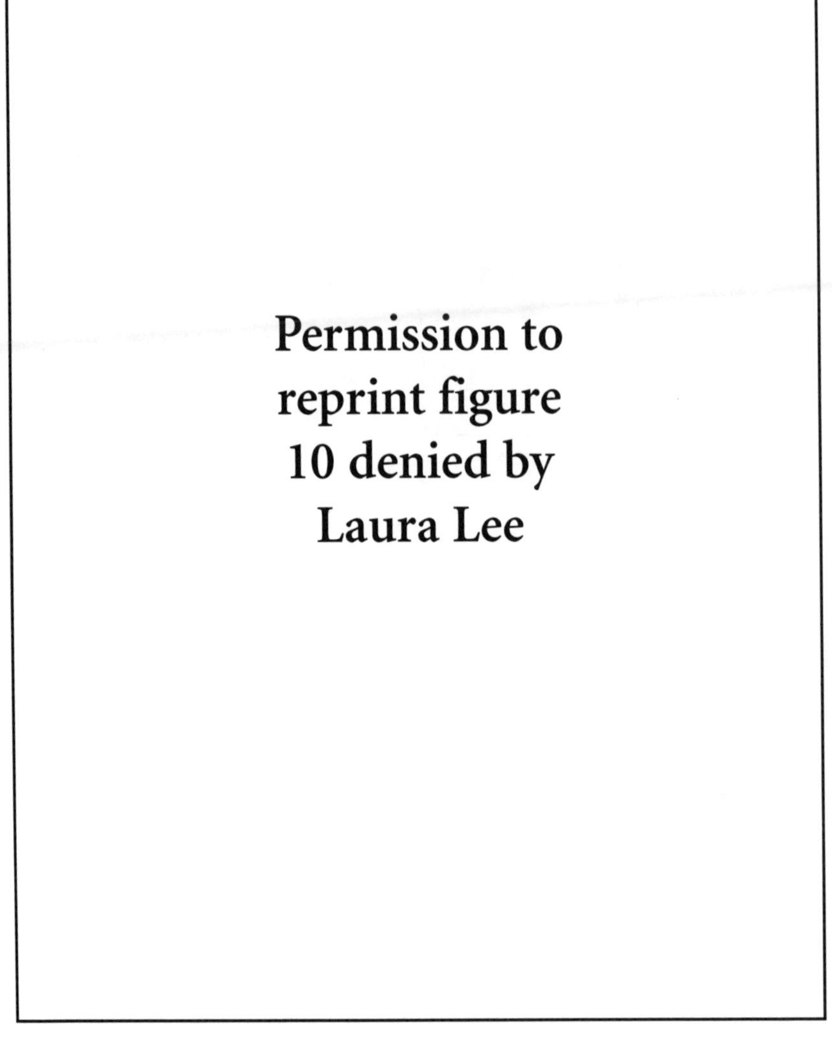

Figure 10. Lee et al.'s failed attempt to reconstruct the Gronlund et al. THAAD footprint shown in *Fig.* 9. From Laura T. Lee, Martin T. Durbin, Rick B. Adleson, Sean K. Collins and Wai T. Lee, *The Abuse of 'Footprints' for Theater Missile Defenses and the ABM Treaty (U)*. Sparta, Inc. Study commissioned by BMDO, September 1994, viewgraph #6. Cleared for public release to Keith Payne, Theodore Postol, and Richard Garwin, February 1995.

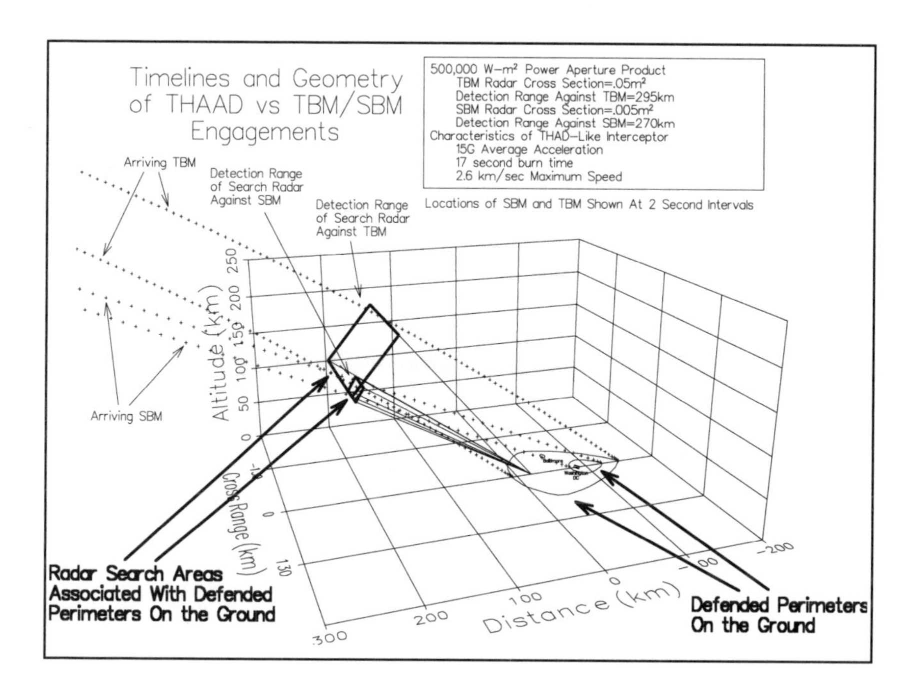

Figure 11. Iterative optimization mechanism. This three dimensional plot shows the defended perimeters on the ground that might be achievable with a THAAD-like defense system and the areas of sky that must be searched by the radar in order to achieve the defended areas. Since the lateral size of the "defended footprint" on the ground is smaller for strategic missiles relative to that of theater missiles, so is the lateral size of the radar sky-search area for a strategic missile. This illustrates that in order to get a correct estimate of the size of a defended footprint, it is necessary to both consider the size of the achievable defended ground region, which is in part a result of interceptor dynamics, and the capabilities of the search radar. Postol alleges that none of these factors were taken into account by Lee et al. Diagram and caption courtesy of Theodore Postol.